M. Nikravesh, L. A. Zadeh, V. Korotkikh (Eds.)

Fuzzy Partial Differential Equations and Relational Equations

Springer-Verlag Berlin Heidelberg GmbH

Studies in Fuzziness and Soft Computing, Volume 142

Editor-in-chief
Prof. Janusz Kacprzyk
Systems Research Institute
Polish Academy of Sciences
ul. Newelska 6
01-447 Warsaw
Poland
E-mail: kacprzyk@ibspan.waw.pl

Further volumes of this series can be found on our homepage: springeronline.com

Vol. 123. V. Torra (Ed.)
Information Fusion in Data Mining, 2003
ISBN 3-540-00676-1

Vol. 124. X. Yu, J. Kacprzyk (Eds.)
Applied Decision Support with Soft Computing, 2003
ISBN 3-540-02491-3

Vol. 125. M. Inuiguchi, S. Hirano and S. Tsumoto (Eds.)
Rough Set Theory and Granular Computing, 2003
ISBN 3-540-00574-9

Vol. 126. J.-L. Verdegay (Ed.)
Fuzzy Sets Based Heuristics for Optimization, 2003
ISBN 3-540-00551-X

Vol 127. L. Reznik, V. Kreinovich (Eds.)
Soft Computing in Measurement and Information Acquisition, 2003
ISBN 3-540-00246-4

Vol 128. J. Casillas, O. Cordón, F. Herrera, L. Magdalena (Eds.)
Interpretability Issues in Fuzzy Modeling, 2003
ISBN 3-540-02932-X

Vol 129. J. Casillas, O. Cordón, F. Herrera, L. Magdalena (Eds.)
Accuracy Improvements in Linguistic Fuzzy Modeling, 2003
ISBN 3-540-02933-8

Vol 130. P.S. Nair
Uncertainty in Multi-Source Databases, 2003
ISBN 3-540-03242-8

Vol 131. J.N. Mordeson, D.S. Malik, N. Kuroki
Fuzzy Semigroups, 2003
ISBN 3-540-03243-6

Vol 132. Y. Xu, D. Ruan, K. Qin, J. Liu
Lattice-Valued Logic, 2003
ISBN 3-540-40175-X

Vol. 133. Z.-Q. Liu, J. Cai, R. Buse
Handwriting Recognition, 2003
ISBN 3-540-40177-6

Vol 134. V.A. Niskanen
Soft Computing Methods in Human Sciences, 2004
ISBN 3-540-00466-1

Vol. 135. J.J. Buckley
Fuzzy Probabilities and Fuzzy Sets for Web Planning, 2004
ISBN 3-540-00473-4

Vol. 136. L. Wang (Ed.)
Soft Computing in Communications, 2004
ISBN 3-540-40575-5

Vol. 137. V. Loia, M. Nikravesh, L.A. Zadeh (Eds.)
Fuzzy Logic and the Internet, 2004
ISBN 3-540-20180-7

Vol. 138. S. Sirmakessis (Ed.)
Text Mining and its Applications, 2004
ISBN 3-540-20238-2

Vol. 139. M. Nikravesh, B. Azvine, I. Yager, L.A. Zadeh (Eds.)
Enhancing the Power of the Internet, 2004
ISBN 3-540-20237-4

Vol. 140. A. Abraham, L.C. Jain, B.J. van der Zwaag (Eds.)
Innovations in Intelligent Systems, 2004
ISBN 3-540-20265-X

Vol. 141. G.C. Onwubolu, B.V. Babu
New Optimzation Techniques in Engineering, 2004
ISBN 3-540-20167-X

Masoud Nikravesh
Lotfi A. Zadeh
Victor Korotkikh (Eds.)

Fuzzy Partial Differential Equations and Relational Equations

Reservoir Characterization and Modeling

Springer

Prof. Masoud Nikravesh
E-mail: nikravesh@cs.berkeley.edu

Prof. Dr. Lotfi A. Zadeh
E-mail: zadeh@cs.berkeley.edu

University of California
Dept. Electrical Engineering and Computer
Science - EECS
94720 Berkeley, CA
USA

Dr. Victor Korotkikh
Central Queensland University
Faculty of Informatic/Communication
4740 Mackeay QLD
Australia
E-mail: v.korotkich@cqu.edu.au

ISSN 1434-9922

DOI 10.1007/978-3-540-39675-8

Library of Congress Cataloging-in-Publication-Data

Fuzzy partial differential equations and relational equations : reservoir characterization and modeling / Masoud Nikravesh, Lotfi A. Zadeh, Victor Korotkikh (eds.).
p. cm. -- (Studies in fuzziness and soft computing, ISSN 1434-9922 ; v. 142)
Includes bibliographical references and index.

1. Petroleum engineering--Mathematics. 2. Fuzzy mathematics. 3. Soft computing.
I. Nikravesh, Masoud, 1959- II. Zadeh, Lotfi Asker. III Korotkikh, Victor. IV. Series.
TN871.F89 2004
665.5'01'51--dc22

Originally published by Springer-Verlag Berlin Heidelberg New York in 2004
MyCopy version of the original edition 2004

Cover design: E. Kirchner, Springer-Verlag, Heidelberg
Printed on acid free paper 62/3020/M - 5 4 3 2 1 0
www.springer.com/mycopy

"So far as the law of mathematics refer to reality they are not certain; and so far as they are certain, they don't refer to reality"

Albert Einstein, 1951

Forewords

Fuzzy Partial Differential Equations and Relational Equations Reservoir Characterization and Modeling

Masoud Nikravesh, with Prof Lotfi Zadeh and Victor Korotkikh should be commanded for compiling an authoritative collection on a subject that has not received enough attention: Partial differential equations are the foundation of physical laws governing two different oil industry applications: Darcy's law and wave equation. The first one has to do with the movement of fluids through porous media. The second one governs propagation of waves (both acoustic and electromagnetic) through the subsurface. How to solve these equations under various assumptions for their respective parameters has been the subject of numerous doctoral dissertation topics, academic research programs, oil and gas company technology development activities and service companies' software packages. They all attempt to make the solution more feasible, assumptions more realistic and approximations more acceptable.

The fact is, given the complex nature of the hydrocarbon bearing reservoirs and considerable heterogeneities rock formation through which seismic waves propagate or fluids (oil, gas and water) flow, parameterization of those equations is a formidable task. The extremely sparse nature of the available data, coupled with very limited direct and accurate measurements (well logs, flow rates, and core samples) makes the modeling and validation job even more difficult. In addition, different types of uncertainties, measurement errors, and various approximations associated with idealistic assumptions of the medium with respect to governing physical laws, makes the theoretical equations less reliable.

Once we recognize the inherent inadequacies of the conventional mathematical techniques and classical equations based on deterministic and crisp parameterization we understand the need for alternatives. Even if we have to rewrite the book on reservoir simulation and geophysical imaging, we will need to resort to stochastic and fuzzy-logic based methods. That is, we need to use wave equations comprised of random or fuzzy coefficients describing subsurface geometry, velocities and densities. This will enable us describe and parameterize the medium through which acoustic waves propagate, particularly when a substantial amount of heterogeneity is present, more effectively.

Likewise, conventional Darcy law, describing permeability of the rocks in terms of "measurable" quantities can be generalized to account for the imprecision, uncertainty and measurement errors. Permeability is of course an important reservoir property. It controls the flow rate and directional movement of different phase fluids (namely gas, water and oil) through the reservoir formations. Darcy's method is based on a partial differential equation is established for an idealized situation dealing with "a horizontal linear flow of an incompressible fluid". Realistically, in a highly heterogeneous and anisotropic, multi-phase fluid environment, these assumptions are too restrictive and a more sophisticated and rigorous treatment of the problem becomes necessary. This however can not be accomplished using conventional mathematics.

The present book is a first step towards introducing fuzzy partial differential and relation equations to the oil industry application problems in general and reservoir simulation in particular. It is envisioned that these methods will begin to find there true relevance and applications in many of the petroleum industry problems, This will enable us to treat model parameterization, inversion and simulation more effectively with closer relevance to the real problems we are facing every day. Prof. Zadeh maintained several years ago: "As complexity increases precise statements lose meaning and meaningful statements lose precision." Using fuzzy differential equations, we may no longer need to find the illusive "precise" equations to describe the "precise physical phenomena", describing our ever increasing complex problems.

Fred Aminzadeh
Houston, Texas
April 2003

Preface

During last decade significant progress has been made in the oil industry by using soft computing technology. Underlying this evolving technology there have been ideas transforming the very language we use to describe problems with imprecision, uncertainty and partial truth.

These developments offer exciting opportunities, but at the same time it is becoming clearer that further advancements are confronted by fundamental problems. The whole idea of how human process information lies at the core of the challenge. There are already new ways of thinking about the problems within theory of perception-based information. This theory aims to understand and harness the laws of human perceptions to dramatically improve the processing of information. A matured theory of perception-based information is likely to be proper positioned to contribute to the solution of the problems and provide all the ingredients for a revolution in science, technology and business.

In this context, Berkeley Initiative in Soft Computing (BISC), University of California, Berkeley from one side and Chevron-Texaco from another formed a Technical Committee to organize a Meeting entitled "State of the Art Assessment and New Directions for Research" to understand the significance of the fields accomplishments, new developments and future directions. The Technical Committee selected and invited 15 scientists (and oil industry experts as technical committee members) from the related disciplines to participate in the Meeting, which took place at the University of California, Berkeley, and March 15-17, 2002.

The chapters of the book are evolved from presentations made by the participants at the Meeting. The papers include reports from the front of soft computing in the oil industry and address the problems of the fields by considering a very important topic of fuzzy partial differential equations, fuzzy relational equations and fuzzy difference equations.

The book starts with the chapter "Soft Computing for Reservoir Characterization" written by Masoud Nikravesh and Fred Aminzadeh. This overview paper highlights the role of soft computing techniques for intelligent reservoir characterization and exploration, seismic data processing and characterization, well logging, reservoir mapping and engineering. Reservoir characterization plays a crucial role in modern reservoir management. It helps to make sound reservoir decisions and improves the asset value of the oil and gas companies. The ultimate product is a reservoir model with realistic tolerance for imprecision and uncertainty. Soft computing aims to exploit such a tolerance for solving practical problems. In reservoir characterization, these intelligent techniques can be used for uncertainty analysis, risk assessment, data fusion and data mining, which are applicable to feature extraction from seismic attributes, well logging, reservoir mapping and engineering. The chapter stresses that future research should focus on the integration of data and disciplinary knowledge for improving our understanding of reservoir data and reducing our prediction uncertainty.

The laws of human information processing may be still far away from our knowledge and experience. It seems that a better position would be to understand them within an irreducible theory, i.e., a theory not allowing a deeper explanatory base. The chapter "An Approach to the Mathematical Theory of Perception-Based Information" by Victor Korotkikh considers an approach to the mathematical theory of perception-based information by using an irreducible theory that captures a new type of processes, i.e., hierarchical formations of integer relations. It is suggested that the optimal functioning of a complex system may be a law of the integer formations and perception-based information plays a key role in the use of this law.

Roberto Tagliaferri, Angelo Ciaramella, Antonio Di Nola and Radim Belohlavek in their chapter "Fuzzy Neural Networks Based on Fuzzy Logic Algebras Valued Relations" propose a method to build a fuzzy neural network based on fuzzy relations with truth values in a suitable algebraic structure. The properties of the network are analyzed in detail and interesting theorems on the stability of parameter variations are proved. Furthermore, in the paper the architecture of the fuzzy network with the fuzzification and defuzzification models is illustrated.

Probability and possibility theories deal with different types of uncertainty. The chapter "Simulating Continuous Dynamical Systems under Conditions of Uncertainty: the Probability and the Possibility Approaches" by Gianluca Bontempi discusses the role of these formalisms in the simulation of continuous dynamical systems where parameters and/or initial conditions are uncertain. The evolution law of possibility distributions in continuous dynamical system is derived and compared with consolidated results in probability theory. The analysis illustrates the limits of conventional Monte Carlo techniques for the simulation of dynamical systems where the uncertainty is expressed in a non probabilistic form. The chapter proposes a new algorithm for the numerical simulation of a differential system, where the uncertainty of parameters and/or initial conditions is represented by fuzzy distributions.

The chapter "Resolution of Min-Max Fuzzy Relational Equations" by Ketty Peeva provides a methodology and polynomial time algorithm for inverse problem resolution for min-max fuzzy relational equations. An exact and universal method is presented for solving min-max fuzzy linear systems of equations. A method and algorithm for solving min-max fuzzy relational equations are also proposed in the chapter.

Vilem Novak develops a systematic theory of the evaluating linguistic expressions and their semantics in the chapter "Fuzzy Relation Equations with

Words". The focus of the paper is on fuzzy relation equations derived from the fuzzy data given by these expressions. It is shown that pure expressions mostly cannot lead to solvable fuzzy relation equations. Thus, the results propose that fuzzy numbers should be used. Moreover, the chapter suggests that if the structure of the data corresponds to a function then the solution can even be in the form of the widely used Mamdani-Assilian formula.

The chapter "A Normative View on Possibility Distributions" by Christer Carlsson, Robert Fuller and Peter Majlender considers fuzzy numbers from a normative point of view and illustrates the concepts of possibilistic mean, covariance, variance and correlation by several examples. It is shown that zero correlation does not always imply non-interactivity. The chapter also presents the limitations of direct definitions of joint possibility distributions from individual fuzzy numbers, for example, when one simply aggregates the membership values of two fuzzy numbers by a triangular norm.

The chapter "FREs: the ODEs and PDEs of the Fuzzy Modelling Paradigm' by Bernard De Baets explains how inverse problems can be solved in the framework of polynomial lattice equations and presents a uniform framework in which various types of image and composition equations can be dealt with. The paper suggests that future work will be to provide a similar framework for approximate solution methods in case of voidness of the solution set.

Fuzzy relations give a unified framework for knowledge representation and processing. The chapter "Equations and Inequalities with BK-Products of Relations" by Ladislav Kohout presents an algebraic setting for dealing with fuzzy relational inequalities and equations. The algebraic apparatus that provides constructive means for formulating the relational algorithms is based on pseudoassociative compositions of fuzzy relations called BK-products. In the paper BK-products are used to develop the theory of generalized morphisms, which is an extension and substantial generalization of crisp and fuzzy relational homomorphisms.

Many general problems on the resolution of composite fuzzy relation equations have been investigated for last three decades. Elie Sanchez in his chapter "Decomposition of Fuzzy Relations and Functional Relations" shows how to solve a class of SUP-min fuzzy relational equations. Such relations are related to systems of truth-qualified propositions. In the paper a concept of the decomposition of a fuzzy relation by a fuzzy set is introduced. The problem of decomposability of a fuzzy relation by a given fuzzy set is solved by combining these methodologies.

Fuzzy systems modeling is expected to become useful in simulating hydrogeologic system behavior. In the chapter "Introduction to Modeling of Hydrogeologic Systems Using Fuzzy Differential Equations" by Boris Faybishenko a hydrogeologic system as a fuzzy system is presented and a fuzzy logic form of Darcy's equation is derived. Based on this equation, a fuzzy logic form of the parabolic-type partial differential equation is derived. The elliptic-type, Laplace equation, and the parabolic-type, Richards's equation, partial differential equations were approximated using fuzzy variables and solved by using the basic principles of fuzzy arithmetic. The results of fuzzy systems modeling are then compared with those obtained by using deterministic models.

The chapter "Construction of Granular Derivatives and Solution of Granular Initial Value Problem" by Ildar Batyrshin considers rule based derivatives and discusses rules construction methods of granular differentials. The main feature of the approach is a definition of granular differential as a fuzzy linear function defined on a fuzzy interval. The set of such differentials together with fuzzy initial value determined by a rule base is used for the construction of a solution of granular initial value problem. The approach may be applied to descriptions of the uncertain dependencies between variables of complex systems and technological processes.

Fuzzy partial differential equations can be applied for modeling of mechanical systems with uncertain parameters. The chapter "Numerical Solutions of Fuzzy Partial Differential Equations and its Applications in Computational Mechanics" by Andrzej Pownuk presents an algorithm for solving fuzzy partial differential equations arising in computational mechanics. The algorithm is based on finite element method and sensitivity analysis. The algorithm is used to solve engineering problems with many thousands of variables.

We would like to thank the authors of the papers and gratefully acknowledge the Chevron-Texaco – specially, David Wilkinson and Adwait Chawathe – for the financial and technical support, which made the Meeting and book possible.

University of California

Berkeley, USA

May 2003

Masoud Nikravesh

Lotfi Zadeh

Victor Korotkikh

Table of Contents

Soft Computing for Reservoir Characterization

Masoud Nikravesh
Berkeley Initiative in Soft Computing (BISC) Program
Computer Science Division- Department of EECS
University of California, Berkeley, CA 94720
Email: nikravesh@cs.berkeley.edu, URL: http://www.cs.berkeley.edu/~nikraves/

F. Aminzadeh,
dGB-USA
Houston, TX

Abstract: As our problems become too complex to rely only on one discipline and as we find ourselves at the midst of information explosion multi-disciplinary analysis methods and data mining approaches in the petroleum industry become more of a necessity than professional curiosity. To tackle difficult problems ahead of us, we need to bring down the walls we have built around traditional disciplines such as petroleum engineering, geology, geophysics and geochemistry, and embark on true muti-disciplinary solutions. Our data, methodologies and workflow will have to cut across different disciplines. As a result, today's "integration" which is based on integration of results will have to give way to a new form of integration, that is, discipline integration. In addition, to solve our complex problems we need to go beyond standard mathematical techniques. Instead, we need to complement the conventional analysis methods with a number of emerging methodologies and soft computing techniques such as Expert Systems, Artificial Intelligence, Neural Network, Fuzzy Logic, Genetic Algorithm, Probabilistic Reasoning, and Parallel Processing techniques. Soft computing differs from conventional (hard) computing in that, unlike hard computing, it is tolerant of imprecision, uncertainty, and partial truth. Soft Computing is also tractable, robust, efficient and inexpensive. In this overview paper, we highlight role of Soft Computing techniques for intelligent reservoir characterization and exploration, seismic data processing and characterization, well logging, reservoir mapping and engineering. Reservoir characterization plays a crucial role in modern reservoir management. It helps to make sound reservoir decisions and improves the asset value of the oil and gas companies. It maximizes integration of multi-disciplinary data and knowledge and improves the reliability of the reservoir predictions. The ultimate product is a reservoir model with realistic tolerance for imprecision and uncertainty. Soft computing aims to exploit such a tolerance for solving practical problems. In reservoir characterization, these intelligent techniques can be used for uncertainty analysis, risk assessment, data fusion and data mining which are applicable to feature extraction from seismic attributes, well logging, reservoir mapping and engineering. The main goal is to integrate soft data such as geological data with hard data such as 3D seismic and production data to build a reservoir and stratigraphic model. While some individual methodologies (esp. neurocomputing) have gained much popularity during the past few years, the true benefit of soft computing lies on the integration of its constituent methodologies rather than use in isolation.

Future research should focus on the integration of data and disciplinary knowledge for improving our understanding of reservoir data and reducing our prediction uncertainty.

1 Introduction

Last decade has witnessed significant advances in transforming geosciences and well data into drillable prospects, generating accurate structural models and creating reservoir models with associated properties. This has been made possible through improvements in data integration, quantification of uncertainties, effective use of geophysical modeling for better describing the relationship between input data and reservoir properties, and use of unconventional statistical methods. Soft computing techniques such as neural networks and fuzzy logic and their appropriate usage in many geophysical and geological problems has played a key role in the progress made in recent years. However there is a consensus of opinion that we have only begun the scratch the surface in realizing full benefits of soft computing technology. Many challenges remain when we are facing with characterization of reservoirs with substantial heterogeneity and fracturing, exploring in the areas with thin-bedded stacked reservoirs and regions with poor data quality or limited well control and seismic coverage and quantifying uncertainty and confidence interval of the estimates. Among the inherent problems we need to overcome are: inadequate and uneven well data sampling, non-uniqueness in cause and effect in subsurface properties versus geosciences data response, different scales of seismic, log and core data and finally how to handle changes in the reservoir as the characterization is in progress.

This paper reviews the recent geosciences applications of soft computing (SC) with special emphasis on exploration. The role of soft computing as an effective method of data fusion will be highlighted. SC is consortium of computing methodologies (Fuzzy Logic (GL), Neuro-Computing (NC), Genetic Computing (GC), and Probabilistic Reasoning (PR) including; Genetic Algorithms (GA), Chaotic Systems (CS), Belief Networks (BN), Learning Theory (LT)) which collectively provide a foundation for the Conception, Design and Deployment of Intelligent Systems. The role model for Soft Computing is the Human Mind. Unlike the conventional or hard computing, it is tolerant of imprecision, uncertainty and partial truth. It is also tractable, robust, efficient and inexpensive. Among main components of soft computing, the artificial neural networks, fuzzy logic and the genetic algorithms in the "exploration domain" will be examined. Specifically, the earth exploration applications of SC in various aspects will be discussed. We outlines the unique roles of the three major methodologies of soft computing – neurocomputing, fuzzy logic and evolutionary computing. We will summarize a number of relevant and documented reservoir characterization applications. We will also provide a list of recommendations for the future use of soft computing. This includes the hybrid of various methodologies (e.g. neural-fuzzy or neuro-fuzzy, neural-genetic, fuzzy-genetic and neural-fuzzy-genetic) and the latest tool of "computing with words" (CW) (Zadeh 1996). CW provides a completely new insight into computing with imprecise, qualitative and linguistic phrases and is a potential tool for geological modeling which is based on words rather than exact numbers.

These applications are divided into two broad categories. One has to do with improving the efficiency in various tasks that are necessary for the processing and manipulation and fusion of different types of data used in exploration. Among these applications are: first arrival picking, noise elimination, structural mapping, horizon picking, event tracking and integration of data from different sources. The other application area is pattern recognition, identification and prediction of different rock properties under the surface. This is usually accomplished by training the system from known rock properties using a number of attributes derived from the properly fused input data (e.g., 2D and 3D seismic, gravity, well log and core data, ground penetrating radar and synthetic aperture radar and other types remote sensing data). Then a similarity measure with certain threshold level is used to determine the properties where no direct measurement is available.

2 The Role of Soft Computing Techniques for Intelligent Reservoir Characterization and Exploration

Soft computing is bound to play a key role in the earth sciences. This is in part due to subject nature of the rules governing many physical phenomena in the earth sciences. The uncertainty associated with the data, the immense size of the data to deal with and the diversity of the data type and the associated scales are important factors to rely on unconventional mathematical tools such as soft computing. Many of these issues are addressed in a recent books, Nikravesh et al. (2002), Wong et al (2001), recent special issues, Nikravesh et al. (2001a and 2001b) and Wong and Nikravesh (2001) and other publications such as Zadeh (1994), Zadeh and Aminzadeh (1995), and Aminzadeh and Jamshidi (1994).

Recent applications of soft computing techniques have already begun to enhance our ability in discovering new reserves and assist in improved reservoir management and production optimization. This technology has also been proven useful in production from low permeability and fractured reservoirs such as fractured shale, fractured tight gas reservoirs and reservoirs in deep water or below salt which contain major portions of future oil and gas resources. Through new technology and data acquisition to processing and interpretation the rate of success in exploration has risen to 40 percent in 1990 from 30 percent in the 1980s. In some major oil companies the overall, gas and oil well drilling success rates have risen to an average of 47 percent in 1996 from 3-30 percent in the early 1990s (SOURCE OF THIS DATA?). For example, in US only, by year 2010, these innovative techniques are expected to contribute over 2 trillion cubic feet (Tcf)/year of additional gas production and 100 million barrels per year of additional oil. This cumulative will be over 30 Tcf of gas reserves and 1.2 billion barrels in oil reserve and will add over $8 billion to revenue in 2010.

Intelligent techniques such as neural computing, fuzzy reasoning, and evolutionary computing for data analysis and interpretation are an increasingly powerful tool for making breakthroughs in the science and engineering fields by transforming the data into information and information into knowledge.

In the oil and gas industry, these intelligent techniques can be used for uncertainty analysis, risk assessment, data fusion and mining, data analysis and interpretation, and knowledge discovery, from diverse data such as 3-D seismic, geological data, well log, and production data. It is important to mention that during 1997, the US industry spent over $3 billion on seismic acquisition, processing and interpretation. In addition, these techniques can be a key to cost effectively locating and producing our remaining oil and gas reserves. Techniques can be used as a tool for:

1) Lowering Exploration Risk
2) Reducing Exploration and Production cost
3) Improving recovery through more efficient production
4) Extending the life of producing wells.

In what follows we will address data processing / fusion / mining, first. Then, we will discuss interpretation, pattern recognition and intelligent data analysis.

2.1 Mining and Fusion of Data

In the past, classical data processing tools and physical models solved many real-world problems. However, with the advances in information processing we are able to further extend the boundaries and complexities of the problems we tackle. This is necessitated by the fact that, increasingly, we are faced with multitude of challenges: On the one hand we are confronted with more unpredictable and complex real-world, imprecise, chaotic, multi-dimensional and multi-domain problems with many interconnected parameters in situations where small variability in parameters can change the solution completely. On the other hand, we are faced with profusion and complexity of computer-generated data. Making sense of large amounts of imprecise and chaotic data, very common in earth sciences applications, is beyond the scope of human ability and understanding. What this implies is that the classical data processing tools and physical models that have addressed many problems in the past may not be sufficient to deal effectively with present and future needs.

In recent years in the oil industry we have witnessed massive explosion in the data volume we have to deal with. As outlined at, Aminzadeh, 1996, this is caused by increased sampling rate, larger offset and longer record acquisition, multi-component surveys, 4-D seismic and, most recently, the possibility of continuous recording in "instrumented oil fields". Thus we need efficient techniques to process such large data volumes. Automated techniques to refine the data (trace editing and filtering), selecting the desired event types (first break picking) or automated interpretation (horizon tracking) are needed for large data volumes. Fuzzy logic and neural networks have been proven to be effective tools for such applications. To make use of large volumes of the field data and multitude of associated data volumes (e.g. different attribute volumes or partial stack or angle gathers), effective data compression methods will be of increasing significance, both for fast data transmission efficient processing, analysis and visualization and economical data storage. Most likely, the biggest impact of advances in data compression techniques will be realized when geoscientists have the ability to fully process and analyze data in the compressed domain. This will make it possible to carry out computer-intensive processing of large volumes of data in a fraction of the time, resulting in tremendous cost reductions. Data mining is another alternative that helps identify the most information rich part of the large volumes of data. Again in many recent reports, it has been demonstrated that neural networks and fuzzy logic, in combination of some of the more conventional methods such as eigenvalue or principal component analysis are very useful.

Figure 1 shows the relationship between Intelligent Technology and Data Fusion/Data Mining. **Tables 1** and **2** show the list of the Data Fusion and Data Mining techniques. **Figure 2** and **Table 3** show the Reservoir Data Mining and Reservoir Data Fusion concepts and techniques. **Table 4** shows the comparison between Geostatistical and Intelligent techniques. In sections II through VIII, we will highlight some of the recent applications of these methods in various earth sciences disciplines.

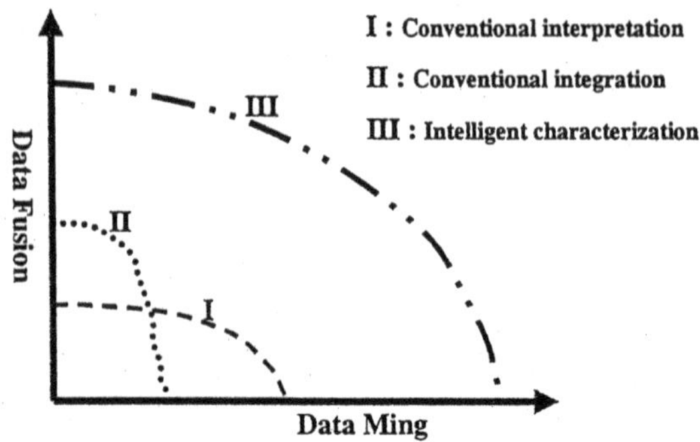

Figure 1. Intelligent Technology

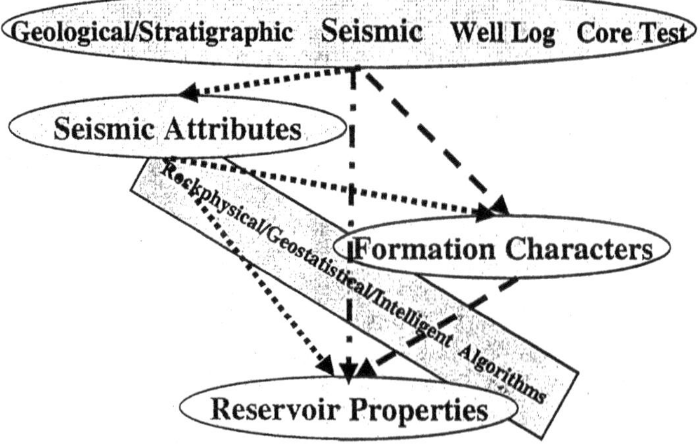

Figure 2. Reservoir Data Mining

Table 1.

Data Mining Techniques

- Deductive Database Client
- Inductive Learning
- Clustering
- Case-based Reasoning
- Visualization
- Statistical Package

Table 2.

Data Fusion Techniques

- Deterministic
  - Transform based *(projections,...)*
  - Functional evaluation based *(vector quantization, ...)*
  - Correlation based *(pattern match, if/then productions)*
  - Optimization based *(gradient-based, feedback, LDP, ...)*
- Non-deterministic
  - Hypothesis testing *(classification, ...)*
  - Statistical estimation *(Maximum likelihood, ...)*
  - Discrimination function *(linear aggregation, ...)*
  - Neural network *(supervised learning, clustering, ...)*
  - Fuzzy Logic *(Fuzzy c-Mean Clustering, ...)*
- Hybrid *(Genetic algorithms, Bayesian network, ...)*

Table 3.

Reservoir Data Fusion

- **Rockphysical**
  - transform seismic data to attributes and reservoir properties
  - formulate seismic/log/core data to reservoir properties
- **Geostatistical**
  - transform seismic attributes to formation characters
  - transform seismic attributes to reservoir properties
  - simulate the 2D/3D distribution of seismic and log attributes
- **Intelligent**
  - clustering anomalies in seismic/log data and attributes
  - ANN layers for seismic attribute and formation characters
  - supervised training model to predict unknown from existing
  - hybrid such as GA and SA for complicated reservoirs

Table 4.

Geostatistical vs. Intelligent

- Geostatistical
  - Data assumption: a certain probability distribution
  - Model: weight functions come from variogram trend, stratigraphic facies, and probability constraints
  - Simulation: Stochastic, not optimized
- Intelligent
  - Data automatic clustering and expert-guided segmentation
  - Classification of relationship between data and targets
  - Model: weight functions come from supervised training based on geological and stratigraphic information
  - Simulation: optimized by GA, SA, ANN, and BN

2.2 Intelligent Interpretation and Data Analysis

Once all the pertinent data is properly integrated (fused) one has to extract the relevant information from the data and draw the necessary conclusions. This can be done either true reliance on human expert or an intelligent system that has the capability to learn and modify its knowledge base as new information become available. For detailed review of various applications of soft computing in intelligent interpretation, data analysis and pattern recognition see Aminzadeh (1989), Aminzadeh (1991) and Aminzadeh and Jamshidi (1995).

Although seismic signal processing has advanced tremendously over the last four decades, the fundamental assumption of a "convolution model" is violated in many practical settings. Sven Treitel, in Aminzadeh (1995) was quoted to pose the question: *What if, mother earth refuses to convolve?* Among such situations are: highly heterogeneous environments, very absorptive media (such as unconsolidated sand and young sediments), fractured reservoirs, and mud volcano, karst and gas chimneys. In such cases we must consider non-linear processing and interpretation methods. Neural networks fractals, fuzzy logic, genetic algorithms, chaos and complexity theory are among such non-linear processing and analysis techniques that have been proven to be effective. The highly heterogeneous earth model that geophysics attempts to quantify is an ideal place for applying these concepts. The subsurface lives in a hyper-dimensional space (the properties can be considered as the additional space dimension), but its actual response to external stimuli initiates an internal coarse-grain and self-organization that results in a low-dimensional structured behavior. Fuzzy logic and other non-linear methods can describe shapes and structures generated by chaos. These techniques will push the boundaries of seismic resolution, allowing smaller-scale anomalies to be characterized.

2.3 Pattern Recognition

In the 1960s and 1970s, pattern recognition techniques were used only by statisticians and were based on statistical theories. Due to recent advances in computer systems and technology, artificial neural networks and fuzzy logic models have been used in many pattern recognition applications ranging from simple character recognition, interpolation, and extrapolation between specific patterns to the most sophisticated robotic applications. To recognize a pattern, one can use the standard multi-layer perceptron with a back-propagation learning algorithm or simpler models such as self-organizing networks (Kohonen, 1997) or fuzzy c-means techniques (Bezdek, 1981; Jang and Gulley, 1995). Self-organizing networks and fuzzy c-means techniques can easily learn to recognize the topology, patterns, or seismic objects and their distribution in a specific set of information. Much of the early applications of pattern recognition in the oil industry were highlighted at Aminzadeh, 1989.

2.4 Clustering

Cluster analysis encompasses a number of different classification algorithms that can be used to organize observed data into meaningful structures. For example, k-means is an algorithm to assign a specific number of centers, k, to represent the clustering of N points (k<N). These points are iteratively adjusted so that each point is assigned to one cluster, and the centroid of each cluster is the mean of its assigned points. In general, the k-means technique will produce exactly k different clusters of the greatest possible distinction.

Alternatively, fuzzy techniques can be used as a method for clustering. Fuzzy clustering partitions a data set into fuzzy clusters such that each data point can belong to multiple clusters. Fuzzy c-means (FCM) is a well-known fuzzy clustering technique that generalizes the classical (hard) c-means algorithm and can be used where it is unclear how many clusters there should be for a given set of data.

Subtractive clustering is a fast, one-pass algorithm for estimating the number of clusters and the cluster centers in a set of data. The cluster estimates obtained from subtractive clustering can be used to initialize iterative optimization-based clustering methods and model identification methods.

In addition, the self-organizing map technique known as Kohonen's self-organizing feature map (Kohonen, 1997) can be used as an alternative for clustering purposes. This technique converts patterns of arbitrary dimensionality (the pattern space) into the response of one- or two-dimensional arrays of neurons (the feature space). This unsupervised learning model can discover any relationship of interest such as patterns, features, correlations, or regularities in the input data, and translate the discovered relationship into outputs. The first application of clustering techniques to combine different seismic attributes was introduced in the mid eighties (Aminzadeh and Chatterjee, 1984)

2.5 Data integration and Reservoir property estimation

Historically, the link between reservoir properties and seismic and log data have been established either through "statistics-based" or "physics-based" approaches. The latter, also known as model based approaches attempt to exploit the changes in seismic character or seismic attribute to a given reservoir property, based on physical phenomena. Here, the key issues are sensitivity and uniqueness. Statistics based methods attempt to establish a heuristic relationship between seismic measurements and prediction values from examination of data only. It can be argued that a hybrid method, combining the strength of statistics and physics based method would be most effective. **Figure 3**, taken from Aminzadeh, 1999 shows the concepts schematically.

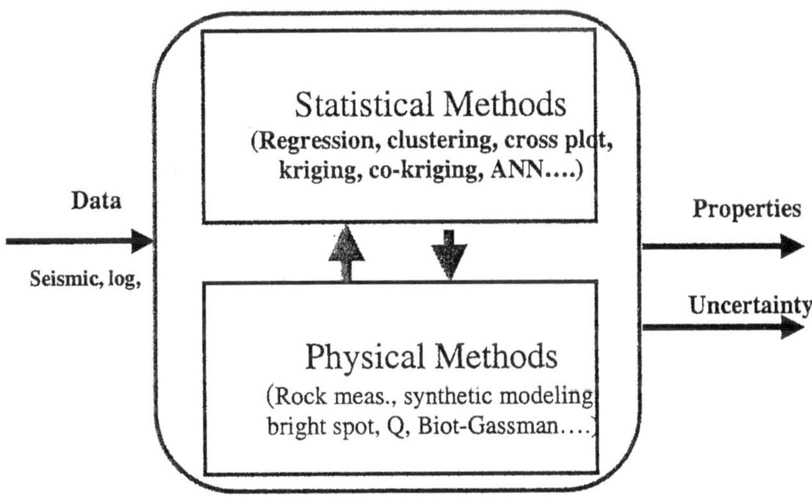

Figure 3. A schematic description of physics-based (blue), statistics-based (red) and hybrid method (green)

Many geophysical analysis methods and consequently seismic attributes are based on physical phenomena. That is, based on certain theoretical physics (wave propagation, Biot-Gassman Equation, Zoeppritz Equation, tuning thickness, shear wave splitting, etc.) certain attributes may be more sensitive to changes in certain reservoir properties. In the absence of a theory, using experimental physics (for example rock property measurements in a laboratory environment such as the one described in the last section of this paper) and/or numerical modeling, one can identify or validate suspected relationships. Although physics-based methods and direct measurements (the ground truth) is the ideal and reliable way to establish such correlations, for various reasons it is not always practical. Those reasons range from lack of known theories, difference between the laboratory environment and field environment (noise, scale, etc.) and the cost for conducting elaborate physical experiments.

Statistics-based methods aim at deriving an explicit or implicit heuristic relationship between measured values and properties to be predicted. Neural-networks and fuzzy-neural networks-based methods are ideally suitable to establish such implicit relationships through proper training. We all attempt to establish a relationship between different seismic attributes, petrophysical measurements, laboratory measurements and different reservoir properties. In such statistics-based method one has keep in mind the impact of noise on the data, data population used for statistical analysis, scale, geologic environment, scale and the correlation between different attributes when performing clustering or regressions. The statistics-based conclusions have to be reexamined and their physical significance explored.

2.6 Quantification of data uncertainty and prediction error and confidence interval

One of the main problems we face is to handle non-uniqueness issue and quantify uncertainty and confidence intervals in our analysis. We also need to understand the incremental improvements in prediction error and confidence range from introduction of new data or a new analysis scheme. Methods such as evidential reasoning and fuzzy logic are most suited for this purpose. **Figure 4** shows the distinction between conventional probability and theses techniques. "Point probability," describes the probability of an event, for example, having a commercial reservoir. The implication is we know exactly what this probability is. Evidential reasoning, provides an upper bound (plausibility) and lower bound (credibility) for the event the difference between the two bounds is considered as the ignorance range. Our objective is to reduce this range through use of all the new information. Given the fact that in real life we may have non-rigid boundaries for the upper and lower bounds and we ramp up or ramp down our confidence for an event at some point, we introduce fuzzy logic to handle and we refer to it as "membership grade". Next-generation earth modeling will incorporate quantitative representations of geological processes and stratigraphic / structural variability. Uncertainty will be quantified and built into the models.

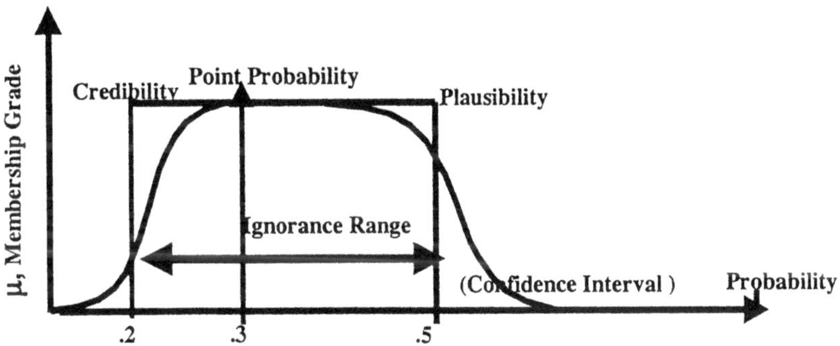

Figure 4. Point probability, evidential reasoning and fuzzy logic

On the issue of non-uniqueness, the more sensitive the particular seismic character to a given change to reservoir property, the easier to predict it. The more unique influence of the change in seismic character to changes in a specific reservoir property, the higher the confidence level in such predictions. Fuzzy logic can handle subtle changes in the impact of different reservoir properties on the wavelet response. Moreover comparison of multitude of wavelet responses (for example near, mid and far offset wavelets) is easier through use of neural networks. As discussed in Aminzadeh and de Groot, 2001, let us assume a seismic pattern for three different lithologies (sand, shaly sand and shale) are compared from different well information and seismic response (both model and field data) and the respective seismic character within the time window or the reservoir interval with four "classes" of wavelets, (w1, w2, w3.and w4). These 4 wavelets (basis wavelets) serve as a segmentation vehicle. The histograms in **Figure 5a** show what classes of wavelets that are likely to be present for given lithologies. In the extreme positive (EP) case we would have one wavelet uniquely representing one lithology. In the extreme negative case (EN) we would have a uniform distribution of all wavelets for all lithologies. In most cases unfortunately we are closer to NP than to EP. The question is how best we can get these distributions move from the EN side to EP side thus improving our prediction capability and increasing confidence level. The common sense is to add enhance information content of the input data.

How about if we use wavelet vectors comprised of pre-stack data (in the simple case, mid, near far offset data) as the input to a neural network to perform the classification? Intuitively, this should lead to a better separation of different lithologies (or other reservoir properties). Likewise, including three component data as the input to the classification process would further improve the confidence level. Naturally, this requires introduction of a new "metric" measuring "the similarity" of these "wavelet vectors". This can be done using the new basis wavelet vectors as input to a neural network applying different weights to mid, near and far offset traces. This is demonstrated conceptually, in **Figure 5** to predict lithology. Compare the sharper histograms of the vector wavelet classification (in this case, mid, near, and far offset gathers) in **Figure 5b**, against those of **Figure 5a** based on scalar wavelet classification.

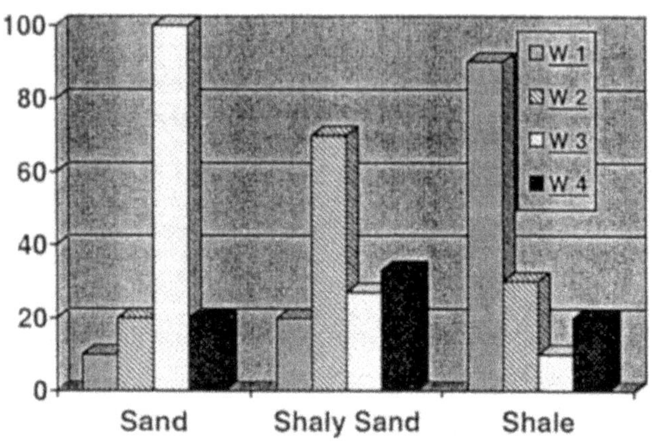

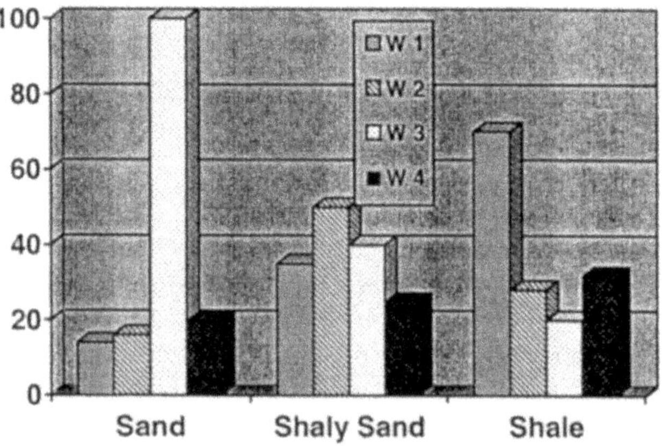

Figure 5- Statistical distribution of different wavelet types versus lithologies, A, -pre-stack data. B, stacked data

3 Artificial Neural Network and Geoscience Applications of ANN for Exploration

Although Artificial neural networks (ANN) were introduced in the late fifties (Rosenblatt, 1962), the interests in them have been increasingly growing in recent years. This has been in part due to new applications fields in the academia and industry. Also, advances in computer technology (both hardware and software) have made it possible to develop ANN capable of tackling practically meaningful problems with a reasonable response time. Simply put, neural networks are computer models that attempt to simulate specific functions of human nervous system. This is accomplished through some parallel structures comprised of non-linear processing nodes that are connected by fixed (Lippmann, 1987), variable (Barhen et al., 1989) or fuzzy (Gupta and Ding, 1994) weights. These weights establish a relationship between the inputs and output of each "Neuron" in the ANN. Usually ANN have several "hidden" layers each layer comprised of several neurons. If the feed-forward (FF) network (FF or concurrent networks are those with unidirectional data flow). The full technical details can be found in Bishop (Bishop, 1995). If the FF network is trained by back propagation (BP) algorithms, they are called BP. Other types of ANN are supervised (self organizing) and auto (hetro) associative networks.

Neurocomputing represents general computation with the use of artificial neural networks. An artificial neural network is a computer model that attempts to mimic simple biological learning processes and simulate specific functions of human nervous system. It is an adaptive, parallel information processing system which is able to develop associations, transformations or mappings between objects or data. It is also the most popular intelligent technique for pattern recognition to date.

The major applications of neurocomputing are seismic data processing and interpretation, well logging and reservoir mapping and engineering.

Good quality seismic data is essential for realistic delineation of reservoir structures. Seismic data quality depends largely on the efficiency of data processing. The processing step is time consuming and complex. The major applications include first arrival picking, noise elimination, structural mapping, horizon picking and event tracking. A detailed review can be found in Nikravesh and Aminzadeh (1999).

For interwell characterization, neural networks have been used to derive reservoir properties by crosswell seismic data. In Chawathé et al. (1997), the authors used a neural network to relate five seismic attributes (amplitude, reflection strength, phase, frequency and quadrature) to gamma ray (GR) logs obtained at two wells in the Sulimar Queen field (Chaves County). Then the GR response was predicted between the wells and was subsequently converted to porosity based on a field-specific porosity-GR transform. The results provided good delineation of various lithofacies.

Feature extraction from 3D seismic attributes is an extremely important area. Most statistical methods are failed due to the inherent complexity and nonlinear information content. **Figure 3** shows an example use of neural networks for segmenting seismic characters thus deducing information on the seismic facies and reservoir properties (lithology, porosity, fluid saturation and sand thickness). A display of the level of confidence (degree of match) between the seismic character at a given point versus the representative wavelets (centers of clusters) is also shown. Combining this information with the seismic model derived from the well logs while perturbing for different properties gives physical meaning of different clusters.

Monson and Pita (1997) applied neural networks to find relationships between 3D seismic attributes and well logs. The study provided realistic prediction of log responses far away from the wellbore. Boadu (1997) also used similar technology to relate seismic attributes to rock properties for sandstones. In Nikravesh et al.[7], the author applied a combination of k-means clustering, neural networks and *fuzzy c-means (1984)* (a clustering algorithm in which each data vector belongs to each of the clusters to a degree specified by a membership grade) techniques to characterize a field that produces from the Ellenburger Dolomite. The techniques were used to perform clustering of 3D seismic attributes and to establish relationships between the clusters and the production log. The production log was established away from wellbore. The production log and the clusters were then superimposed at each point of a 3D seismic cube. They also identified the optimum locations for new wells based on the connectivity, size and shape of the clusters related to the pay zones.

The use of neural networks in well logging has been popular for nearly one decade. Many successful applications have been documented (Wong et al., 1998; Brus et al., 1999; Wong et al., 2000). The most recent work by Bruce et al. (2000) presented a state-of-the-art review of the use of neural networks for predicting permeability from well logs. In this application, the network is used as a nonlinear regression tool to develop transformation between well logs and core permeability. Such a transformation can be used for estimating permeability in uncored intervals and wells. In this work, the permeability profile was predicted by a Bayesian neural network. The network was trained by a training set with four well logs (GR, NPHI, RHOB and RT) and core permeability. The network also provided a measure of confidence (the standard deviation of a Gaussian function): the higher the standard deviation ("sigma"), the lower the prediction reliability. This is very useful for understanding the risk of data extrapolation. The same tool can be applied to estimate porosity and fluid saturations. Another important application is the clustering of well logs for the recognition of lithofacies (Rogers et al., 1992). This provides useful information for improved petrophysical estimates and well correlation.

Neurocomputing has also been applied to reservoir mapping. In Wong et al.(1997) and Wang et al.(1998, 1999, 1999), the authors applied a radial basis function neural network to relate the conceptual distribution of geological facies (in the form of hand drawings) to reservoir porosity. It is able to incorporate the general property trend provided by local geological knowledge and to simulate fine-scaled details when used in conjunction with geostatistical simulation techniques. In Caers (1999) and Caers and Journel (1998), the authors trained a neural network to recognize the local conditional probability based on multiple-point information retrieved from a "training image," which can be any densely populated image (e.g. outcrop data, photographs, hand drawings, seismics, etc.). The conditional probability was used in stochastic simulation with a Markov Chain sampler (e.g. Markov Chain Monte Carlo). These methodologies can be applied to produce 3D model of petrophysical properties using multiple seismic attributes and conceptual geological maps. This is a significant advantage compared to the conventional geostatistical methods which are limited to two-point statistics (e.g. variograms) and simple objects (e.g. channels).

In Nikravesh et al.(1996), the authors used neural networks to predict field performance and optimize oil recovery by water injection in the Lost Hill Diatomite (Kern County). They constructed several neural networks to model individual well behavior (wellhead pressure and injection-production history) based on data obtained from the surrounding wells. The trained networks were used to predict future fluid production. The results matched well with the actual data. The study also led to the best oil recovery with the minimum water injected.

In what follows we will review the geoscience applications in these broad areas: Data Processing and Prediction. We will not address other geoscience applications such as: classification of multi- source remote sensing data Bennediktson et. al., (1990), earthquake prediction, Aminzadeh et. al. (1994), and ground water remediation, Johnson and Rogers (1995).

3.1 Data Processing

Various types of geoscience data are used in the oil industry to ultimately locate the most prospective locations for oil and gas reservoirs. These data sets go through extensive amount of processing and manipulation before they are analyzed and interpreted. The processing step is very time consuming yet a very important one. ANN have been utilized to help improve the efficiency of operation in this step. Under this application area we will examine: First seismic arrival picking, and noise elimination problems. Also, see Aminzadeh (1991) and McCormack (1991) and Zadeh Aminzadeh (1995) and Aminzadeh et al (1999) for other related applications.

3.1.1 First Arrival Picking

Seismic data are the response of the earth to any disturbance (compressional waves or shear waves). The seismic source can be generated either artificially (petroleum seismology, PS) or, naturally, (earthquake seismology, ES). The recorded seismic data are then processed and analyzed to make an assessment of the subsurface (both the geological structures and rock properties) in PS and the nature of the source (location or epicenter and magnitude, for example, in Richter scale) in ES. Conventional PS relies heavily on compressional (P- wave) data while ES is essentially based on the shear (S- wave) data.

The first arrivals of P and S waves on a seismic record contain useful information both in PS and ES. However one should make sure that the arrival is truly associated with a seismically generated event not a noise generated due to various factors. Since we usually deal with thousands of seismic records, their visual inspection for distinguishing FSA from noise, even if reliable, could be quite time consuming.

One of the first geoscience applications of ANN has been to streamline the operation of identifying the FSA in an efficient and reliable manner. Among the recent publications in this area are: McCormack (1990) and Veezhinathan et al. (1991). Key elements of the latter (V91) are outlined below:

Here, the FSA picking is treated as a pattern recognition problem. Each event is classified either as an FSA or non-FSA. A segment of the data within a window is used to obtain four "Hilbert" attributes of the seismic signal. The Hilbert attributes of seismic data were introduced by Taner et al. (1979). In V91, these attributes are derived from seismic signal using a sliding time window. Those attributes are: 1) Maximum Amplitude, 2) Mean Power Level; MPL, 3) Power Ratios, and 4) Envelop Slop Peak. These types of attributes have been used by Aminzadeh and Chatterjee (1984) for predicting gas sands using clustering and discernment analysis technique.

In V91, the network processes three adjacent peaks at a time to decide whether the center peak is an FSA or a non-FSA. A BPN with five hidden layers combined with a post processing scheme accomplished correct picks of 97%. Adding a fifth attribute, Distance from Travel Time Curve, generated satisfactory results without the need for the post processing step.

McCormack (1990) created a binary image from the data and used it to train the network to move up and down across the seismic record to identify the FSA. This image-based approach captures space-time information in the data but requires a large number of input units, thus necessitating a large network. Some empirical schemes are used to ensure its stability.

3.1.2 Noise Elimination

A related problem to FSA is editing noise from the seismic record. The objective here is to identify events with non-seismic origin (the reverse of FSA) and then remove them from the original data in order to increase the signal to noise ratio. Liu et al. (1989), McCormack (1990) and Zhang and Li (1994) are Some of the publications in this area.

Zhang and Li (1994) handled the simpler problem, to edit out the whole noisy trace from the record. They initiate the network in the "learning" phase by "scanning " over the whole data set. The weights are adapted in the learning phase either with some human input as the distinguishing factors between "good" and "bad" traces or during an unsupervised learning phase. Then in the "recognizing" phase the data are scanned again and depending upon whether the output of the network is less than or greater than a threshold level the trace is either left alone or edited out as a bad trace.

3.2 Identification and Prediction

Another major application area for ANN in the oil industry is to predict various reservoir properties. This ultimately is used a decision tool for exploration and development drilling and redevelopment or extension of the existing fields. The input data to this prediction problem is usually processed and interpreted seismic and log data and/or a set of attributes derived from the original data set. Historically, many "hydrocarbon indicators" have been proposed to make such predictions. Among them are: the bright spot analysis Sherif and Geldart (1982), amplitude versus offset analysis, Ostrander (1982), seismic clustering analysis, Aminzadeh and Chatterjee (1984),fuzzy pattern recognition, Griffiths (1987) and other analytical methods, Agterberg and Griffiths (1991). Many of the ANN developed for this purpose are built around the earlier techniques either for establishing a relationship between the raw data and physical properties of the reservoirs and/or to train the network using the previously established relationships.

Huang and Williamson (1994) have developed a general regression neural network, GRNN to predict rock's total organic carbon (TOC) using well log data. First, they model the relationship between the resistivity log and TOC with a GRNN, using published data. After training the ANN in two different modes, the GRNN found optimum values of sigma. Sigma is an important smoothing parameter used in GRNN. They have established the superiority of GRNN over BP-ANN in determining the architecture of the network. After completing the training phase a predictive equation for determining TOC was derived. Gogan et al (1995) used

ANN for determining lithology and fluid saturation from well log and pre-stack seismic data. Various seismic attributes from partial stacks (mid, near and far offsets) as an input to ANN. The network was calibrated using synthetic (theoretical) data with pre stack seismic response of known lithologies and saturation from the well log data. The output of the network was a set of classes of lithologies and saturations.

3.3 Neural Network and Nonlinear Mapping

In this section, a series of neural network models will be developed for nonlinear mapping between wireline logs. A series of neural network models will also be developed to analyze actual welllog data and seismic information and the nonlinear mapping between wireline logs and seismic attributes will be recognized.

In this study, wireline logs such as travel time (DT), gamma ray (GR), and density (RHOB) will be predicted based on SP, and resistivity (RILD) logs. In addition, we will predict travel time (DT) based on induction resistivity and vice versa. In this study, all logs are scaled uniformly between -1 and 1 and results are given in scaled domain.

Figures 6A through **6E** show typical behavior of SP, RILD, DT, GR, and RHOB logs in scaled domain. The design of a neural network to predict DT, GR, and RHOB based on RILD and SP logs starts with filtering, smoothing, and interpolating values (in a small horizon) for missing information in the data set. A first-order filter and a simple linear recursive parameter estimator for interpolating were used to filter and reconstruct the noisy data. The available data were divided into three data sets: training, testing, and validation. The network was trained based on the training data set and continuously tested using a test data set during the training phase. The network was trained using a backpropagation algorithm and modified Levenberge-Marquardt optimization technique. Training was stopped when prediction deteriorated with step.

3.3.1 Travel time (DT) prediction based on SP and resistivity (RILD) logs.

The neural network model to predict the DT has 14 input nodes (two windows of data each with 7 data points) representing SP (7 points or input nodes) and RILD (7 points or input nodes) logs. The hidden layer has 5 nodes. The output layer has 3 nodes representing the prediction of the DT (a window of 3 data point). Typical performance of the neural network for training, testing, and validation data sets is

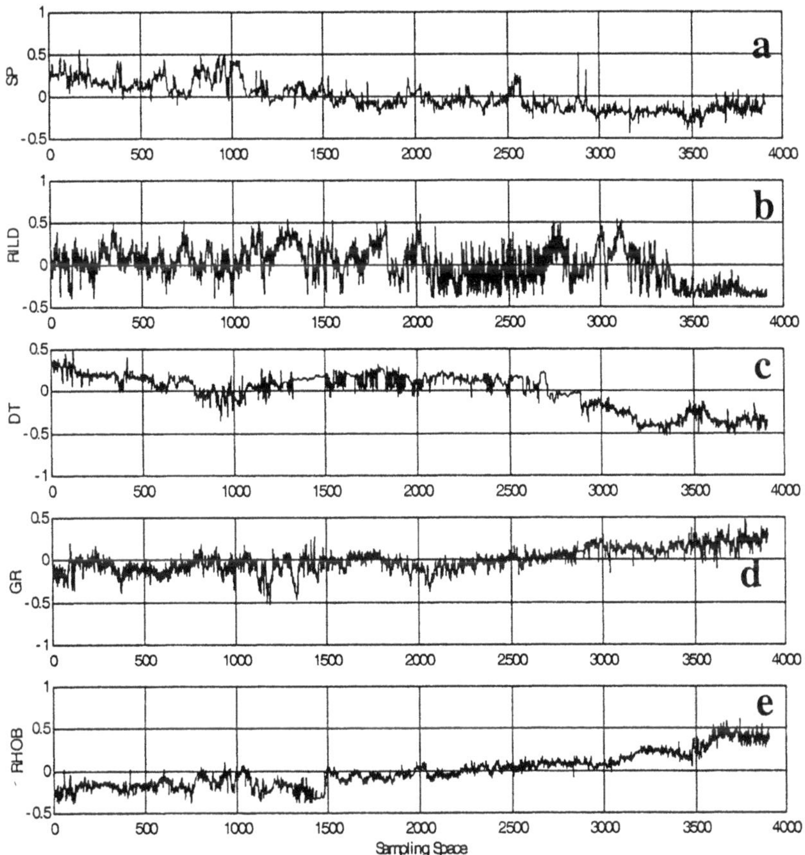

Figure 6. Typical behavior of SP, RILD, DT, GR, and RHOB logs.

shown in **Figures 7A, 7B, 7C**, and **7D.** The network shows good performance for prediction of DT for training, testing, and validation data sets. However, there is not a perfect match between actual and predicted values for DT is the testing and validation data sets. This is due to changes of the lithology from point to point. In other words, some of the data points in the testing and validation data sets are in a lithological layer which was not presented in the training phase. Therefore, to have perfect mapping, it would be necessary to use the layering information (using other types of logs or linguistic information) as input into the network or use a lar

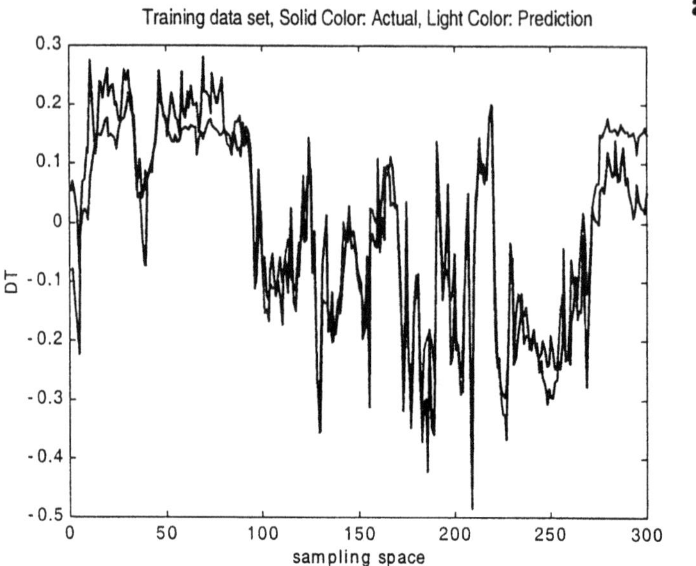
a
Training data set, Solid Color: Actual, Light Color: Prediction
DT
sampling space
0.3
0.2
0.1
0
-0.1
-0.2
-0.3
-0.4
-0.5
0
50
100
150
200
250
300

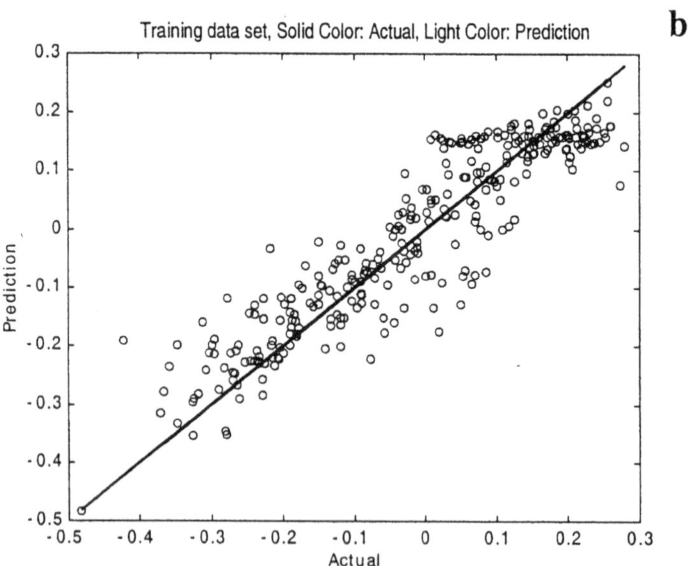
b
Training data set, Solid Color: Actual, Light Color: Prediction
Prediction
Actual
0.3
0.2
0.1
0
-0.1
-0.2
-0.3
-0.4
-0.5
-0.5
-0.4
-0.3
-0.2
-0.1
0
0.1
0.2
0.3

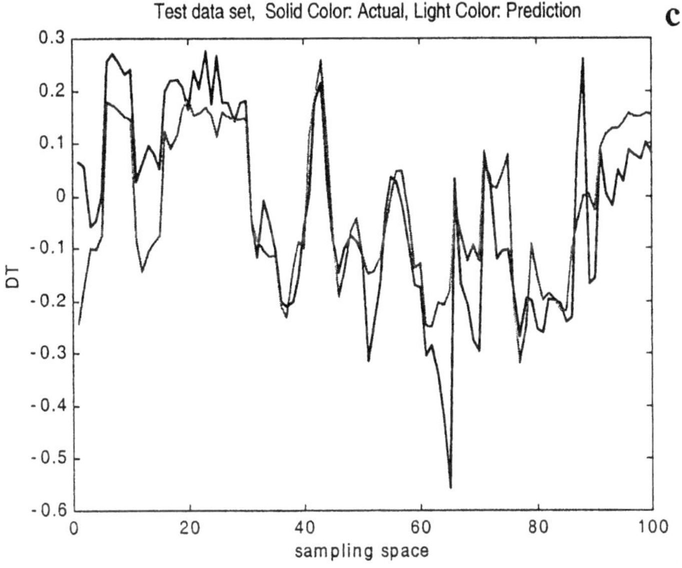

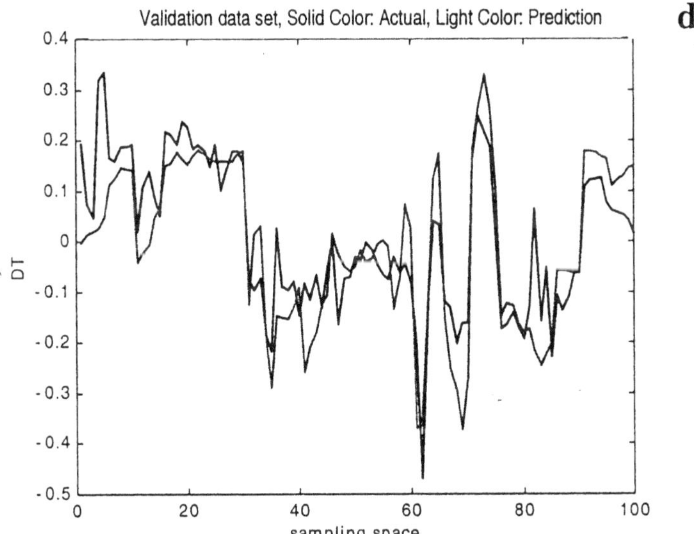

Figure 7. Typical neural network performance for prediction of DT based on SP and RILD.

ger data set for the training data set which represent all the possible behaviors in the data.

3.3.2 Gamma ray (GR) prediction based on SP and resistivity (RILD) logs.

In this study, a neural network model is developed to predict GR based on SP and RILD log. The network has 14 input nodes (two windows of data, each with 7 data points) representing SP (7 points or input nodes) and RILD (7 points or input nodes) logs. The hidden layer has 5 nodes. The output layer has 3 nodes representing the prediction of the GR. **Figures 8A** through **8D** show the performance of the neural network for training, testing, and validation data. The neural network model shows a good performance for prediction of GR for training, testing, and validation data sets. In comparison with previous studies (DT prediction), this study shows that the GR is not as sensitive as DT to noise in the data. In addition, a better global relationship exits between SP-resisitivity-GR rather than SP-resisitivity-DT. However, the local relationship is in the same order of complexity. Therefore, two models have the same performance for training (excellent performance). However, the model for prediction of GR has a better generalization property. Since the two models have been trained based on the same criteria, it is unlikely that this lack of mapping for generalization is due to over fitting during the training phase.

3.3.3 Density (RHOB) prediction based on SP and resistivity (RILD) logs.

To predict density based on SP and RILD logs, a neural network model with 14 input nodes representing SP (7 points or input nodes) and RILD (7 points or input nodes) logs, 5 nodes in the hidden layer, and 3 nodes in the output layer representing the prediction of the RHOB is developed. **Figures 9A** through **9D** show a typical performance of the neural network for the training, testing, and validation data sets. The network shows excellent performance for the training data set as shown in **Figures 9A** and **9B**. The model shows a good performance for the testing data set as shown in **Figure 9C**. **Figure 9D** shows the performance of the neural network for the validation data set. The model has relatively good performance for the validation data set. Therefore, there is not a perfect match between the actual and predicted values for RHOB for the testing and validation data set. Since RHOB is directly related to lithology and layering, and to have perfect mapping, it would be necessary to use the layering information (using other types of logs or linguistic information) as an input into the network or use a larger data set for the training

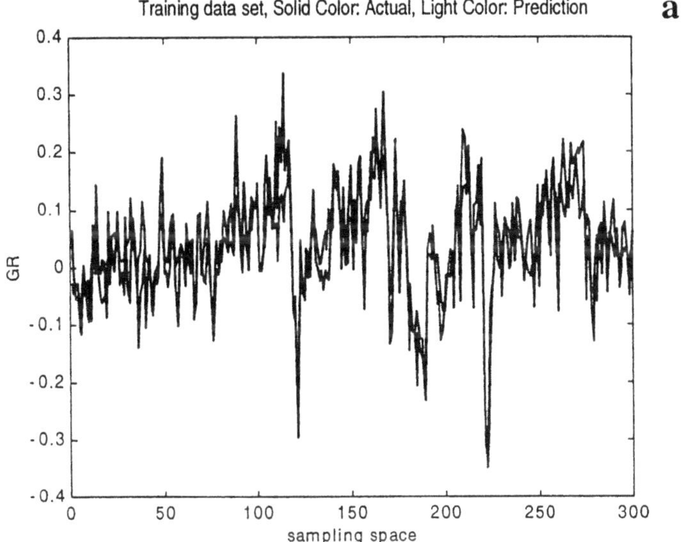

Training data set, Solid Color: Actual, Light Color: Prediction
a
0.4
0.3
0.2
0.1
0
-0.1
-0.2
-0.3
-0.4
GR
0
50
100
150
200
250
300
sampling space

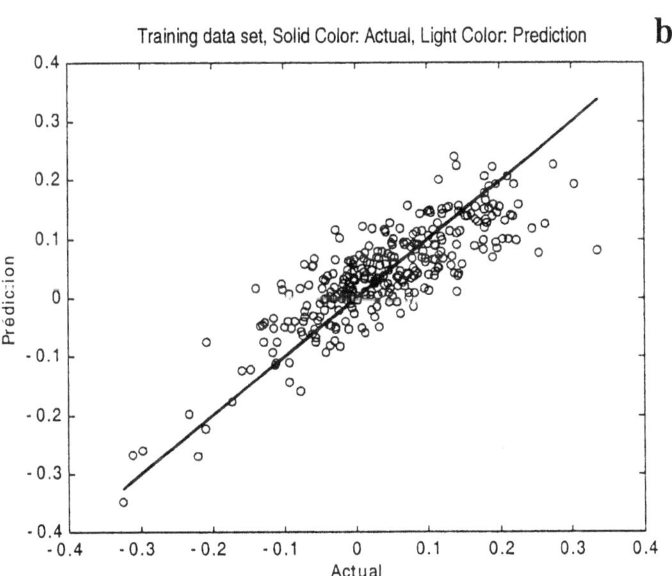

Training data set, Solid Color: Actual, Light Color: Prediction
b
0.4
0.3
0.2
0.1
0
-0.1
-0.2
-0.3
-0.4
Prediction
-0.4
-0.3
-0.2
-0.1
0
0.1
0.2
0.3
0.4
Actual

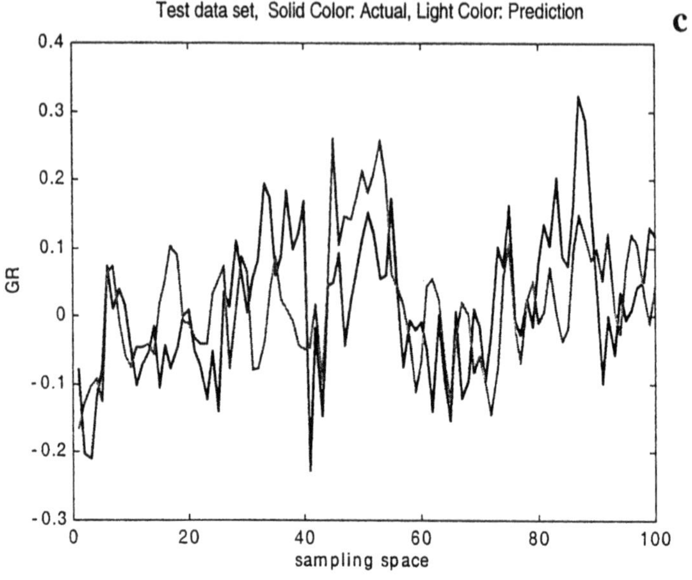

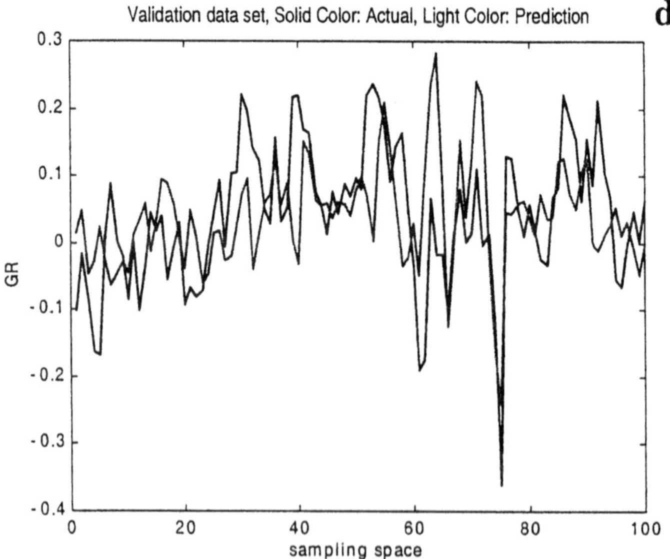

Figure 8. Typical neural network performance for prediction of GR based on SP and RILD.

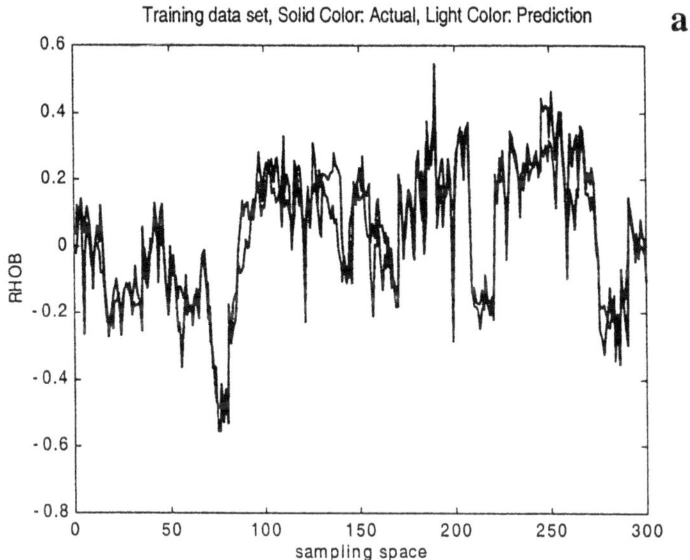
Training data set, Solid Color: Actual, Light Color: Prediction
a
0.6
0.4
0.2
0
-0.2
-0.4
-0.6
-0.8
RHOB
0
50
100
150
200
250
300
sampling space

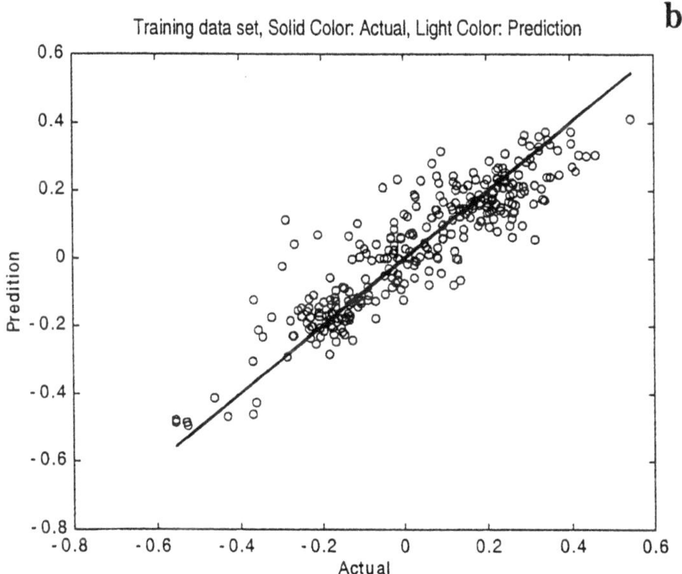
Training data set, Solid Color: Actual, Light Color: Prediction
b
0.6
0.4
0.2
0
-0.2
-0.4
-0.6
-0.8
Predition
-0.8
-0.6
-0.4
-0.2
0
0.2
0.4
0.6
Actual

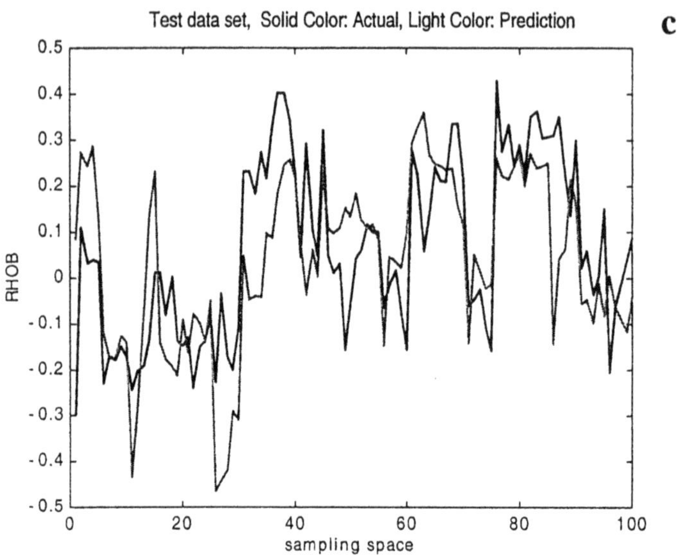

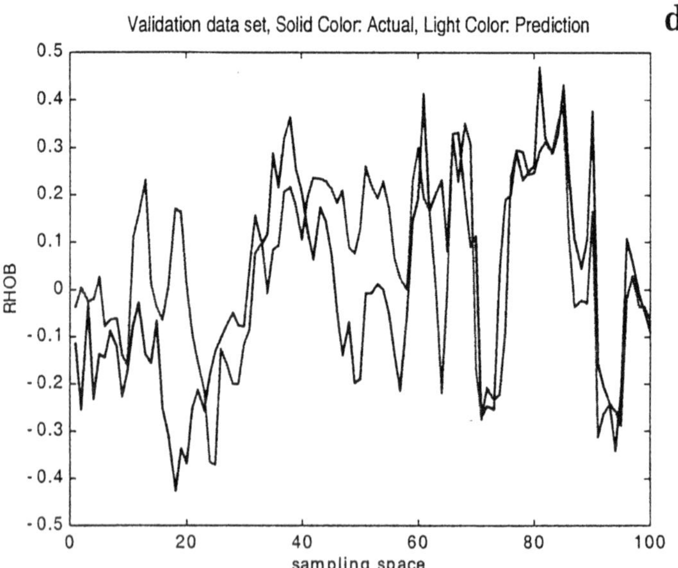

Figure 9. Typical neural network performance for prediction of RHOB based on SP and RILD.

data set which represent all the possible behaviors in the data. In these cases, one can use a knowledge-based approach using knowledge of an expert and select more diverse information which represent all different possible layering as a training data set. Alternatively, one can use an automated clustering technique to recognize the important clusters existing in the data and use this information for selecting the training data set.

3.3.4 Travel time (DT) prediction based on resistivity (RILD).

The neural network model to predict the DT has 11 input nodes representing a RILD log. The hidden layer has 7 nodes. The output layer has 3 nodes representing the prediction of the DT. Using engineering knowledge, a training data set is carefully selected so as to represent all the possible layering existing in the data. The typical performance of neural network for the training, testing, and validation data sets is shown in **Figures 10A** through **10D**. As expected, the network has excellent performance for prediction of DT. Even though only RILD logs are used for prediction of DT, the network model has better performance than when SP and RILD logs used for prediction of DT (comparing **Figures 7A** through **7D** with **5A** through **5D**). However, in this study, knowledge of an expert was used as extra information. This knowledge not only reduced the complexity of the model, but also better prediction was achieved.

3.3.5 Resistivity (RILD) prediction based on travel time (DT)

In this section, to show that the technique presented in the previous section is effective, the performance of the inverse model is tested. The network model has 11 input nodes representing DT, 7 nodes in the hidden layer, and 3 nodes in the output layer representing the prediction of the RILD. **Figures 11A** through **11D** show the performance of the neural network model for the training, testing, and validation data sets. **Figures 11A** and **11B** show that the neural network has excellent performance for the training data set. **Figures 6C** and **6D** show the performance of the network for the testing and validation data set. The network shows relatively excellent performance for testing and validation purposes. As was mentioned in the previous section, using engineering knowledge the complexity of the model was reduced and better performance was achieved.

In addition, since the network model (prediction of DT from resisitivity) and its inverse (prediction of resisitivity based on DT) have relatively excellent

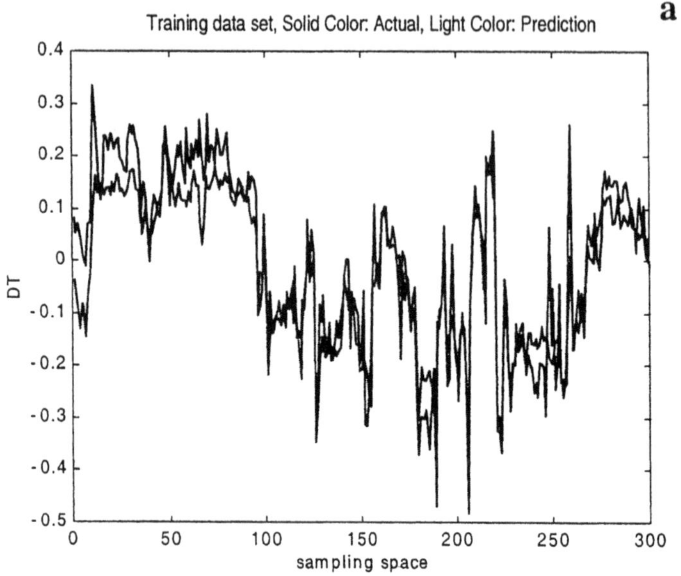
a
Training data set, Solid Color: Actual, Light Color: Prediction
DT
sampling space
0.4
0.3
0.2
0.1
0
-0.1
-0.2
-0.3
-0.4
-0.5
0
50
100
150
200
250
300

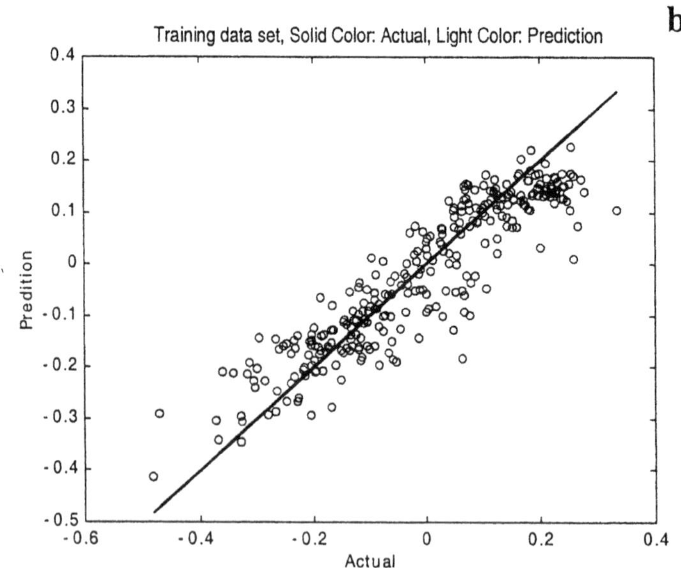
b
Training data set, Solid Color: Actual, Light Color: Prediction
Predition
Actual
0.4
0.3
0.2
0.1
0
-0.1
-0.2
-0.3
-0.4
-0.5
-0.6
-0.4
-0.2
0
0.2
0.4

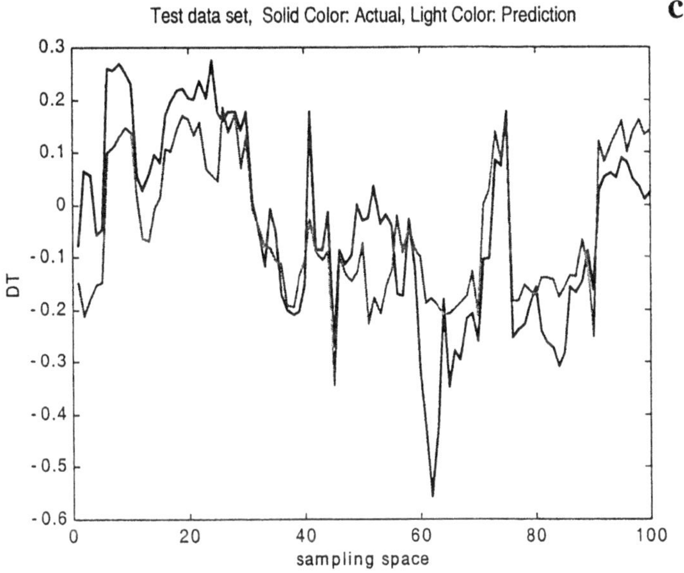

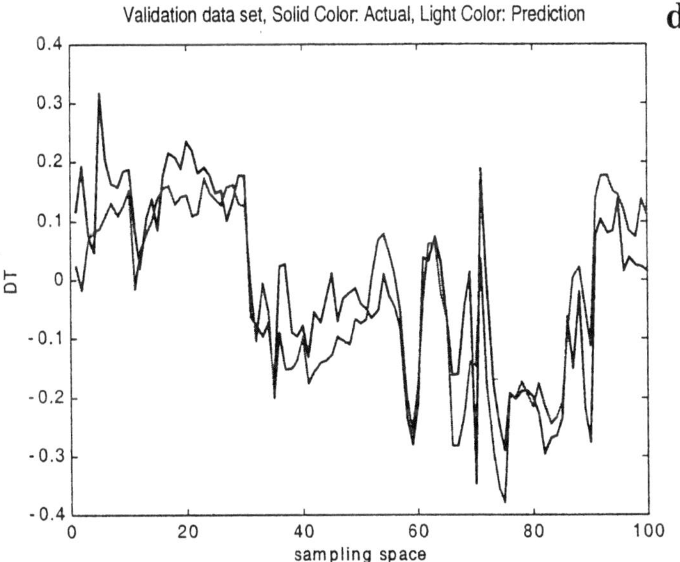

Figure 10. Typical neural network performance for prediction of DT based on RILD.

a

Training data set, Solid Color: Actual, Light Color: Prediction
RESISTIVITY
sampling space
0.6
0.4
0.2
0
-0.2
-0.4
-0.6
-0.8
0
50
100
150
200
250
300

b

Training data set, Solid Color: Actual, Light Color: Prediction
Prediction
Actual
0.6
0.4
0.2
0
-0.2
-0.4
-0.6
-0.8
-0.8
-0.6
-0.4
-0.2
0
0.2
0.4
0.6

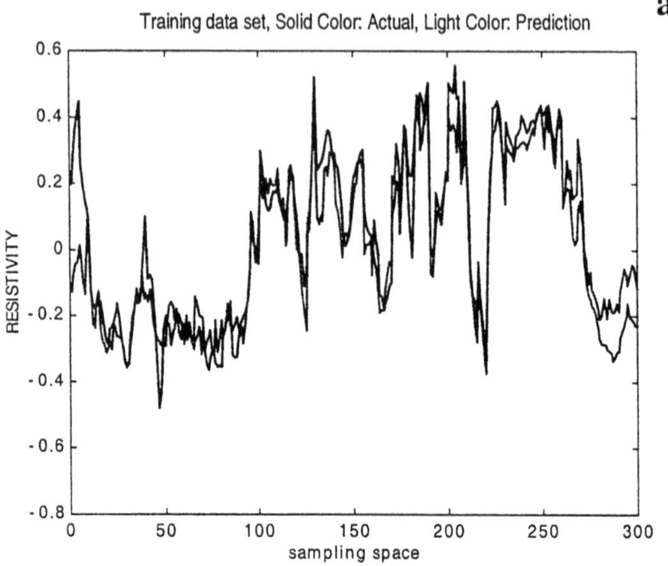

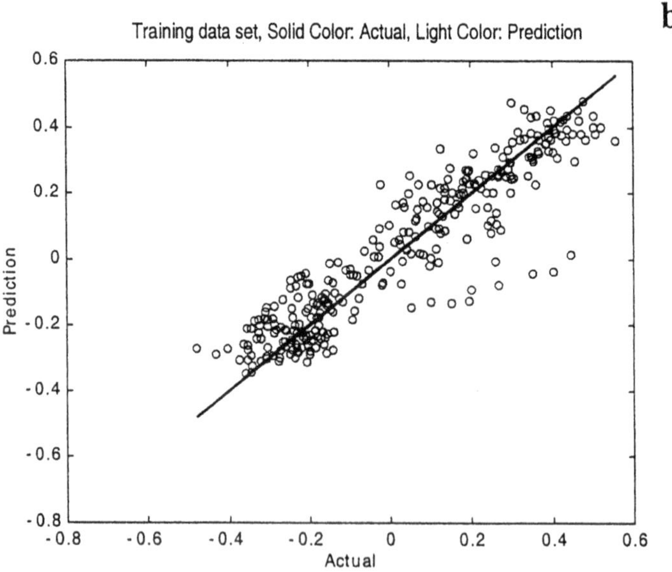

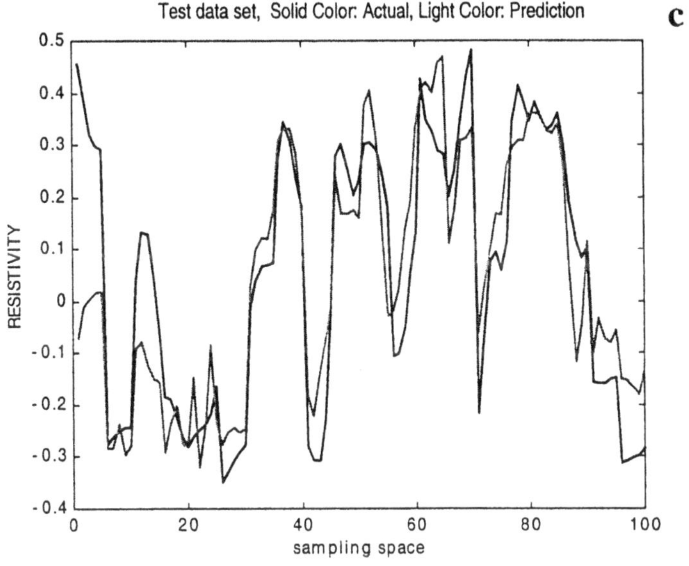

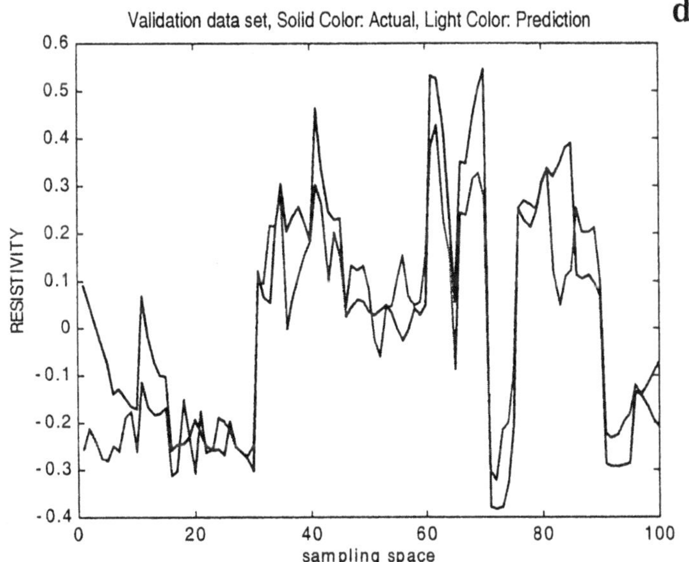

Figure 11. Typical neural network performance for prediction of RILD based on DT.

performance and generalization properties, a one-to-one mapping was achieved. Therefore, this implies that a good representation of layering was selected based on knowledge of an expert.

4 Fuzzy Logic

In recent years, it has been shown that uncertainty may be due to fuzziness rather than chance. Fuzzy logic is considered to be appropriate to deal with the nature of uncertainty in system and human error, which are not included in current reliability theories. The basic theory of fuzzy sets was first introduced by Zadeh (1965). Unlike classical logic which is based on crisp sets of "true and false", fuzzy logic views problems as a degree of "truth", or "fuzzy sets of true and false" (Zadeh 1965). Despite the meaning of the word "fuzzy", fuzzy set theory is not one that permits vagueness. It is a methodology that was developed to obtain an approximate solution where the problems are subject to vague description. In addition, it can help engineers and researchers to tackle uncertainty, and to handle imprecise information in a complex situation. During the past several years, the successful application of fuzzy logic for solving complex problems subject to uncertainty has greatly increased and today fuzzy logic plays an important role in various engineering disciplines. In recent years, considerable attention has been devoted to the use of hybrid neural network-fuzzy logic approaches as an alternative for pattern recognition, clustering, and statistical and mathematical modeling. It has been shown that neural network models can be use to construct internal models that capture the presence of fuzzy rules. However, determination of the input structure and number of membership functions for the inputs has been one of the most important issues of fuzzy modeling.

Fuzzy logic provides a completely new way of modeling complex and ill-defined systems. The major concept of fuzzy logic is the use of a *linguistic variable*, that is a variable whose values are words or sentences in a natural or synthetic language. This also leads to the use of *fuzzy if-then rules*, in which the antecedent and consequents are propositions containing linguistic variables.

In recent years, fuzzy logic, or more generally, fuzzy set theory, has been applied extensively in many reservoir characterization studies. This is mainly due to the fact that reservoir geology is mainly a descriptive science which uses mostly uncertain, imprecise, ambiguous and linguistic information (Bois 1984). Fuzzy set theory has the ability to deal with such information and to combine them with the quantitative observations. The applications are many, including seismic and stratigraphic modeling and formation evaluation.

In Bois (1984), he proposed to use fuzzy set theory as a pattern recognition tool to interpret a seismic section. The algorithm produced a synthetic seismic section by convoluting a geological model with a representative impulse (by deconvolution or signature of the source), which were both of subjective and fuzzy nature. The synthetic section was then compared with the original seismic section in a fuzzy context. In essence, the algorithm searches for the appropriate geological model from the observed seismic section by an iterative procedure, which is a popular way for solving inverse problems.

In Baygun et al.(1985), the authors used fuzzy logic as a classifier for the delineation of geological objects in a mature hydrocarbon reservoir with many wells. The authors showed that fuzzy logic can be used to extract dimensions and orientation of geological bodies, and geologists can use such a technique for reservoir characterization in a practical way which bypassed many tedious steps.

In Nordlund (1996), the author presented a study on dynamic stratigraphic modeling using fuzzy logic. In stratigraphic modeling, it is possible to model several processes simultaneously in space and time. Although many processes can be modeled using conventional mathematics, modeling the change of deposition and erosion on surfaces is often difficult. Formalizing geological knowledge is a difficult exercise as it involves handling of several independent and complex parameters. In addition, most information is qualitative and imprecise, which are unacceptable for direct numerical treatment. In the paper, the author showed a successful use of fuzzy rules to model erosion and deposition. The fuzzified variables included the depth of the reservoirs, the distances to source and shore, a sinusoidal sea-level curve, tectonic subsidence rate and simulation time with depositional surface. From the study, the author demonstrated that a few (10) fuzzy rules could produce stratigraphies with realistic geometry and facies.

In Cuddy (1997), the author applied fuzzy logic to solve a number of petrophysical problems in several North Sea fields. The work included lithofacies and permeability from well logs. Lithofacies prediction was based on the use of a possibility value (Gaussian function with a specific mean and variance) to represent a well log belonging to a certain lithofacies. The lithofacies that was associated with the highest combined fuzzy possibility (multiplication of all values) was taken as the most likely lithofacies for that set of logs. A similar methodology was applied to predict permeability by dividing the core permeability values into ten equal bin sizes on a log scale. The problem was converted into a classification problem. All the results suggested that the fuzzy approach had given better petrophysical estimates compared to the conventional techniques.

Fang and Chen (1997) also applied fuzzy rules to predict porosity and permeability from five compositional and textural characteristics of sandstone in the Yacheng Field (South China Sea). The five input attributes were the relative amounts of rigid grains, ductile grains and detrital matrix, grain size and the Trask sorting coefficient. All the porosity and permeability data were firstly divided into certain number of clusters by fuzzy c-means. The corresponding sandstone characteristics for each cluster were used to general the fuzzy linguistic rules. Each fuzzy cluster produced one fuzzy if-then rule with five input statements. The formulated rules were then used to make linguistic prediction by combining individual conclusion from each rule. If a numerical output was desired, a *defuzzification algorithm (1997)* could be used to extract a crisp output from a fuzzy set. The results showed that the fuzzy modeling gave better results compared to those presented in Bloch (1991).

In Huang et al.(1999), the authors presented a simple but practical fuzzy interpolator for predicting permeability from well logs in the North West Shelf (offshore Australia). The basic idea was to simulate local fuzzy reasoning. When a new input vector (well logs) was given, the system would select two training vectors which were nearest to the new input vector and build a set of piece-wise linear inference rules with the training values, in which the membership value of the training values was one. The study used well log and core data from two wells and the performance was tested at a third well, where actual core data were available for comparison. The accuracy of the permeability predictions at the test well was although similar to that obtained from the authors' previous neural-fuzzy technique, the fuzzy approach was >7,000 times faster in terms of CPU time.

In Nikravesh and Aminzadeh (2000), the authors applied a neural-fuzzy approach to develop an optimum set of rules for nonlinear mapping between porosity, grain size, clay content, P-wave velocity, P-wave attenuation and permeability. The rules developed from a training set were used to predict permeability in another data set. The prediction performance was very good. The study also showed that the integrated technique discovered clear relationships between P-wave velocity and porosity, and P-wave attenuation and clay content, which were useful to geophysicists.

4.1 Geoscience Applications of Fuzzy Logic

The uncertain, fuzzy, and linguistic nature of geophysical and geological data makes it a good candidate for interpretation through fuzzy set theory. The main advantage of this technique is in combining the quantitative data and qualitative information and subjective observation. The imprecise nature of the information available for interpretation (such as seismic data, wirelin logs, geological and lithological data) makes fuzzy sets theory an appropriate tool to utilize. For example, Chappaz (1977) and Bois (1983, 1984) proposed to use fuzzy sets theory in the interpretation of seismic sections. Bois used fuzzy logic as pattern recognition tool for seismic interpretation and reservoir analysis. He concluded that fuzzy set theory, in particular, can be used for interpretation of seismic data which are imprecise, uncertain, and include human error. He maintained these type of error and fuzziness cannot be taken into consideration by conventional mathematics. However, they are perfectly seized by fuzzy set theory. He also concluded that using fuzzy set theory one can determine the geological information using seismic data. Therefore, one can predict the boundary of reservoir in which hydrocarbon exists. B. Baygun et. al (1985) used fuzzy logic as classifier for delineation of geological objects in a mature hydrocarbon reservoir with many wells. B. Baygun et. al have shown that fuzzy logic can be used to extract dimensions and orientation of geological bodies and the geologist can use such a technique for reservoir characterization in a very quick way through bypassing several tedious steps. H. C. Chen et al. in their study used fuzzy set theory as fuzzy regression analysis for extraction of parameter for Archie equation. Bezdek et. al (1981) also reported a series of the applications of fuzzy sets theory in geostatical analysis. Tamhane et al (2002,) show how to integrate linguistic descriptions in petroleum reservoirs using fuzzy logic.

Many of our geophysical analysis techniques such as migration, DMO, wave equation modeling as well as the potential methods (gravity, magnetic, electrical methods) use conventional partial differential wave equations with deterministic coefficients. The same is true for the partial differential equations used in reservoir simulation. For many practical and physical reasons deterministic parameters for the coefficients of these PDEs leads unrealistic (for example medium velocities for seismic wave propagation or fluid flow for Darcy equation). Stochastic parameters in theses cases can provide us with a more practical characterization. Fuzzy coefficients for PDEs can prove to be even more realistic and easy to parameterize. Today's deterministic processing and interpretation ideas will give way to stochastic methods, even if the industry has to rewrite the book on geophysics. That is, using wave equations with random and fuzzy coefficients to describe subsurface velocities and densities in statistical and membership grade terms, thereby enabling a better description of wave propagation in the subsurface–particularly when a substantial amount of heterogeneity is present. More generalized applications of geostatistical techniques will emerge, making it possible to introduce risk and uncertainty at the early stages of the seismic data processing and interpretation loop.

5 Neuro-Fuzzy Techniques

In recent years, considerable attention has been devoted to the use of hybrid neural-network/fuzzy-logic approaches as an alternative for pattern recognition, clustering, and statistical and mathematical modeling. It has been shown that neural network models can be used to construct internal models that recognize fuzzy rules.

Neuro-fuzzy modeling is a technique for describing the behavior of a system using fuzzy inference rules within a neural network structure. The model has a unique feature in that it can express linguistically the characteristics of a complex nonlinear system. As a part of any future research opportunities, we will use the neuro-fuzzy model originally presented by Sugeno and Yasukawa (1993). The neuro-fuzzy model is characterized by a set of rules. The rules are expressed as follows:

R^i : if x_1 is A_1^i and x_2 is A_2^i ... and x_n is A_n^i (Antecedent) [1]

then $y^* = f_i(x_1, x_2, \dots, x_n)$ (Consequent)

where $f_i(x_1, x_2,\dots, x_n)$ can be constant, linear, or a fuzzy set.

For the linear case: $f_i(x_1, x_2,\dots, x_n) = a_{i0} + a_{i1}\, x_1 + a_{i2}\, x_2 + \dots + a_{in}\, x_n$ [2]

Therefore, the predicted value for output y is given by:

$$y = \Sigma_i\, \mu_i\, f_i(x_1, x_2,\dots, x_n) \,/\, \Sigma \mu_i \quad [3]$$

with

$$\mu_i = \Pi_j\, A_j^i\,(x_j) \quad [4]$$

where R_i is the ith rule, x_j are input variables, y is output, A_j^i are fuzzy membership functions (fuzzy variables), a_{ij} constant values.

As a part of any future research opportunities, we will use the Adaptive Neuro-Fuzzy Inference System (ANFIS) technique originally presented by Jang (1992). The model uses neuro-adaptive learning techniques, which are similar to those of neural networks. Given an input/output data set, the ANFIS can construct a fuzzy inference system (FIS) whose membership function parameters are adjusted using the back-propagation algorithm or other similar optimization techniques. This allows fuzzy systems to learn from the data they are modeling.

5.1 Neural-fuzzy model for rule extraction

In this section, a neuro-fuzzy model will be developed for model identification and knowledge extraction (rule extraction) purposes. The model is characterized by a set of rules which can be further used for representation of data in the form of linguistic variables. Therefore, in this situation the fuzzy variables become linguistic variables. The neuro-fuzzy technique is used to implicitly cluster the data while finding the nonlinear mapping. The neuro-fuzzy model developed in this study is an approximate fuzzy model with triangular and Gaussian membership functions originally presented by Sugeno and Yasukawa (1993). K-Mean technique is used for clustering and the network is trained using a backpropagation algorithm and modified Levenberge-Marquardt optimization technique.

In this study, the effect of rock parameters and seismic attenuation on permeability will be analyzed based on soft computing techniques and experimental data. The software will use fuzzy logic techniques because the data and our requirements are imperfect. In addition, it will use neural network techniques, since the functional structure of the data is unknown. In particular, the software will be used to group data into important data sets; extract and classify dominant and interesting patterns that exist between these data sets; and discover secondary, tertiary and higher-order data patterns. The objective of this section is to predict the permeability based on grain size, clay content, porosity, P-wave velocity, and P-wave attenuation.

5.2 Prediction of permeability based on porosity, grain size, clay content, P-wave velocity, and P-wave attenuation.

In this section, a neural-fuzzy model will be developed for nonlinear mapping and rule extraction (knowledge extraction) between porosity, grain size, clay content, P-wave velocity, P-wave attenuation and permeability. **Figure 12** shows typical data, which has been used in this study. In this study, permeability will be predicted based on the following rules and equations. (1) through (4):

IF Rock Type= Sandstones
[5]
AND Porosity=[p1,p2]
AND Grain Size =[g1,g2]
AND Clay Content =[c1,c2]
AND P-Wave Vel.=[pwv1,pwv2]
AND P-Wave Att.=[pwa1,pwa2]
THEN Y*= a0+a1*P+a2*G+a3*C+a4*PWV+a5*PWA.

Where, P is %porosity, G is grain size, C is clay content, PWV is P-wave velocity, and P-wave attenuation, Y*, is equivalent to f in equation 1.

Data are scaled uniformly between -1 and 1 and the result is given in the scaled domain. The available data were divided into three data sets: training, testing, and validation. The neuro-fuzzy model was trained based on a training data set and continuously tested using a test data set during the training phase. Training

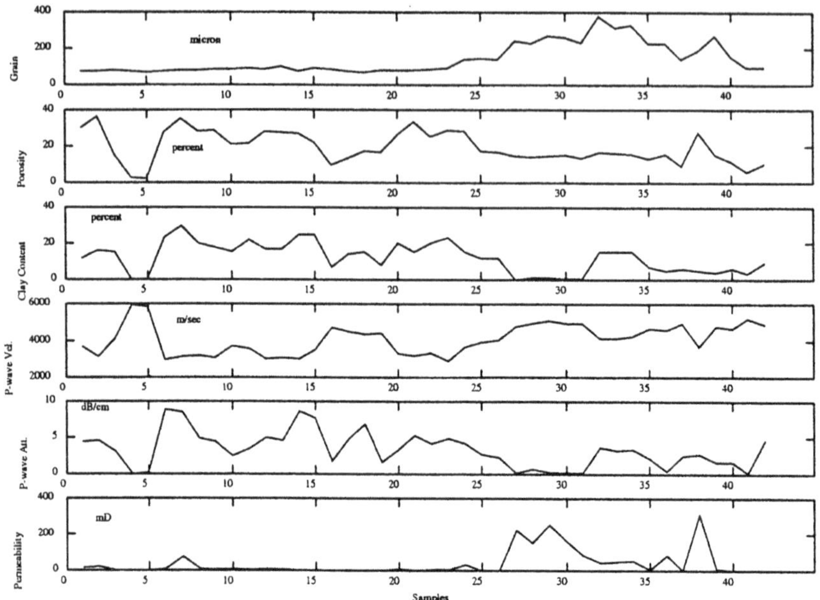

Figure 12. Actual data (Boadu 1997).

was stopped when it was found that the model's prediction suffered upon continued training. Next, the number of rules was increased by one and training was repeated. Using this technique, an optimal number of rules were selected. **Figures 13A** through **13E** and **Table 5** show typical rules extracted from the data. In **Table 1**, Column 1 through 5 show the membership functions for porosity, grain size, clay content, P-wave velocity, and P-wave attenuation respectively. Using the model defined by equations 1 through 4 and membership functions defined in **Figures 13A** through **13E** and **Table 5**, permeability was predicted as shown in **Figure 14A**. In this study, 7 rules were identified for prediction of permeability based on porosity, grain size, clay content, P-wave velocity, and P-wave attenuation (**Figures 13A** through **13E** and **Figure 14A**). In addition, 8 rules were identified for prediction of permeability based on porosity, clay content, P-wave velocity, and P-wave attenuation (**Figure 14B**). Ten rules were identified for prediction of permeability based on porosity, P-wave velocity, and P-wave attenuation (**Figure 14C**). Finally, 6 rules were identified for prediction of permeability based on grain size, clay content, P-wave velocity, and P-wave attenuation (**Figure 14D**). The neural network model shows very good performance for prediction of permeability. In this situation, not only a nonlinear mapping and relationship was identified between porosity, grain size, clay content, P-wave velocity, and P-wave attenuation, and permeability, but the rules existing between data were also

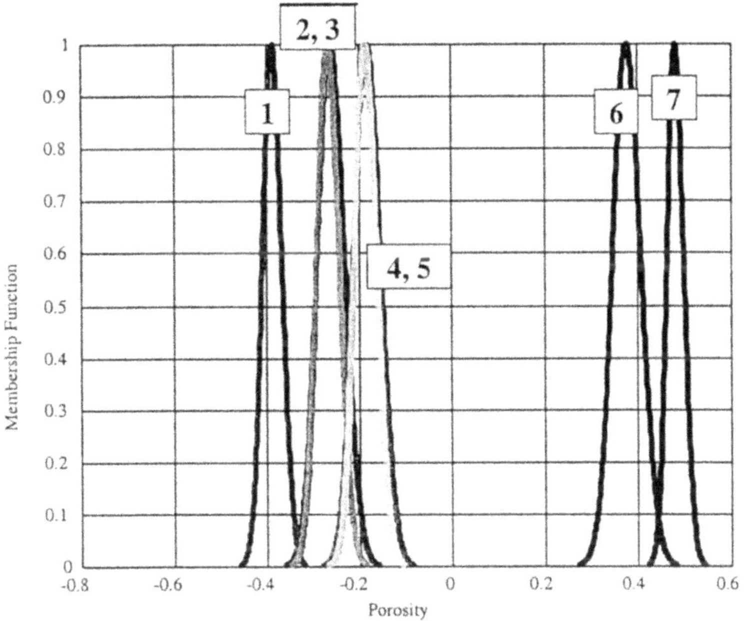

Figure 13A. Typical rules extracted from data, 7 Rules (Porosity).

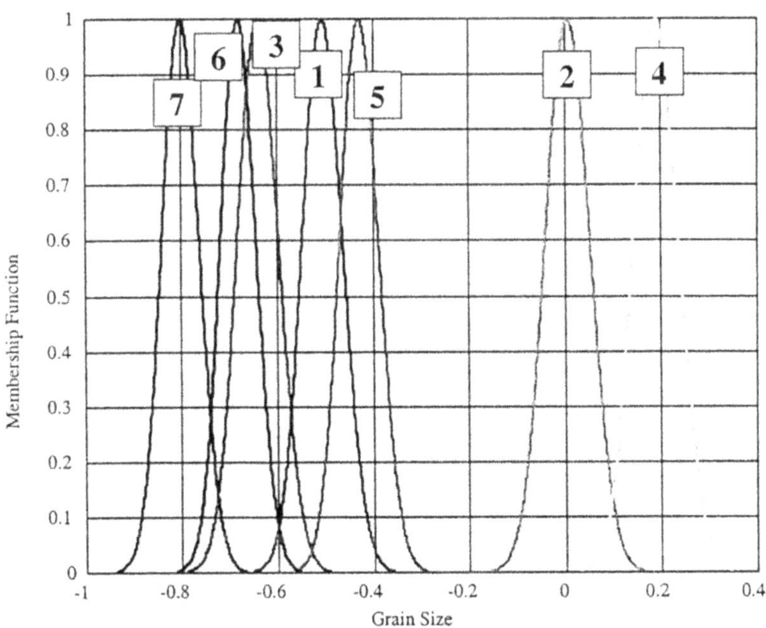

Figure 13B. Typical rules extracted from data, 7 Rules (Grain Size).

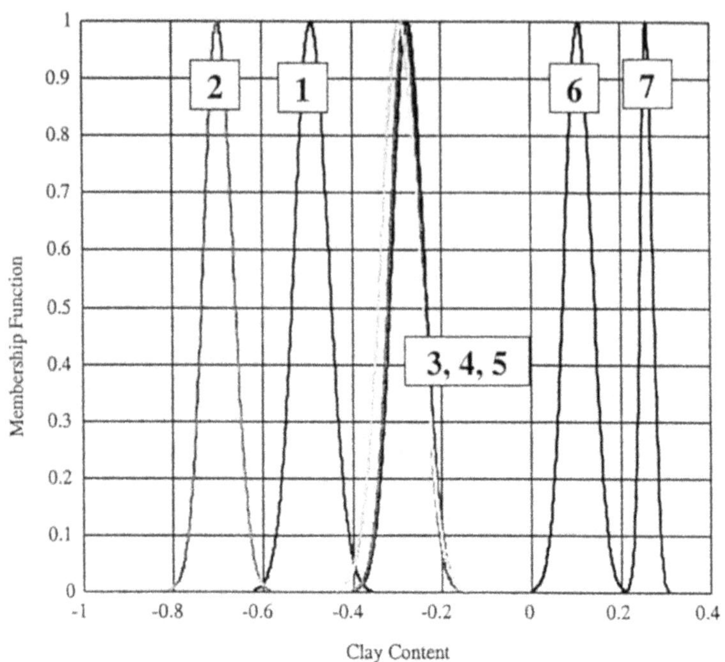

Figure 13C. Typical rules extracted from data, 7 Rules (Clay Content).

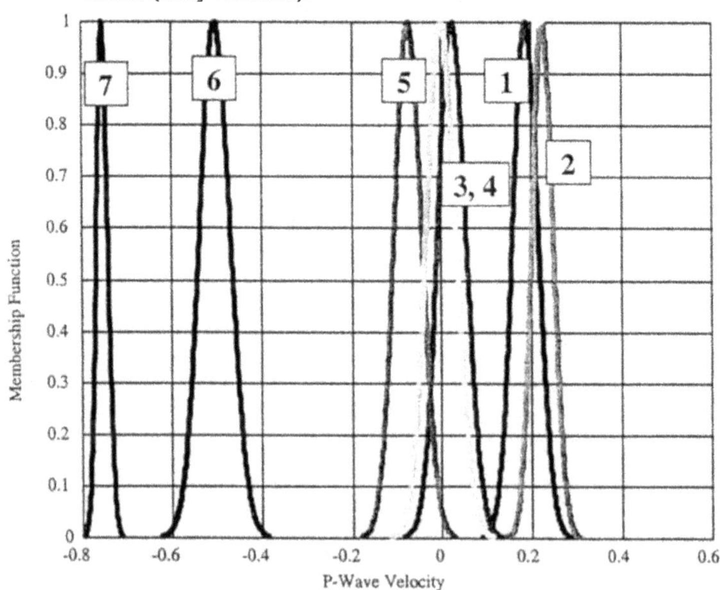

Figure 13D. Typical rules extracted from data, 7 Rules (P-Wave Velocity).

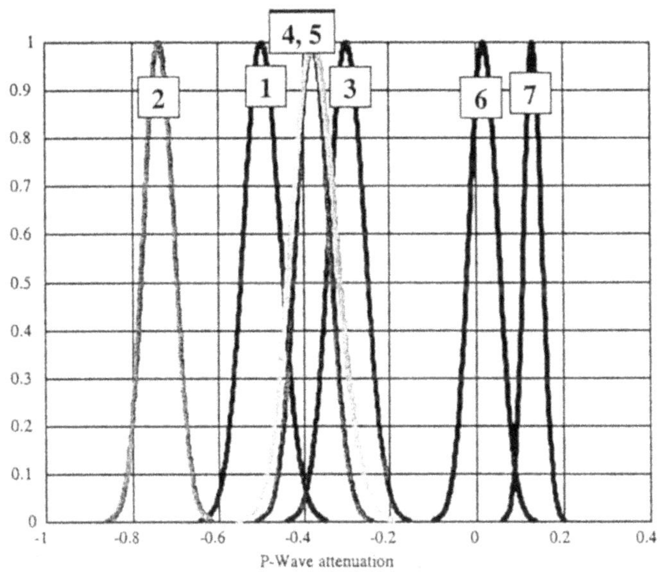

Figure 13E. Typical rules extracted from data, 7 Rules (P-Wave Attenuation).

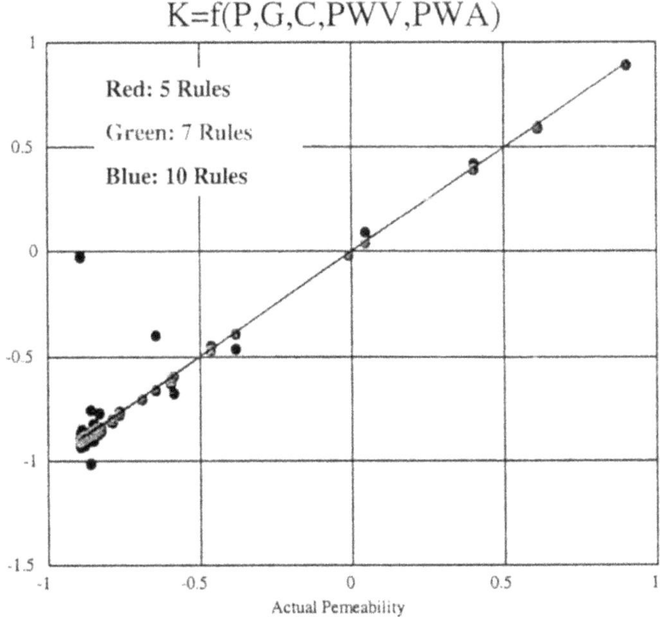

Figure 14A. Performance of Neural-Fuzzy model for prediction of permeability [K= f (P, G, C, PWV, PWA)].

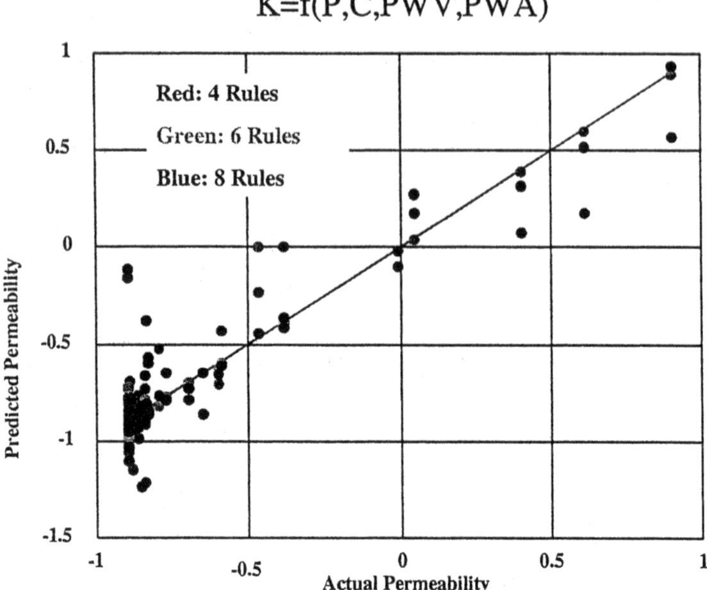

Figure 14B. Performance of Neural-Fuzzy model for prediction of permeability [K= f (P, C, PWV, PWA)].

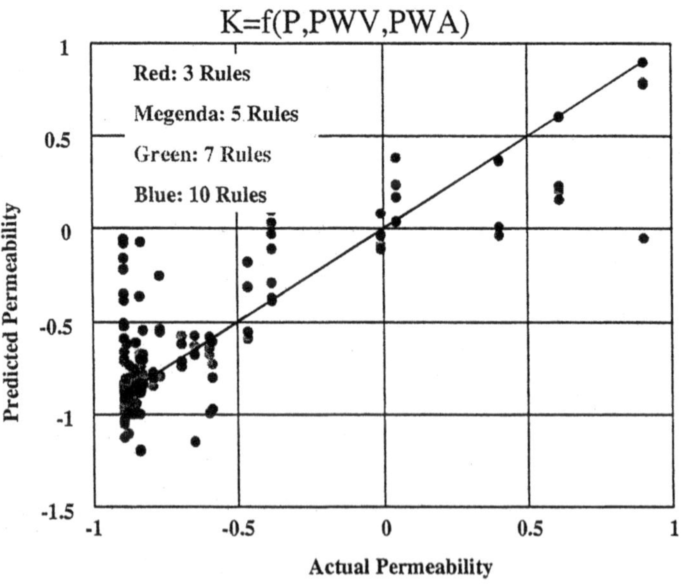

Figure 14C. Performance of Neural-Fuzzy model for prediction of permeability [K= f (P, PWV, PWA)].

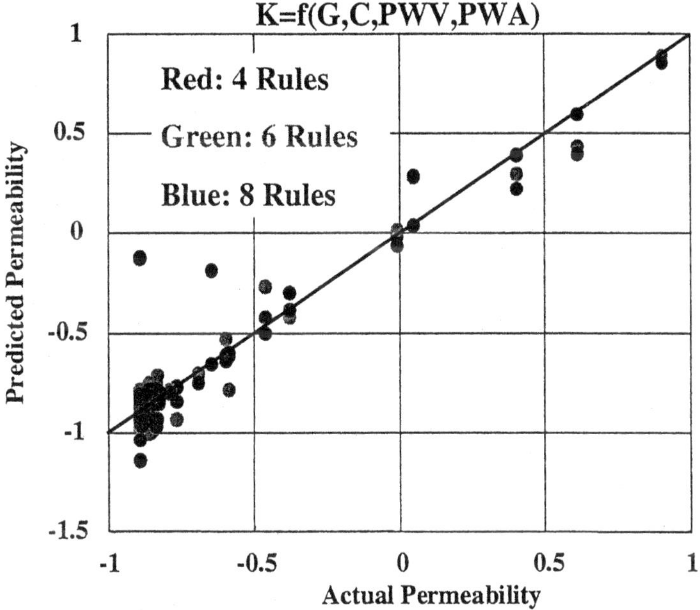

Figure 14D. Performance of Neural-Fuzzy model for prediction of permeability [K= f (G, C, PWV, PWA)].

Table 5. Boundary of rules extracted from data.

Porosity	Grain Size	Clay Content	P_Wave Velocity	P_Wave Attenuation
[-0.4585, -0.3170]	[-0.6501, -0.3604]	[-0.6198, -0.3605]	[0.0893, 0.2830]	[-0.6460 -0.3480]
[0.4208, 0.5415]	[-0.9351, -0.6673]	[0.2101, 0.3068]	[-0.7981, -0.7094]	[0.0572 0.2008]
[-0.3610, -0.1599]	[-0.7866, -0.4923]	[-0.3965, -0.1535]	[-0.0850, 0.1302]	[-0.4406 -0.1571]
[-0.2793, -0.0850]	[-0.5670, -0.2908]	[-0.4005, -0.1613]	[-0.1801, 0.0290]	[-0.5113 -0.2439]
[-0.3472, -0.1856]	[-0.1558, 0.1629]	[-0.8093, -0.5850]	[0.1447, 0.3037]	[-0.8610 -0.6173]
[0.2700, 0.4811]	[-0.8077, -0.5538]	[-0.0001, 0.2087]	[-0.6217, -0.3860]	[-0.1003 0.1316]
[-0.2657, -0.1061]	[0.0274, 0.3488]	[-0.4389, -0.1468]	[-0.1138, 0.1105]	[-0.5570 -0.1945]

identified. For this case study, our software clustered the parameters as grain size, P-wave velocity/porosity (as confirmed by **Figure 15** since a clear linear relationship exists between these two variables), and P-wave attenuation/clay content (as it is confirmed by **Figure 16** since an approximate linear relationship exists between these two variables). In addition, using the rules extracted, it was shown that P-wave velocity is closely related to porosity and P-wave attenuation is closely related to clay content. Boadu (1997) also indicated that the most influential rock parameter on the attenuation is the clay content. In addition our software ranked the variables in the order grain size, p-wave velocity, p-wave attenuation and clay content/porosity (since clay content and porosity can be predicted from p-wave velocity and p-wave attenuation).

6 Genetic Algorithms

Evolutionary computing represents computing with the use of some known mechanisms of evolution as key elements in algorithmic design and implementation. A variety of algorithms have been proposed. They all share a common conceptual base of simulating the evolution of individual structures via processes of parent selection, mutation, crossover and reproduction. The major one is the genetic algorithms (GAs) (Holland, 1975).

Genetic algorithm (GA) is one of the stochastic optimization methods which is simulating the process of natural evolution. GA follows the same principles as those in nature (survival of the fittest, Charles Darwin). GA first was presented by John Holland as an academic research. However, today GA turn out to be one of the most promising approaches for dealing with complex systems which at first nobody could imagine that from a relative modest technique. GA is applicable to multi-objectives optimization and can handle conflicts among objectives. Therefore, it is robust where multiple solution exist. In addition, it is highly efficient and it is easy to use.

Another important feature of GA is its ability to extract knowledge in terms of fuzzy rules. GA is now widely used and applied to discovery of fuzzy rules. However, when the data sets are very large, it is not easy to extract the rules. To overcome such a limitation, a new coding technique has been presented recently. The new coding method is based on biological DNA. The DNA coding method and the mechanism of development from artificial DNA are suitable for knowledge extraction from large data set. The DNA can have many redundant parts which is important for extraction of knowledge. In addition, this technique allows overlapped representation of genes and it has no constraint on crossover points. Also, the same type of mutation can be applied to every locus. In this technique, the length of chromosome is variable and it is easy to insert and/or delete any part

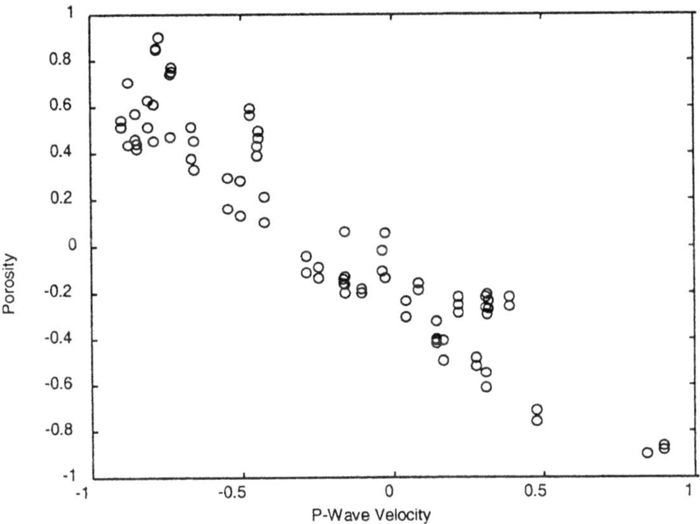

Figure 15. Relationship between P-Wave Velocity and Porosity.

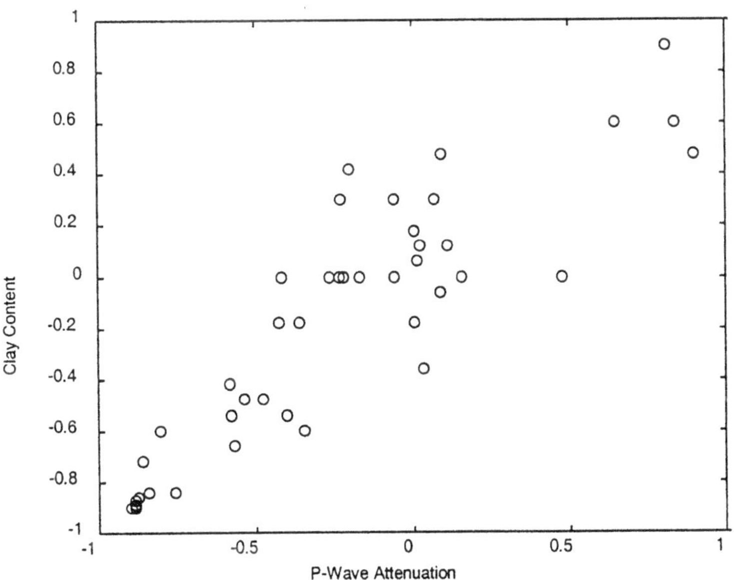

Figure 16. Relationship between P-Wave Attenuation and Clay Content.

of DNA. Today, genetic algorithm can be used in a hierarchical fuzzy model for pattern extraction and to reduce the complexity of the neuro-fuzzy models. In addition, GA can be use to extract the number of the membership functions required for each parameter and input variables, and for robust optimization along the multidimensional, highly nonlinear and non-convex search hyper-surfaces.

GAs work by firstly encoding the parameters of a given estimator as chromosomes (binary or floating-point). This is followed by populating a range of potential solutions. Each chromosome is evaluated by a fitness function. The better parent solutions are reproduced and the next generation of solutions (children) is generated by applying the genetic operators (crossover and mutation). The children solutions are evaluated and the whole cycle repeats until the best solution is obtained.

The methodology is in fact general and can be applied to optimizing parameters in other soft computing techniques, such as neural networks. In Yao (1999), the author gave an extensive review of the use of evolutionary computing in neural networks with more than 300 references. Three general areas are: evolution of connection weights; evolution of neural network architectures; and evolution of learning rules.

Most geoscience applications began in early 1990s. Gallagher and Sambridge (1994) presented an excellent overview on the use of GAs in seismology. Other applications include geochemical analysis, well logging and seismic interpretation.

Fang et al. first used GAs to predict porosity and permeability from compositional and textural information and the Archie parameters in petrophysics. The same authors later used the same method to map geochemical data into a rock's mineral composition (1996). The performance was much better than the results obtained from linear regression and nonlinear least-squares methods.

In Huang et al.(1998), the authors used GAs to optimize the connection weights in a neural network for permeability prediction from well logs. The study showed that the GA-trained networks (neural-genetic model) gave consistently smaller errors compared to the networks trained by the conventional gradient descent algorithm (backpropagation). However, GAs were comparatively slow in convergence. In Huang et al.(2000), the same authors initialized the connection weights in GAs using the weights trained by backpropagation. The technique was also integrated with fuzzy reasoning, which gave a hybrid system of neural-fuzzy-genetic (Huang, 1998). This improved the speed of convergence and still obtained better results.

Another important feature of GAs is its capability of extracting fuzzy rules. However, this becomes unpractical when the data sets are large in size. To overcome this, a new encoding technique has been presented recently, which is based on the understanding of biological DNA. Unlike the conventional chromosomes, the length of chromosome is variable and it is flexible to insert new parts and/or delete redundant parts. In Yashikawa et al. (1998) and Nikravesh et al.(1998), the authors used a hybrid system of neural-fuzzy-DNA model for knowledge extraction from seismic data, mapping the well logs into seismic data and reconstruction of porosity based on multi-attributes seismic mapping.

6.1 Geoscience Applications of Genetic Algorithms

Most of the applications of the GA in the area of petroleum reservoir or in the area of geoscience are limited to inversion techniques or used as optimization technique. While in other filed, GA is used as a powerful tool for extraction of knowledge, fuzzy rules, fuzzy membership, and in combination with neural network and fuzzy-logic. Recently, Nikravesh et. al, (?) proposed to use a neuro-fuzzy-genetic model for data mining and fusion in the area of geoscience and petroleum reservoirs. In addition, it has been proposed to use neuro-fuzzy-DNA model for extraction of knowledge from seismic data and mapping the wireline logs into seismic data and reconstruction of porosity (and permeability if reliable data exist for permeability) based on multi-attributes seismic mapping. Seismic inversion was accomplished using genetic algorithms by Mallick (1999). Potter et al (1999) used GA for stratigraphic analysis. For an overview of GA in exploration problems see McCormack et al (1999)

7 Principal Component Analysis and Wavelet

Some of the data Fusion and data mining methods used in exploration applications are as follows.

First we need to reduce the space to make the data size more manageable as well as reducing the time required for data processing. We can use Principal Component Analysis. Using the eigen value and vectors, we can reduce the space domain. We choose the eigenvector corresponding to the largest eigenvalues. Then in the eigenvector space we use Fuzzy K-Mean or Fuzzy C-Mean technique. For details of Fuzzy C-Means algorithm see Cannon et al (1986). Also, see Lashgari (1991), Aminzadedh (1989) and Aminzadeh (1994) for the application of Fuzzy Logic and Fuzzy K-Means algorithm in several earth exploration problems.

We can also use Wavelet and extract the patterns and Wavelets describing different geological settings and the respective rock properties. Using the Wavelet and neural network, we can fuse the data for nonlinear modeling. For clustering purposes, we can use the output from Wavelet and use Fuzzy C-Mean or Fuzzy

K-Mean. To use uncertainty and see the effect of the uncertainty, it is easy to add the distribution to each point or some weight for importance of the data points. Once we assign some weight to each point, then we can correspond each weight to number of points in a volume around each point.

Of course the techniques based on principal component analysis has certain limitations. One of the limitations is when SNR is negative or zero causing the technique to fail. The reason for this is the singularity of the variance and covariance matrices. Therefore, an important step is to use KF or some sort of Fuzzy set theory for noise reduction and extraction of Signal.

8 Intelligent Reservoir Characterization

In reservoir engineering, it is important to characterize how 3-D seismic information is related to production, lithology, geology, and logs (e.g. porosity, density, gamma ray, etc.) (Boadu 1997; Nikravesh 1998a-b; Nikravesh et al., 1998; Chawathe et al. 1997; Yoshioka et al 1996; Schuelke et al. 1997; Monson and Pita 1997, Aminzadeh and Chatterjee, 1985). Knowledge of 3-D seismic data will help to reconstruct the 3-D volume of relevant reservoir information away from the well bore. However, data from well logs and 3-D seismic attributes are often difficult to analyze because of their complexity and our limited ability to understand and use the intensive information content of these data. Unfortunately, only linear and simple nonlinear information can be extracted from these data by standard statistical methods such as ordinary Least Squares, Partial Least Squares, and nonlinear Quadratic Partial Least-Squares. However, if *a priori* information regarding nonlinear input-output mapping is available, these methods become more useful.

Simple mathematical models may become inaccurate because several assumptions are made to simplify the models in order to solve the problem. On the other hand, complex models may become inaccurate if additional equations, involving a more or less approximate description of phenomena, are included. In most cases, these models require a number of parameters that are not physically measurable. Neural networks (Hecht-Nielsen 1989) and fuzzy logic (Zadeh 1965) offer a third alternative and have the potential to establish a model from nonlinear, complex, and multi-dimensional data. They have found wide application in analyzing experimental, industrial, and field data (Baldwin et al. 1990; Baldwin et al. 1989; Pezeshk et al. 1996; Rogers et al. 1992; Wong et al. 1995a, 1995b; Nikravesh et al. 1996; Nikravesh and Aminzadeh, 1997). In recent years, the utility of neural network and fuzzy logic analysis has stimulated growing interest among reservoir engineers, geologists, and geophysicists (Nikravesh et al. 1998; Nikravesh 1998a; Nikravesh 1998b; Nikravesh and Aminzadeh 1998; Chawathe et al. 1997; Yoshika et al. 1996; Schuelke et al. 1997; Monson and Pita 1997; Boadu 1997; Klimentos and McCann 1990; Aminzadeh and Katz 1994). Boadu (1997) and Nikravesh et al. (1998) applied artificial neural networks and neuro-fuzzy successfully to find relationships between seismic data and rock properties of sandstone. In a recent study, Nikravesh and Aminzadeh (1999) used an artificial neural network to further analyze data published by Klimentos and McCann (1990) and analyzed by Boadu (1997). It was concluded that to find nonlinear relationships, a neural network model provides better performance than does a multiple linear regression model. Neural network, neuro-fuzzy, and knowledge-based models have been successfully used to model rock properties based on well log databases (Nikravesh, 1998b).

Monson and Pita (1997), Chawathe et al. (1997) and Nikravesh (1998b) applied artificial neural networks and neuro-fuzzy techniques successfully to find the relationships between 3-D seismic attributes and well logs and to extrapolate mapping away from the well bore to reconstruct log responses.

Adams et al. (1999a and 1999b), Levey et al. (1999), Nikravesh et al. (1999a and 1999b) showed schematically the flow of information and techniques to be used for intelligent reservoir characterization (IRESC) (**Figure 17**). The main goal will be to integrate soft data such as geological data with hard data such as 3-D seismic, production data, etc. to build a reservoir and stratigraphic model. Nikravesh et al. (1999a and 1999b) were developed a new integrated methodology to identify a nonlinear relationship and mapping between 3-D seismic data and production-log data and the technique was applied to a producing field. This advanced data analysis and interpretation methodology for 3-D seismic and production-log data uses conventional statistical techniques combined with modern soft-computing techniques. It can be used to predict: 1. mapping between production-log data and seismic data, 2. reservoir connectivity based on multi-attribute analysis, 3. pay zone recognition, and 4. optimum well placement (**Figure 18**). Three criteria have been used to select potential locations for infill drilling or recompletion (Nikravesh et al., 1999a and 1999b): 1. continuity of the selected cluster, 2. size and shape of the cluster, and 3. existence of high Production-Index values inside a selected cluster with high Cluster-Index values. Based on these criteria, locations of the new wells were selected, one with high continuity and potential for high production and one with low continuity and potential for low production. The neighboring wells that are already in production confirmed such a prediction (**Figure 18**).

Although these methodologies have limitations, the usefulness of the techniques will be for fast screening of production zones with reasonable accuracy. This new methodology, combined with techniques presented by Nikravesh (1998a, 1998b), Nikravesh and Aminzadeh (1999), and Nikravesh et al. (1998), can be used to reconstruct well logs such as DT, porosity, density, resistivity, etc. away from the well bore. By doing so, net-pay-zone thickness, reservoir models, and geological representations will be accurately identified. Accurate reservoir characterization through data integration is an essential step in reservoir modeling, management, and production optimization.

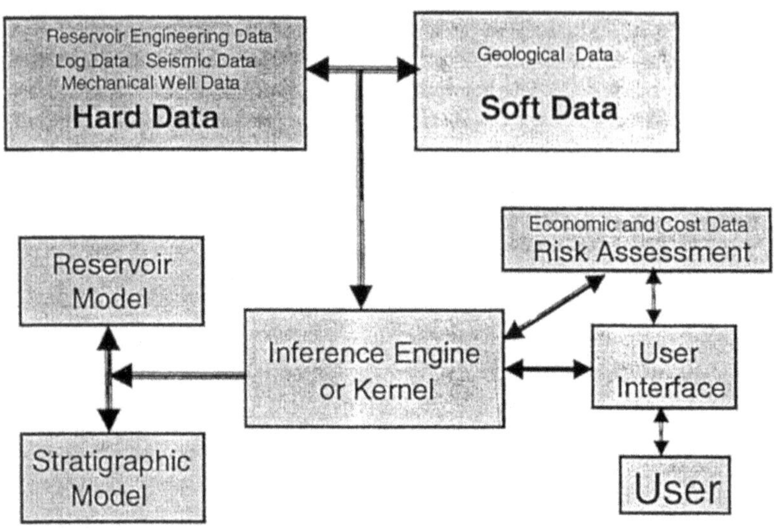

Figure 17. Integrated Reservoir Characterization (IRESC).

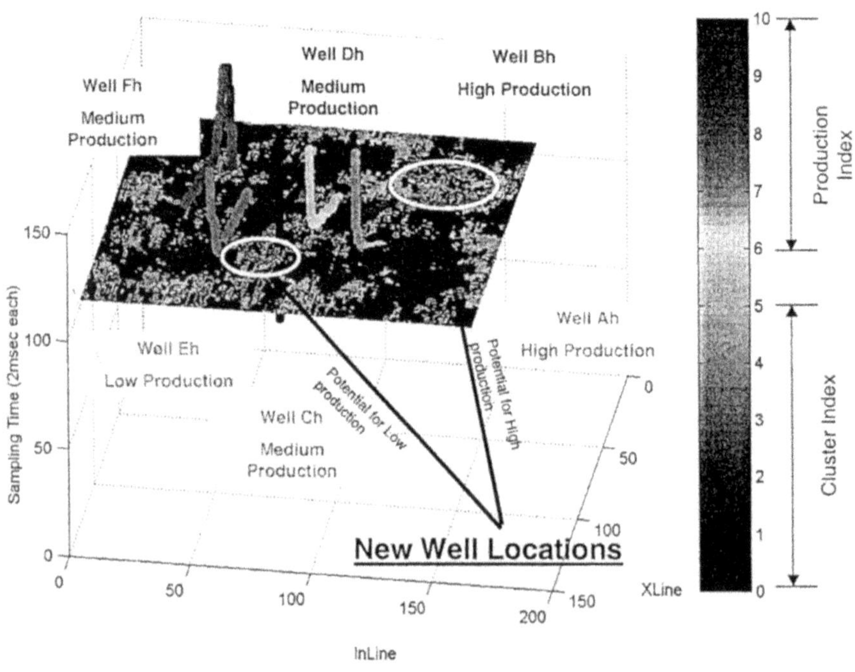

Figure 18. Optimal well placement (Nikravesh et al., 1999a and 1999b).

8.1 Reservoir Characterization

Figure 17 shows schematically the flow of information and techniques to be used for intelligent reservoir characterization (IRESC). The main goal is to integrate soft data such as geological data with hard data such as 3-D seismic, production data, etc. to build reservoir and stratigraphic models. In this case study, we analyzed 3-D seismic attributes to find similarity cubes and clusters using three different techniques: 1. k-means, 2. neural network (self-organizing map), and 3. fuzzy c-means. The clusters can be interpreted as lithofacies, homogeneous classes, or similar patterns that exist in the data. The relationship between each cluster and production-log data was recognized around the well bore and the results were used to reconstruct and extrapolate production-log data away from the well bore. The results from clustering were superimposed on the reconstructed production-log data and optimal locations to drill new wells were determined.

8.1.1 Examples

Our example are from fields that produce from the Ellenburger Group. The Ellenburger is one of the most prolific gas producers in the conterminous United States, with greater than 13 TCF of production from fields in west Texas. The Ellenburger Group was deposited on an Early Ordovician passive margin in shallow subtidal to intertidal environments. Reservoir description indicates the study area is affected by a karst-related, collapsed paleocave system that acts as the primary reservoir in the field studied (Adams et al., 1999; Levey et al., 1999).

8.1.2 Area 1

The 3-D seismic volume used for this study has 3,178,500 data points (**Table 6**). Two hundred, seventy-four well-log data points intersect the seismic traces. Eighty-nine production log data points are available for analysis (19 production and 70 non-production). A representative subset of the 3-D seismic cube, production log data, and an area of interest were selected in the training phase for clustering and mapping purposes. The subset (150 samples, with each sample equal to 2 msec of seismic data or approximately 20 feet of Ellenburger dolomite) was designed as a section (670 seismic traces) passing through all the wells as shown in **Figure 19** and has 100,500 (670*150) data points. However, only 34,170 (670*51) data points were selected for clustering purposes, representing the main Ellenburger focus area. This subset covers the horizontal boreholes of producing wells, and starts approximately 15 samples (300 feet) above the Ellenburger, and

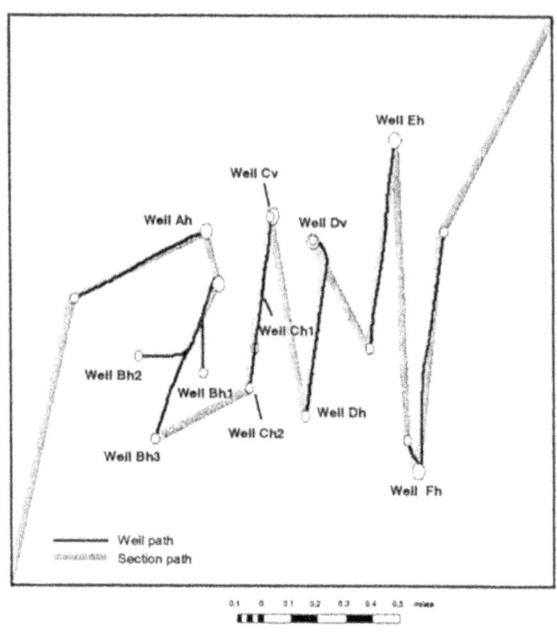

Figure 19. Seismic section passing through all the wells, Area 1.

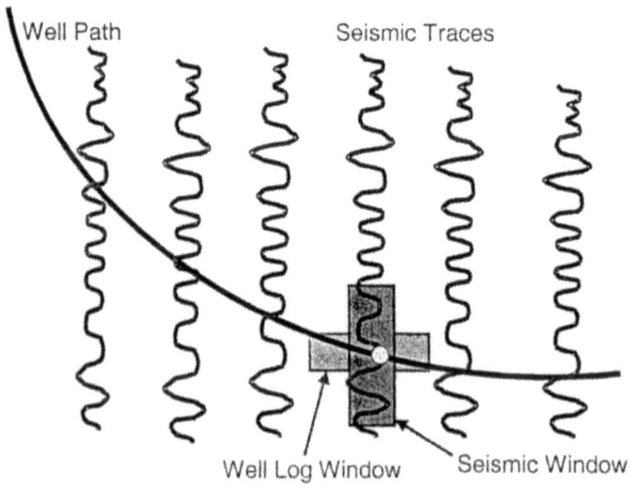

Figure 20. Schematic diagram of how the well path intersects the seismic traces.

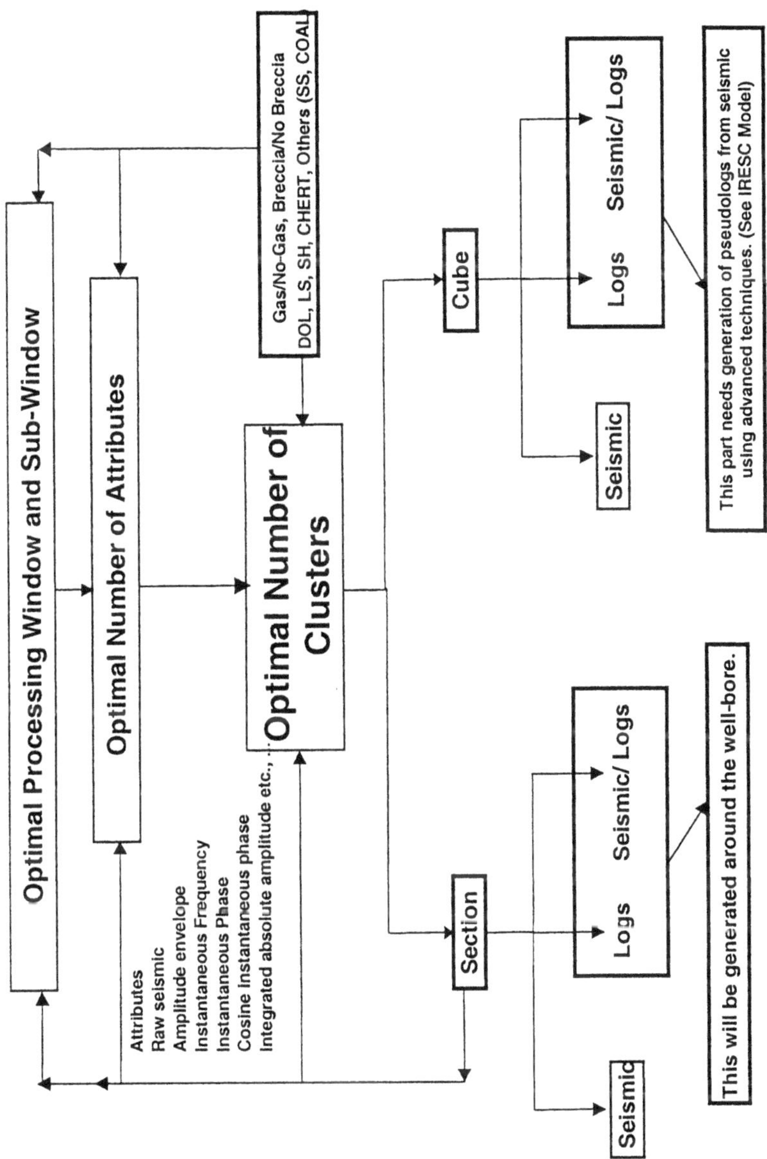

Figure 21. Iterative technique to select an optimal number of clusters, seismic attributes, and optimal processing windows.

Table 6. Typical statistics for main focus area, Area 1, and Ellenburger.

Data	
Cube	**Section**
InLine: 163	Total Number of Traces: 670
Xline: 130	Time Sample: 150
Time Sample: 150	Total Number of Points: 100,500
Total Number of Points: 3,178,500	Used for Clustering: 34,170
	Section/Cube=%3.16
	For Clustering: %1.08
Well Data	**Production Data**
Total Number of Points: 274	Total Number of Points: 89
Well Data/Section: %0.80	Production: 19
Well Data/Cube:%0.009	No Production: 70
	Production Data/Section:%0.26
	Production Data/Cube:%0.003

Table 7. List of the attributes calculated in this study.

Attribute No.	abbrev.	Attribute
1.	ampenv	Amplitude Envelope
2.	ampwcp	Amplitude Weighted Cosine Phase
3.	ampwfr	Amplitude Weighted Frequency
4.	ampwph	Amplitude Weighted Phase
5.	aprpol	Apparent Polarity
6.	avgfre	Average Frequency
7.	cosiph	Cosine Instantaneous Phase
8.	deriamp	Derivative Instantaneous Amplitude
9.	deriv	Derivative
10.	domfre	Dominant Frequency
11.	insfre	Instantaneous Frequency
12.	inspha	Instantaneous Phase
13.	intaamp	Integrated Absolute Amplitude
14.	integ	Integrate
15.	raw	Raw Seismic
16.	sdinam	Second Derivative Instantaneous Amplitude
17.	secdev	Second Derivative

ends 20 samples (400 feet) below the locations of the horizontal wells. In addition, the horizontal wells are present in a 16-sample interval, for a total interval of 51 samples (102 msec or 1020 feet). **Table 6** shows typical statistics for this case study. **Figure 20** shows a schematic diagram of how the well path intersects the seismic traces. For clustering and mapping, there are two windows that must be optimized, the seismic window and the well log window. Optimal numbers of seismic attributes and clusters need to be determined, depending on the nature of the problem. **Figure 21** shows the iterative technique that has been used to select an optimal number of clusters, seismic attributes, and optimal processing windows for the seismic section shown in **Figure 19**. Expert knowledge regarding geological parameters has also been used to constrain the maximum number of clusters to be selected. In this study, six attributes have been selected (Raw Seismic, Instantaneous Amplitude, Instantaneous Phase, Cosine Instantaneous Phase, Instantaneous Frequency, and Integrate Absolute Amplitude) out of 17 attributes calculated (**Table 7**).

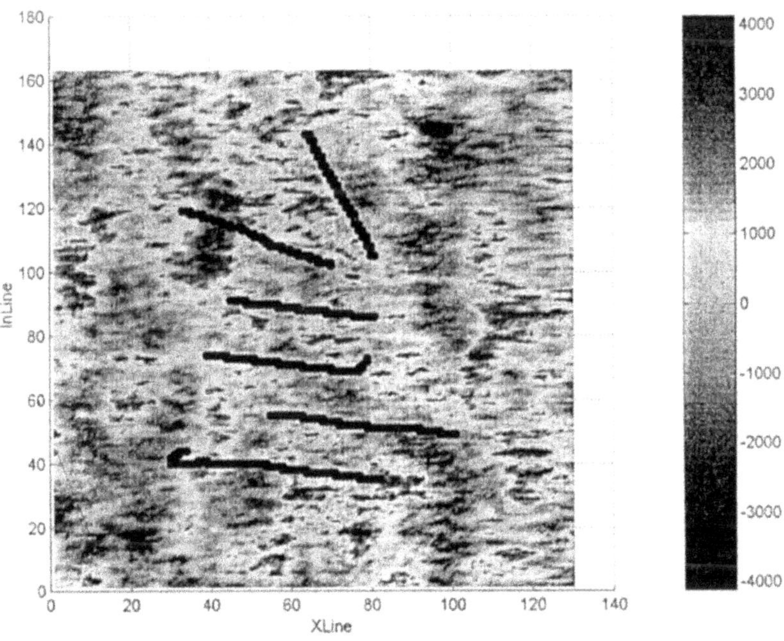

Figure 22. Typical time slice of Raw Seismic in Area 1 with Rule 1.

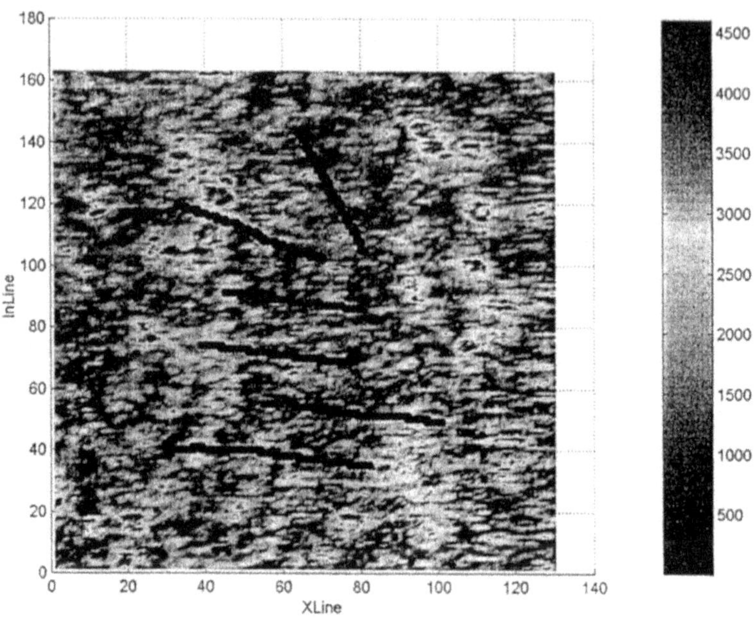

Figure 23. Typical time slice of Amplitude envelope in Area 1 with Rule 1.

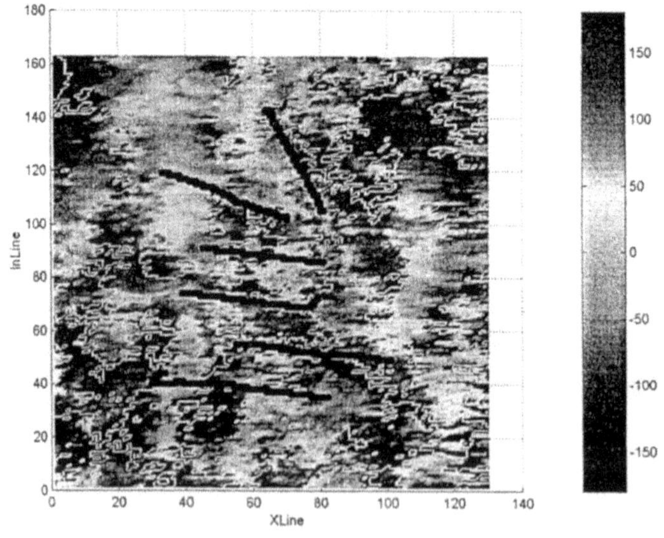

Figure 24. Typical time slice of Instantaneous Phase in Area 1 with Rule 1.

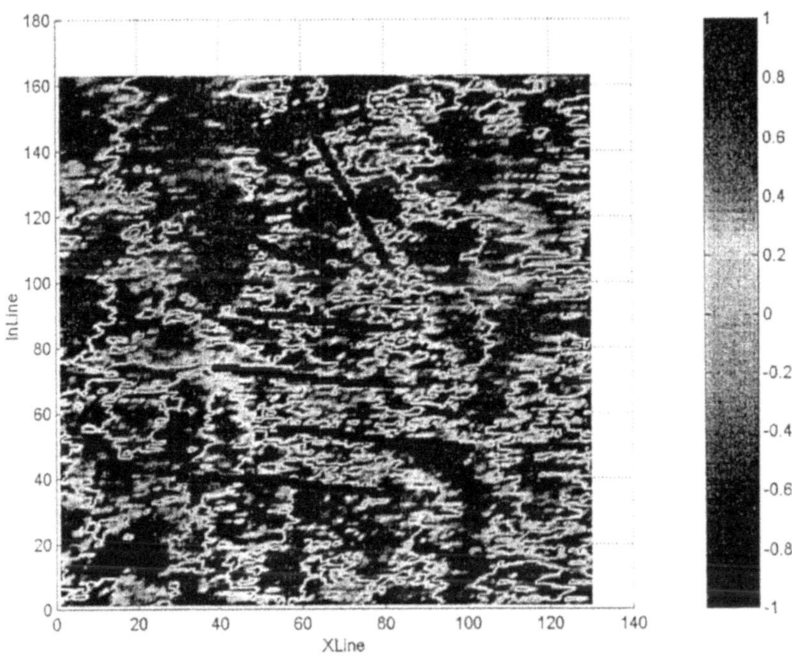

Figure 25. Typical time slice of Cosine Instantaneous Phase in Area 1 with Rule 1.

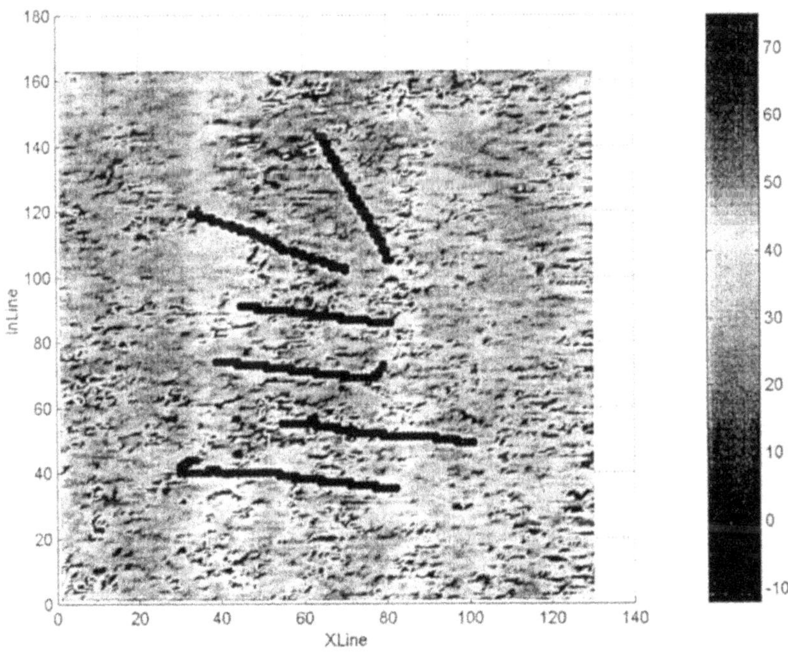

Figure 26. Typical time slice of Instantaneous Frequency in Area 1 with Rule 1.

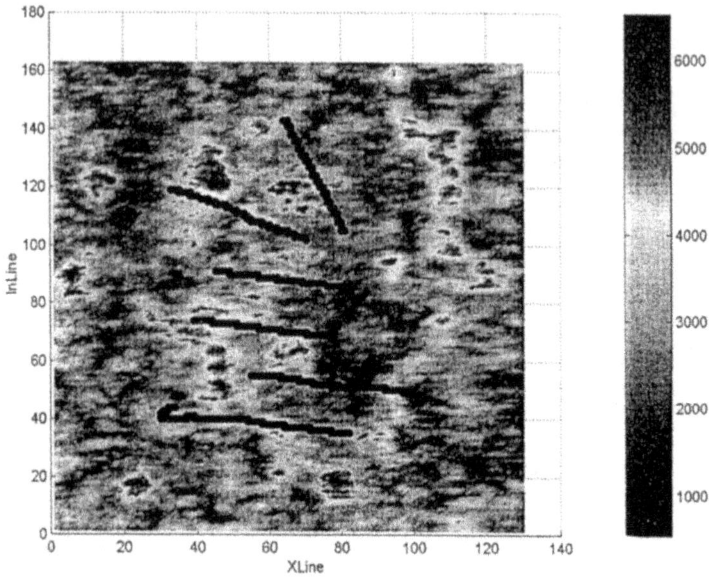

Figure 27. Typical time slice of Integrated Absolute Amplitude in Area 1 with Rule 1.

Figures 22 through **27** show typical representations of these attributes in our case study. Ten clusters were recognized, a window of one sample was used as the optimal window size for the seismic, and a window of three samples was used for the production log data. Based on qualitative analysis, specific clusters with the potential to be in producing zones were selected. Software was developed to do the qualitative analysis and run on a personal computer using Matlab™ software. **Figure 28** shows typical windows and parameters of this software. Clustering was based on three different techniques, k-means (statistical), neural network , and fuzzy c-means clustering. Different techniques recognized different cluster patterns as shown by the cluster distributions (**Figures 29A** through **31**). **Figures 29-31** through **14** show the distribution of clusters in the section passing through the wells as shown in **Figure 19**. By comparing k-mean (**Figure 12A**) and neural network clusters (**Figure 29**) with fuzzy clusters (**Figure 14**), one can conclude that the neural network predicted a different structure and patterns than did the other techniques. **Figures 29B and 29C** show a typical time-slice from the 3-D seismic cube that has been reconstructed with the extrapolated k-means cluster data. Finally, based on a qualitative analysis, specific clusters that have the potential to include producing zones were selected. Each clustering technique produced two clusters that included most of the production data. Each of these three pairs of clusters is equivalent.

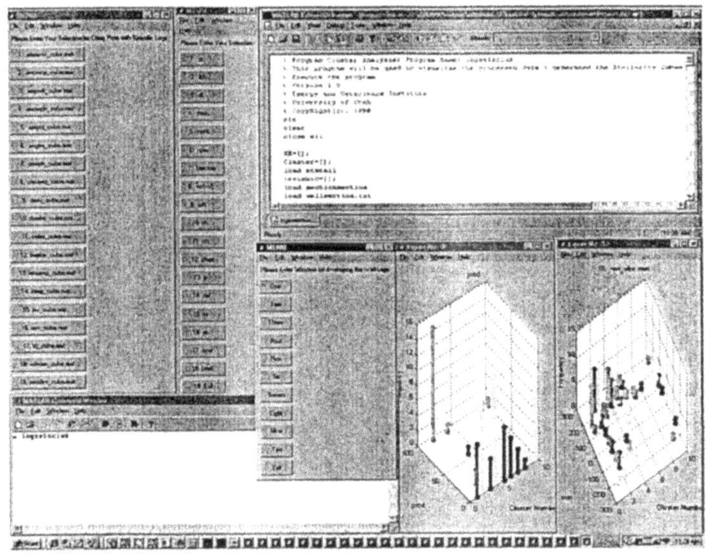

Figure 28. MatLab ™ software to do the qualitative analysis.

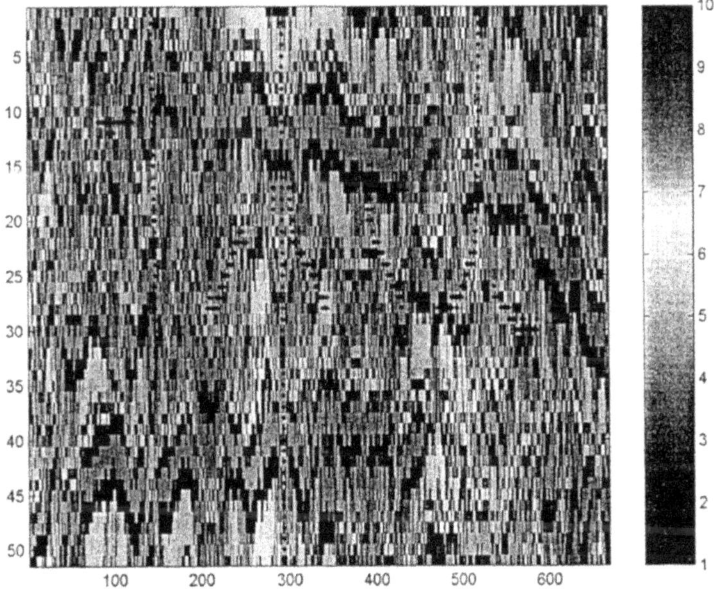

Figure 29A. Typical k-means distribution of clusters in the section passing through the wells as shown in Figure 19.

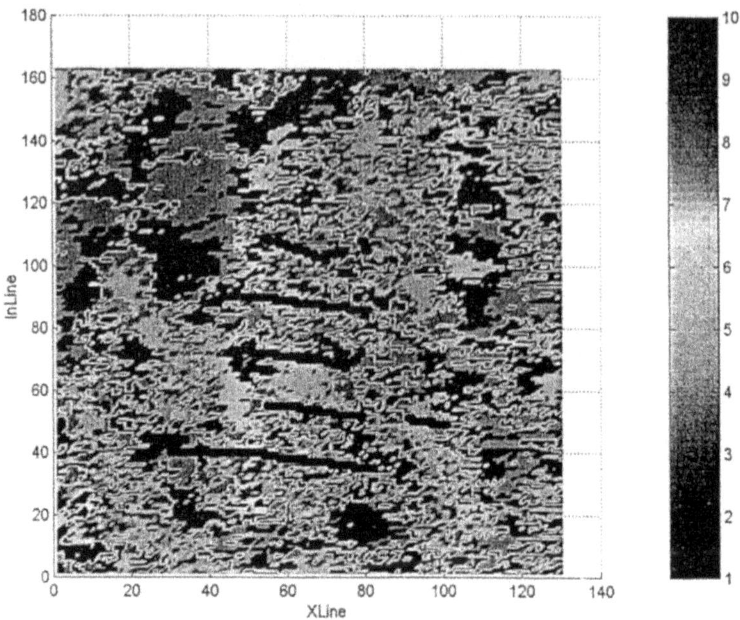

Figure 29B. Typical 2-D presentation of time-slice of k-means distribution of clusters.

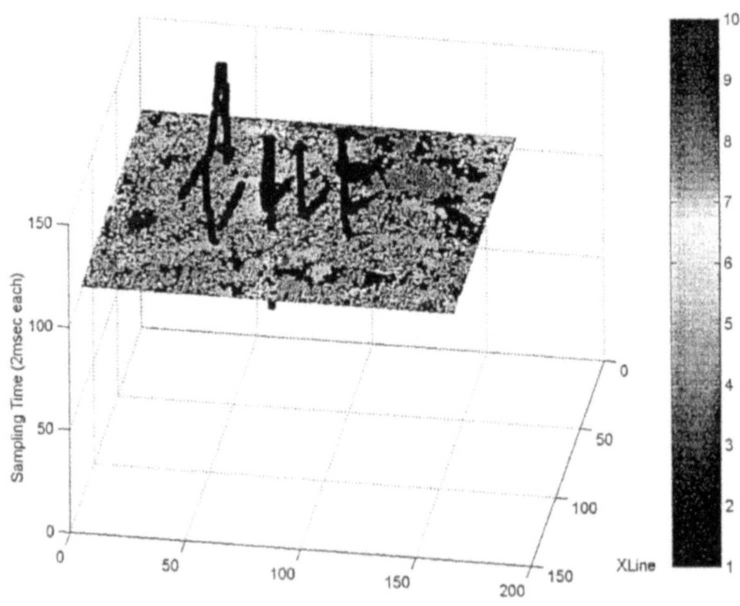

Figure 29C. Typical 3-D presentation of time-slice of k-means distribution of clusters.

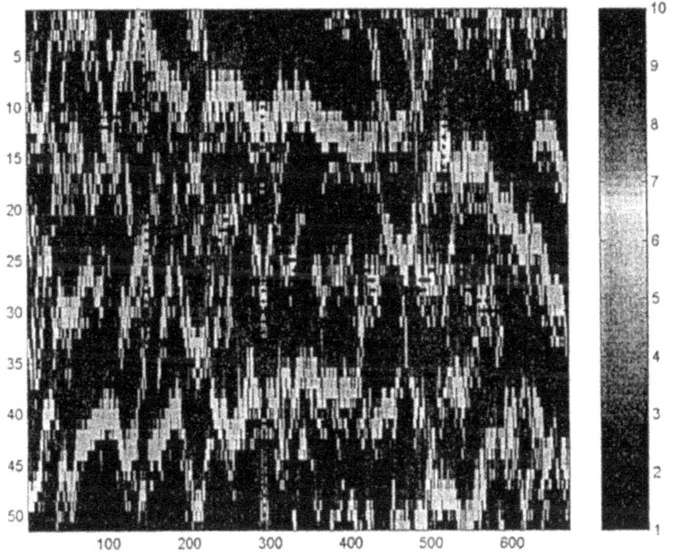

Figure 30. Typical neural network distribution of clusters in the section passing through the wells as shown in Figure 19.

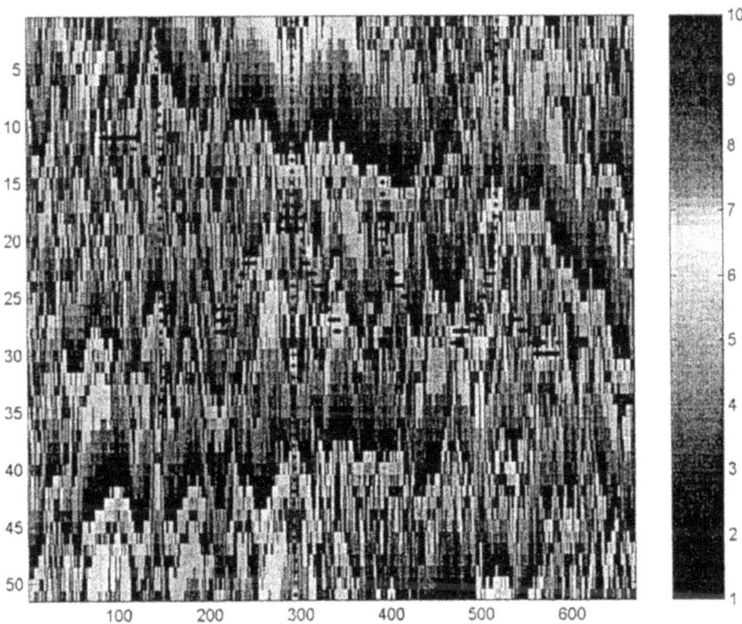

Figure 31. Typical fuzzy c-means distribution of clusters in the section passing through the wells as shown in Figure 19.

To confirm such a conclusion, cluster patterns were generated for the section passing through the wells as shown in **Figure 19**. **Figures 32 through 34** show the two clusters from each technique that correlate with production: clusters one and four from k-means clustering (**Figure 32**); clusters one and six from neural network clustering (**Figure 33**); and clusters six and ten from fuzzy c-means clustering (**Figure 34**). By comparing these three cross sections, one can conclude that, in the present study, all three techniques predicted the same pair of clusters based on the objective of predicting potential producing zones. However, this may not always be the case because information that can be extracted by the different techniques may be different. For example, clusters using classical techniques will have sharp boundaries whereas those generated using the fuzzy technique will have fuzzy boundaries.

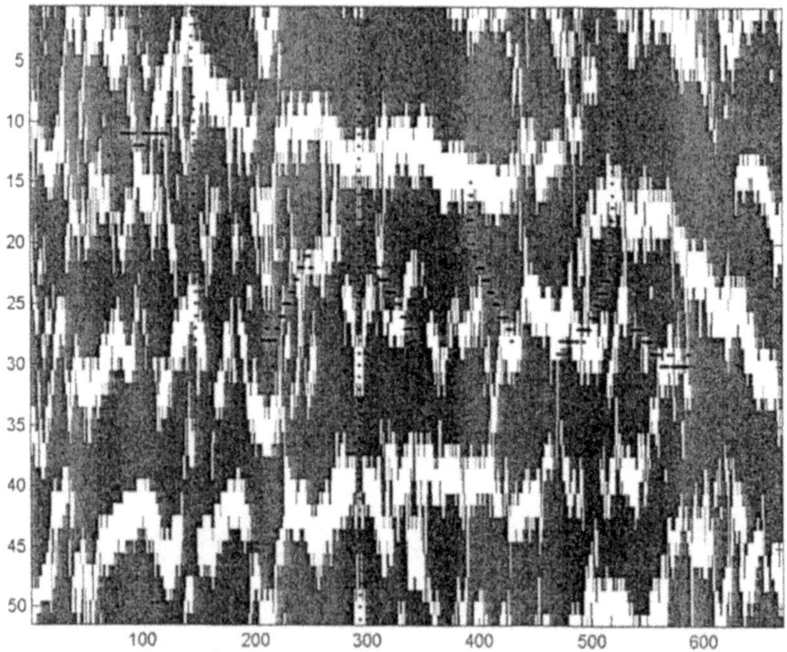

Figure 32. Clusters one and four that correlate with production using the k-means clustering technique on the section passing through the wells as shown in Figure 19.

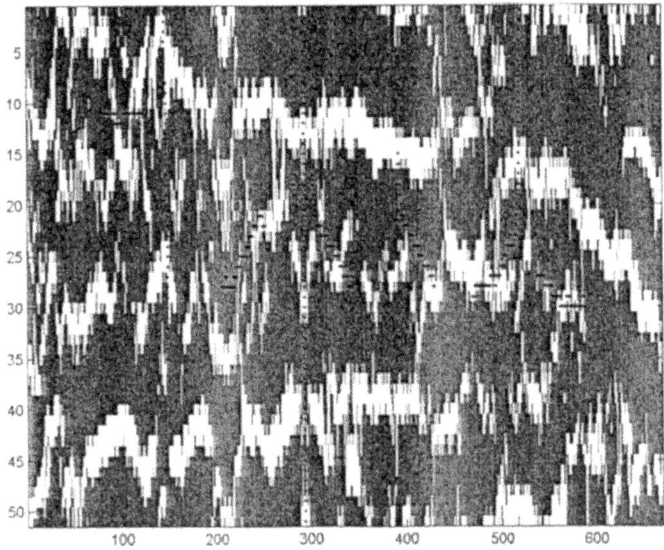

Figure 33. Clusters one and six that correlate with production using the neural network clustering technique on the section passing through the wells as shown in Figure 19.

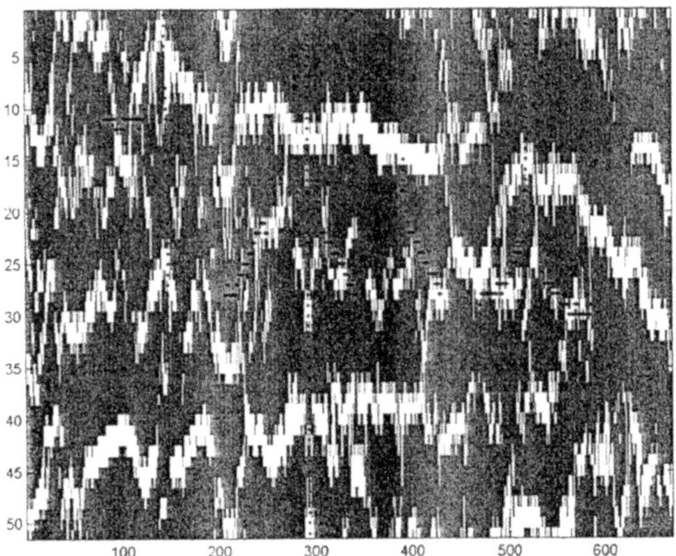

Figure 34. Clusters six and ten that correlate with production using the fuzzy c-means clustering technique on the section passing through the wells as shown in Figure 19.

Based on the clusters recognized in **Figures 32** through **34** and the production log data, a subset of the clusters has been selected and assigned as cluster 11 as shown in **Figures 35 and 36.** In this sub-cluster, the relationship between production-log data and clusters has been recognized and the production-log data has been reconstructed and extrapolated away from the well bore. Finally, the production-log data and the cluster data were superimposed at each point in the 3-D seismic cube. **Figures 37A and 37B (Figure 18)** show a typical time-slice of a 3-D seismic cube that has been reconstructed with the extrapolated production-log data and cluster data. The color scale in **Figures 37A and 20B** is divided into two indices, Cluster Index and Production Index. Criteria used to define Cluster Indices for each point are expressed as a series of dependent **IF-THEN** statements.

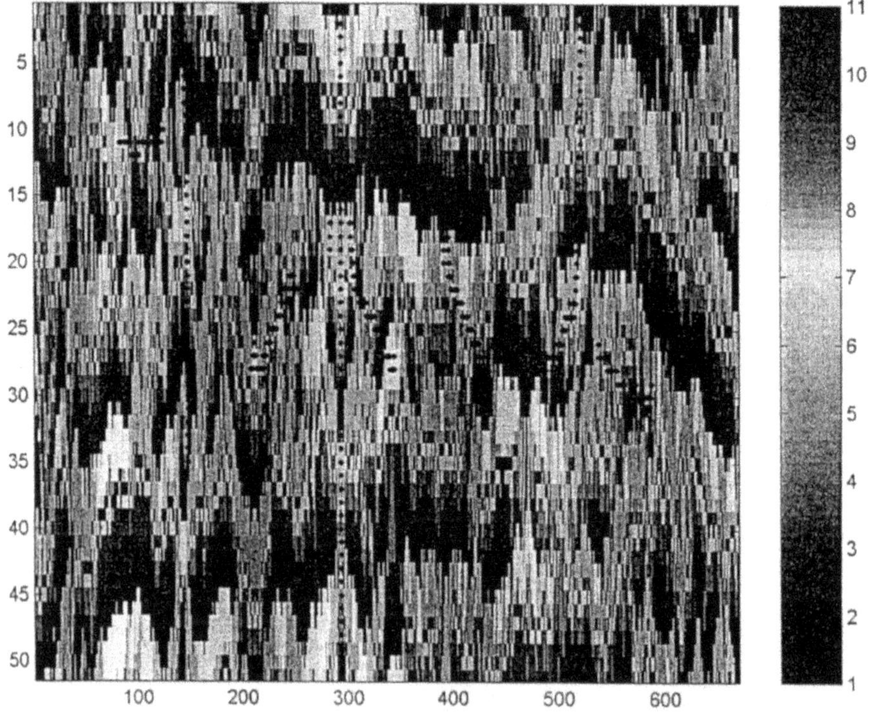

Figure 35. The section passing through all wells showing a typical distribution of clusters generated by combining the three sets of 10 clusters created by each clustering technique (k-means, neural network, and fuzzy c-means). Note that eleven clusters are displayed rather than the ten clusters originally generated. The eleventh cluster is a subset of the three pairs of clusters that contain most of the production.

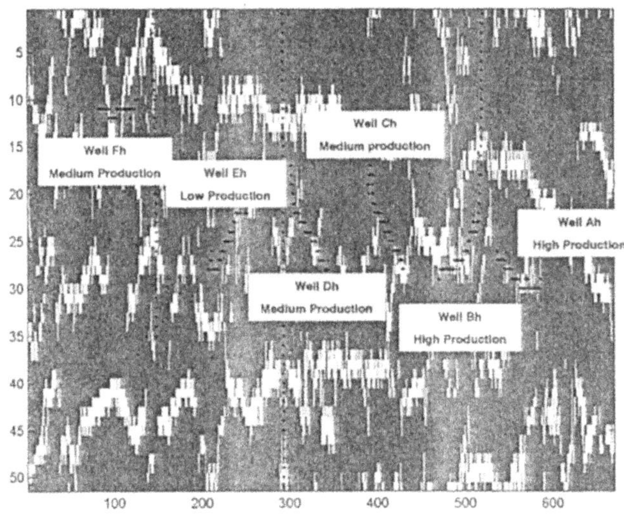

Figure 36. The section passing through all wells showing only the eleventh cluster, a subset of the three pairs of clusters that contains most of the production.

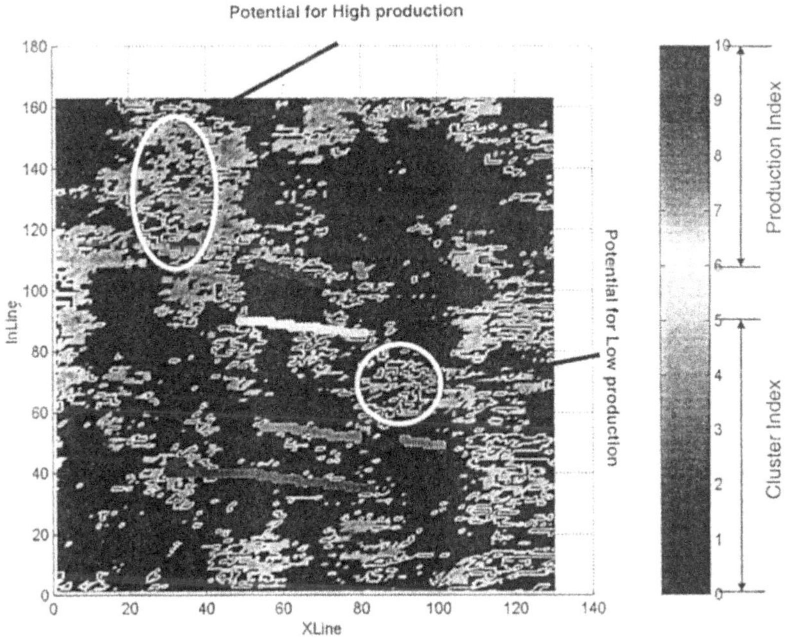

Figure 37A. A 2-D time slice showing clusters and production, with areas indicated for optimal well placement.

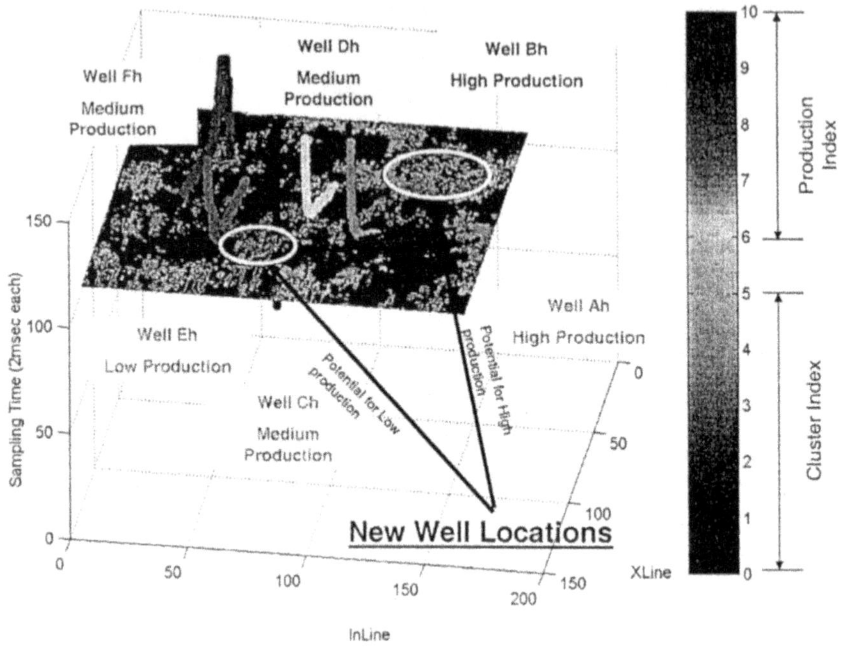

Figure 37B. An oblique view of the 2-D time slice in Figure 20A showing clusters and production, with areas indicated for optimal well placement.

Before defining Production Indices for each point within a specified cluster, a new cluster must first be defined based only on the seismic data that represents production-log data with averaged values greater than 0.50. Averaged values are determined by assigning a value to each sample of a horizontal borehole (two feet/sample). Sample intervals that are producing gas are assigned values of one and non-producing sample intervals are assigned values of zero. The optimum window size for production-log data is three samples and the averaged value at any point is the average of the samples in the surrounding window. After the new cluster is determined, a series of **IF-THEN** statements is used to define the Production Indices.

Three criteria have been used to select potential locations for infill drilling or recompletion: 1. continuity of the selected cluster, 2. size and shape of the cluster, and 3. existence of high Production-Index values inside a selected cluster with high Cluster-Index values. Based on these criteria, locations of the new wells were selected and two such locations are shown in **Figure 37B (Figure 18)**, one with high continuity and potential for high production and one with low continuity and potential for low production. The neighboring wells that are already in production confirm such a prediction as shown in **Figure 37B (Figure 18).**

9. Fractured Reservoir Characterization

In particular when we faced with fractured reservoir characterization, an efficient method of data entry, compiling, and preparation becomes important. Not only the initial model requires considerable amount of data preparation, but also subsequent stages of model updating will require a convenient way to input the new data to the existing data stream. Well logs suites provided by the operator will be supplied to the project team. We anticipate a spectrum of resistivity, image logs, cutting and core where available. A carefully designed data collection phase will provide the necessary input to develop a 3-D model of the reservoir. An optimum number of test wells and training wells needs to be identified. In addition, a new technique needs to be developed to optimize the location and the orientation of each new well to be drilled based on data gathered from previous wells. If possible, we want to prevent clustering of too many wells at some locations and undersampling in other locations thus maintaining a level of randomness in data acquisition. The data to be collected will be dependent on the type of fractured reservoir

The data collected will also provide the statistics to establish the trends, variograms, shape, and distribution of the fractures in order to develop a nonlinear and non-parametric statistical model and various possible realizations of this model. For Example, one can use Stochastic models techniques and Alternative Conditional Expectation (ACE) model developed by Breiman and Friedman [1985] for initial reservoir model prediction This provides crucial information on the variability of the estimated models. Significant changes from one realization to the other indicate a high level of uncertainty, thus the need for additional data to reduce the standard deviation. In addition, one can use our neuro-fuzzy approach to better quantify and perhaps reduce the uncertainties in the characterization of the reservoir.

Samples from well cuttings (commonly available) and cores (where available) from the focus area can also be analyzed semi-quantitatively by XRD analysis of clay mineralogy to determine vertical variability. Calibration to image logs needs to be performed to correlate fracture density to conventional log signature and mineralogical analysis.

Based on the data obtained and the statistical representation of the data, an initial 3-D model of the boundaries of the fractures and its distribution can be developed. The model is represented by a multi-valued parameter, which reflects different subsurface properties to be characterized. This parameter is derived through integration of all the input data using a number of conventional statistical approaches.

A novel "neuro-fuzzy" based algorithm that combines the training and learning capabilities of the conventional neural networks with the capabilities of fuzzy logic to incorporate subjective and imprecise information can be refined for this application. Nikravesh [1998a, b] showed the significant superiority of the neuro-fuzzy approach for data integration over the conventional methods for characterizing the boundaries. Similar method with minor modifications can be implemented and tested for fractured reservoirs.

Based on this information, an initial estimate for distribution of reservoir properties including fracture shape and distribution in 2-D and 3-D spaces can be predicted. Finally, the reservoir model is used as an input to this step to develop an optimum strategy for management of the reservoir. As data collection continues in the observation wells, using new data the model parameters will be updated. These models are then continually evaluated and visualized to assess the effectiveness of the production strategy. The wells chosen in the data collection phase will be designed and operated through a combination of an intelligent advisor.

10 Future Trends and Conclusions

We have discussed the main areas where soft computing can make a major impact in geophysical, geological and reservoir engineering applications in the oil industry. These areas include facilitation of automation in data editing and data mining. We also pointed out applications in non-linear signal (geophysical and log data) processing. And better parameterization of wave equations with random or fuzzy coefficients both in seismic and other geophysical wave propagation equations and those used in reservoir simulation. Of significant importance is their use in data integration and reservoir property estimation. Finally, quantification and reduction of uncertainty and confidence interval is possible by more comprehensive use of fuzzy logic and neural networks. The true benefit of soft computing, which is to use the intelligent techniques in combination (hybrid) rather than isolation, has not been demonstrated in a full extent. This section will address two particular areas for future research: hybrid systems and computing with words.

10.1 Hybrid Systems

So far we have seen the primary roles of neurocomputing, fuzzy logic and evolutionary computing. Their roles are in fact unique and complementary. Many hybrid systems can be built. For example, fuzzy logic can be used to combine results from several neural networks; GAs can be used to optimize the number of fuzzy rules; linguistic variables can be used to improve the performance of GAs; and extracting fuzzy rules from trained neural networks. Although some hybrid systems have been built, this topic has not yet reached maturity and certainly requires more field studies.

In order to make full use of soft computing for intelligent reservoir characterization, it is important to note that the design and implementation of the hybrid systems should aim to improve prediction and its reliability. At the same time, the improved systems should contain small number of sensitive user-definable model parameters and use less CPU time. The future development of hybrid systems should incorporate various disciplinary knowledge of reservoir geoscience and maximize the amount of useful information extracted between data types so that reliable extrapolation away from the wellbores could be obtained.

10.2 Computing with Words

One of the major difficulties in reservoir characterization is to devise a methodology to integrate qualitative geological description. One simple example is the core descriptions in standard core analysis. These descriptions provide useful and meaningful observations about the geological properties of core samples. They may serve to explain many geological phenomena in well logs, mud logs and petrophysical properties (porosity, permeability and fluid saturations). Yet, these details are not utilized due to the lack of a suitable computational tool. Gedeon et al.(1999) provided one of the first attempts to relate these linguistic descriptions (grain size, sorting, matrix, roundness, bioturbation and lamina) to core porosity levels (very poor, poor, fair and good) using intelligent techniques. The results were promising and drawn a step closer to Zadeh's idea on computing with words (Zadeh, 1996).

Computing with words (CW) aims to perform computing with objects which are propositions drawn from a natural language or having the form of mental perceptions. In essence, it is inspired by remarkable human capability to manipulate words and perceptions and perform a wide variety of physical and mental tasks without any measurement and any computations. It is fundamentally different from the traditional expert systems which are simply tools to "realize" an intelligent system, but are not able to process natural language which is imprecise, uncertain and partially true. CW has gained much popularity in many engineering disciplines (Zadeh, 1999a and 1999b). In fact, CW plays a pivotal role in fuzzy logic and vice-versa. Another aspect of CW is that it also involves a fusion of natural languages and computation with fuzzy variables.

In reservoir geology, natural language has been playing a very crucial role for a long time. We are faced with many intelligent statements and questions on a daily basis. For example: "if the porosity is high then permeability is likely to be high"; "most seals are beneficial for hydrocarbon trapping, a seal is present in reservoir A, what is the probability that the seal in reservoir A is beneficial?"; and "high resolution log data is good, the new sonic log is of high resolution, what can be said about the goodness of the new sonic log?"

CW has much to offer in reservoir characterization because most available reservoir data and information are too imprecise. There is a strong need to exploit the tolerance for such imprecision, which is the prime motivation for CW. Future research in this direction will surely provide a significant contribution in bridging reservoir geology and reservoir engineering.

Given the level of interest and the number of useful networks developed for the earth science applications and specially oil industry, it is expected soft computing techniques will play a key role in this field. Many commercial packages based on soft computing are emerging. The challenge is how to explain or "sell" the concepts and foundations of soft computing to the practicing explorationist and convince them of the value of the validity, relevance and reliability of results based on the intelligent systems using soft computing methods.

References

1. Adams, R.D., J.W. Collister, D.D. Ekart, R.A. Levey, and M. Nikravesh, 1999a, Evaluation of gas reservoirs in collapsed paleocave systems, Ellenburger Group, Permian Basin, Texas, AAPG Annual Meeting, San Antonio, TX, 11-14 April.
2. Adams, R.D., M. Nikravesh, D.D. Ekart, J.W. Collister, R.A. Levey, and R.W. Siegfried, 1999b, Optimization of Well Locations in Complex Carbonate Reservoirs, Summer 1999b, Volume 5, Number 2, GasTIPS, GRI 1999b.
3. Aminzadeh, F. and Jamshidi, M.: *Soft Computing: Fuzzy Logic, Neural Networks, and Distributed Artificial Intelligence*, PTR Prentice Hall, Englewood Cliffs, NJ (1994).
4. Adams, R.D., Nikravesh, M., Ekart, D.D., Collister, J.W., Levey, R.A. and Seigfried, R.W.: "Optimization of Well locations in Complex Carbonate Reservoirs," GasTIPS, vol. 5, no. 2, GRI (1999).
5. Aminzadeh, F. 1989, Future of Expert Systems in the Oil Industry, Proceedings of HARC Conference on Geophysics and Computers, A Look at the Future.
6. Aminzadeh, F. and Jamshidi,M., 1995, Soft Computing, Eds., PTR Prentice Hall,
7. Aminzadeh, F. and S. Chatterjee, 1984/85, Applications of clustering in exploration seismology, Geoexploration, v23, p.147-159.
8. Aminzadeh, F., 1989, Pattern Recognition and Image Processing, Elsevier Science Ltd. Aminzadeh, F., 1994, Applications of Fuzzy Expert Systems in Integrated Oil Exploration, Computers and Electrical Engineering, No. 2, pp 89-97

9. Aminzadeh, F., 1991, Where are we now and where are we going?, in Expert Systems in Exploration, Aminzadeh, F. and M. Simaan, Eds., SEG publications, Tulsa, 3-32.
10. Aminzadeh, F., S. Katz, and K. Aki, 1994, Adaptive Neural Network for Generation of Artificial Earthquake Precursors, IEEE Trans. On Geoscience and Remote Sensing, 32 (6)
11. Aminzadeh, F. 1996, Future Geoscience Technology Trends in , Stratigraphic Analysis, Utilizing Advanced Geophysical, Wireline, and Borehole Technology For Petroleum Exploration and Production, GCSEPFM pp 1-6,
12. Aminzadeh, F. Barhen, J. Glover, C. W. and Toomanian, N. B , 1999, Estimation of Reservoir Parameters Using a Hybrid Neural Network, Journal of Science and Petroleum Engineering, Vol. 24, pp 49-56,..
13. Aminzadeh, F. Barhen, J. Glover, C. W. and Toomanian, N. B., 2000, Reservoir parameter estimation using hybrid neural network, Computers & Geoscience, Vol. 26 pp 869-875.
14. Aminzadeh, F., and de Groot, P., Seismic characters and seismic attributes to predict reservoir properties, 2001, Proceedings of SEG-GSH Spring Symposium.
15. Baldwin, J.L., A.R.M. Bateman, and C.L. Wheatley, 1990, Application of Neural Network to the Problem of Mineral Identification from Well Logs, The Log Analysis, 3, 279.
16. Baldwin, J.L., D.N. Otte, and C.L. Wheatley, 1989, Computer Emulation of human mental process: Application of Neural Network Simulations to Problems in Well Log Interpretation, Society of Petroleum Engineers, SPE Paper # 19619, 481.
17. Baygun et. al, 1985, Applications of Neural Networks and Fuzzy Logic in Geological Modeling of a Mature Hydrocarbon Reservoir, Report# ISD-005-95-13a, Schlumberger-Doll Research.
18. Benediktsson, J. A. Swain, P. H., and Erson, O. K., 1990, Neural networks approaches versus statistical methods in classification of multisource remote sensing data, IEEE Geoscience and Remote Sensing, Transactions, Vol. 28, NO. 4, 540-552.
19. Bezdek, J.C., 1981, Pattern Recognition with Fuzzy Objective Function Algorithm, Plenum Press, New York.
20. Bezdek, J.C., and S.K. Pal, *eds*., 1992, Fuzzy Models for Pattern Recognition: IEEE Press, 539p.
21. Boadu, F.K., 1997, Rock properties and seismic attenuation: Neural network analysis, Pure Applied Geophysics, v149, pp. 507-524.
22. Bezdek, J.C., Ehrlich, R. and Full, W.: The Fuzzy C-Means Clustering Algorithm," *Computers and Geosciences* (1984) 10, 191-203.
23. Bois, P., 1983, Some Applications of Pattern Recognition to Oil and Gas Exploration, IEEE Transaction on Geoscience and Remote Sensing, Vol. GE-21, No. 4, PP. 687-701.
24. Bois, P., 1984, Fuzzy Seismic Interpretation, IEEE Transactions on Geoscience and Remote Sensing, Vol. GE-22, No. 6, pp. 692-697.
25. Breiman, L., and J.H. Friedman, Journal of the American Statistical Association, 580, 1985.
26. Cannon, R. L., Dave J. V. and Bezdek, J. C., 1986, Efficient implementation of Fuzzy C- Means algorithms, IEEE Trans. On Pattern Analysis and Machine Intelligence, Vol. PAMI 8, No. 2.
27. Bruce, A.G., Wong, P.M., Tunggal, Widarsono, B. and Soedarmo, E.: "Use of Artificial Neural Networks to Estimate Well Permeability Profiles in Sumatera,

Indonesia," Proceedings of the 27th Annual Conference and Exhibition of the Indonesia Petroleum Association (1999).
28. Bruce, A.G., Wong, P.M., Zhang, Y., Salisch, H.A., Fung, C.C. and Gedeon, T.D.: "A State-of-the-Art Review of Neural Networks for Permeability Prediction," *APPEA Journal* (2000), in press.
29. Caers, J.: "Towards a Pattern Recognition-Based Geostatistics," Stanford Center for Reservoir Forecasting Annual Meeting (1999).
30. Caers, J. and Journel, A.G.: "Stochastic Reservoir Simulation using Neural Networks Trained on Outcrop Data," SPE 49026, SPE Annual Technical Conference and Exhibition, New Orleans (1998), 321-336.
31. Chawathe, A., A. Quenes, and W.W. Weiss, 1997, Interwell property mapping using crosswell seismic attributes, SPE 38747, SPE Annual Technical Conference and Exhibition, San Antonio, TX, 5-8 Oct.
32. Chen, H. C. ,Fang, J. H. ,Kortright, M. E. ,Chen, D. C Novel Approaches to the Determination of Archie Parameters II: Fuzzy Regression Analysis,. SPE 26288-P
33. Cybenko, G., 1989, Approximation by superposition of a sigmoidal function, Math. Control Sig. System, v2, p. 303.
34. Cuddy, S.: "The Application of Mathematics of Fuzzy Logic to Petrophysics," SPWLA Annual Logging Symposium (1997), paper S.
35. Bloch, S.: "Empirical Prediction of Porosity and Permeability in Sandstones," *AAPG Bulletin* (1991) 75, 1145-1160.
36. Fang, J.H., Karr, C.L. and Stanley, D.A.: "Genetic Algorithm and Application to Petrophysics," SPE 26208, unsolicited paper.
37. Fang, J.H., Karr, C.L. and Stanley, D.A.: "Transformation of Geochemical Log Data to Mineralogy using Genetic Algorithms," *The Log Analyst* (1996) 37, 26-31.
38. Fang, J.H. and Chen, H.C.: "Fuzzy Modelling and the Prediction of Porosity and Permeability from the Compositional and Textural Attributes of Sandstone," *Journal of Petroleum Geology* (1997) 20, 185-204.
39. Gallagher, K. and Sambridge, M.: "Genetic Algorithms: A Powerful Tool for Large-Scale Nonlinear Optimization Problems," *Computers and Geosciences* (1994) 20, 1229-1236.
40. Gedeon, T.D., Tamhane, D., Lin, T. and Wong, P.M.: "Use of Linguistic Petrographical Descriptions to Characterise Core Porosity: Contrasting Approaches," submitted to *Journal of Petroleum Science and Engineering* (1999).
41. Gupta, M. M. and H. Ding, 1994, Foundations of fuzzy neural computations, in Soft Computing, Aminzadeh, F. and Jamshidi,M., Eds., PTR Prentice Hall, 165-199.
42. Hecht-Nielsen, R., 1989, Theory of backpropagation neural networks, presented at IEEE Proc., Int. Conf. Neural Network, Washington DC.
43. Hecht-Nielsen, R., 1990, Neurocomputing: Addison-Wesley Publishing, p. 433.Jang, J.S.R., 1992, Self-learning fuzzy controllers based on temporal backpropagation, IEEE Trans. Neural Networks, 3 (5).
44. Holland, J.: *Adaptation in Natural and Artificial Systems*, University of Michigan Press, Ann Harbor (1975).
45. Horikawa, S, T. Furuhashi, A. Kuromiya, M. Yamaoka, and Y. Uchikawa, Determination of Antecedent Structure for Fuzzy Modeling Using Genetic Algo-

rithm, Proc. ICEC'96, IEEE International onference on Evolutionary Computation, Nagoya, Japan, May 20-22, 1996.

46. Horikawa, S, T. Furuhashi, and Y. Uchikawa, On Fuzzy Modeling Using Fuzzy Neural Networks with the Backpropagation Algorithm, IEEE Trans. On Neural Networks, 3 (5), 1992.
47. Horikawa, S, T. Furuhashi, Y. Uchikawa, and T. Tagawa, "A study on Fuzzy Modeling Using Fuzzy Neural Networks", Fuzzy Engineering toward human Friendly Systems, IFES'91.
48. Huang, Z. and Williamson, M. A. , 1994, Geological pattern recognition and modeling with a general regression neural network, Canadian Jornal of Exploraion Geophysics, Vol. 30, No. 1, 60-68
49. Huang, Y., Gedeon, T.D. and Wong, P.M.: "A Practical Fuzzy Interpolator for Prediction of Reservoir Permeability," Proceedings of IEEE International Conference on Fuzzy Systems, Seoul (1999).
50. Huang, Y., Wong, P.M. and Gedeon, T.D.: "Prediction of Reservoir Permeability using Genetic Algorithms," *AI Applications* (1998) 12, 67-75.
51. Huang, Y., Wong, P.M. and Gedeon, T.D.: "Permeability Prediction in Petroleum Reservoir using a Hybrid System," In: *Soft Computing in Industrial Applications*, Suzuki, Roy, Ovaska, Furuhashi and Dote (eds.), Springer-Verlag, London (2000), in press.
52. Huang, Y., Wong, P.M. and Gedeon, T.D.: "Neural-fuzzy-genetic Algorithm Interpolator in Log Analysis," EAGE Conference and Technical Exhibition, Leipig (1998), paper P106.
53. Jang, J.S.R., and N. Gulley, 1995, Fuzzy Logic Toolbox, The Math Works Inc., Natick, MA.
54. Jang, J.S.R., 1992, Self-learning fuzzy controllers based on temporal backpropagation, IEEE Trans. Neural Networks, 3 (5).
55. Jang, J.S.R., 1991, Fuzzy modeling using generalized neural networks and Kalman filter algorithm, Proc. of the Ninth National Conference on Artificial Intelligence, pp. 762-767.
56. Jang, J.-S.R., Sun, C.-T. and Mizutani, E.: *Neuro-Fuzzy and Soft Computing*, Prentice-Hall International Inc., NJ (1997).
57. Johnson, V. M. and Rogers, L. L., Location analysis in ground water remediation using artificial neural networks, Lawrence Livermore National Laboratory Report, UCRL-JC-17289.
58. Klimentos, T., and C. McCann, 1990, Relationship among Compressional Wave Attenuation, Porosity, Clay Content and Permeability in Sandstones, Geophys., v55, p. 991014.
59. Kohonen, T., 1997, Self-Organizing Maps, Second Edition, Springer.Berlin.
60. Kohonen, T., 1987, Self-Organization and Associate Memory, 2nd Edition, Springer Verlag., Berlin.
61. Lashgari, B, 1991, Fuzzy Classification with Applications, in Expert Systems in Exploration, Aminzadeh, F. and M. Simaan, Eds., SEG publications, Tulsa, 3-32.
62. Levey, R., M. Nikravesh, R. Adams, D. Ekart, Y. Livnat, S. Snelgrove, J. Collister, 1999, Evaluation of fractured and paleocave carbonate reservoirs, AAPG Annual Meeting, San Antonio, TX, 11-14 April.
63. Lin, C.T., and Lee, C.S.G., 1996, Neural fuzzy systems, Prentice Hall, Englewood Cliffs, New Jersey.
64. Lippmann, R. P., 1987, An introduction to computing with neural networks, ASSP Magazine, April, 4-22.

65. Liu, X., Xue, P., Li, Y., 1989, Neural networl method for tracing seismic events, 59th annual SEG meeting, Expanded Abstracts, 716-718.
66. MacQueen, J., 1967, Some methods for classification and analysis of multivariate observation: in LeCun, L. M. and Neyman, J., eds., The Proceedings of the 5th Berkeley Symposium on Mathematical Statistics and Probability: University of California Press, v.1, 281-297.
67. Mallick, S., 1999, Some practical aspects of prestack waveform inversion using a genetic algorithm: An example from the east Texas Woodbine gas sand: Geophysics, Soc. of Expl. Geophys., **64,** 326-336
68. Monson, G.D. and Pita, J.A.: "Neural Network Prediction of Pseudo-logs for Net Pay and Reservoir Property Interpretation: Greater Zafiro Field Area, Equatorial Guinea," SEG Annual Meeting (1997).
69. Matsushita, S., A. Kuromiya, M. Yamaoka, T. Furuhashi, and Y. Uchikawa, "A study on Fuzzy GMDH with Comprehensible Fuzzy Rules", 1994 IEEE Symposium on Emerging Tehnologies and Factory Automation.Gaussian Control Problem, Mathematics of operations research, 8 (1) 1983.
70. The Math Works™, 1995, Natick.
71. McCormack, M. D., 1990, Neural computing in geophysics, Geophysics: The Leading Edge of Exploration, 10, 1, 11-15.
72. McCormack, M. D., 1991, Seismic trace editing and first break piking using neural networks, 60th annual SEG meeting, Expanded Abstracts, 321-324.
73. McCormack, M. D., Stoisits, R. F., Macallister, D. J. and Crawford, K. D., 1999, Applications of genetic algorithms in exploration and production`: The Leading Edge, **18,** no. 6, 716-718.
74. Medsker, L.R., 1994, Hybrid Neural Network and Expert Systems: Kluwer Academic Publishers, p. 240.
75. Monson G.D., and J.A. Pita, 1997, Neural network prediction of pseudo-logs for net pay and reservoir property interpretation: Greater Zafiro field area, Equatorial Guinea, SEG 1997 meeting, Dallas, TX.
76. Nikravesh, M, and F. Aminzadeh, 1997, Knowledge discovery from data bases: Intelligent data mining, FACT Inc. and LBNL Proposal, submitted to SBIR-NASA.
77. Nikravesh, M. and F. Aminzadeh, 1999, Mining and Fusion of Petroleum Data with Fuzzy Logic and Neural Network Agents, To be published in Special Issue, Computer and Geoscience Journal, 1999-2000.
78. Nikravesh, M., 1998a, "Mining and fusion of petroleum data with fuzzy logic and neural network agents", CopyRight© Report, LBNL-DOE, ORNL-DOE, and DeepLook Industry Consortium.
79. Nikravesh, M., 1998b, "Neural network knowledge-based modeling of rock properties based on well log databases," SPE 46206, 1998 SPE Western Regional Meeting, Bakersfield, California, 10-13 May.
80. Nikravesh, M., A.E. Farell, and T.G. Stanford, 1996, Model Identification of Nonlinear Time- Variant Processes Via Artificial Neural Network, Computers and Chemical Engineering, v20, No. 11, 1277p.
81. Nikravesh, M., B. Novak, and F. Aminzadeh, 1998, Data mining and fusion with integrated neuro-fuzzy agents: Rock properties and seismic attenuation, JCIS 1998, The Fourth Joint Conference on Information Sciences, North Carolina, USA, October 23-28. .
82. Nikravesh, M., Adams, R.D., Levey, R.A. and Ekart, D.D.: "Soft Computing: Tools for Intelligent Reservoir Characterization (IRESC) and Optimum Well Placement (OWP)," ?????.

83. Nikravesh, M. and Aminzadeh, F.: "Opportunities to Apply Intelligent Reservoir Characterizations to the Oil Fields in Iran: Tutorial and Lecture Notes for Intelligent Reservoir Characterization," unpublished.
84. Nikravesh, M., Kovscek, A.R., Murer, A.S. and Patzek, T.W.: "Neural Networks for Field-Wise Waterflood Management in Low Permeability, Fractured Oil Reservoirs," SPE 35721, SPE Western Regional Meeting, Anchorage (1996), 681-694.
85. Nikravesh, M.: "Artificial Intelligence Techniques for Data Analysis: Integration of Modeling, Fusion and Mining, Uncertainty Analysis, and Knowledge Discovery Techniques," 1998 BISC-SIG-ES Workshop, Lawrence Berkeley National Laboratory, University of California, Berkeley (1998).
86. Nikravesh, M., R.A. Levey, and D.D. Ekart, 1999, Soft Computing: Tools for Reservoir Characterization (IRESC), To be presented at 1999 SEG Annual Meeting.
87. Nikravesh, M. and Aminzadeh, F., 2001, Mining and Fusion of Petroleum Data with Fuzzy Logic and Neural Network Agents, Journal of Petroleum Science and Engineering, Vol. 29, pp 221-238,
88. Nordlund, U.: "Formalizing Geological Knowledge – With an Example of Modeling Stratigraphy using Fuzzy Logic," *Journal of Sedimentary Research* (1996) 66, 689-698.
89. Pezeshk, S, C.C. Camp, and S. Karprapu, 1996, Geophysical log interpretation using neural network, Journal of Computing in Civil Engineering, v10, 136p.
90. Rogers, S.J., J.H. Fang, C.L. Karr, and D.A. Stanley, 1992, Determination of Lithology, from Well Logs Using a Neural Network, AAPG Bulletin, v76, 731p.
91. Potter, D., Wright, J. and Corbett, P., 1999, A genetic petrophysics approach to facies and permeability prediction in a Pegasus well, 61st Mtg.: Eur. Assn. Geosci. Eng., Session:2053.
92. Rosenblatt, F., 1962 Principal of neurodynamics, Spartan Books.
93. Schuelke, J.S., J.A. Quirein, J.F. Sarg, D.A. Altany, and P.E. Hunt, 1997, Reservoir architecture and porosity distribution, Pegasus field, West Texas-An integrated sequence stratigraphic-seismic attribute study using neural networks, SEG 1997 meeting, Dallas, TX.
94. *Selva, C., Aminzadeh, F., Diaz B., and Porras J. M., 200,* Using Geostatistical Techniques For Mapping A Reservoir In Eastern Venezuela Proceedings of 7TH INTERNATIONAL CONGRESS OF THE BRAZILIAN GEOPHYSICAL SOCIETY, OCTOBER 2001, SALVADOR, BRAZIL
95. Sheriff, R. E., and Geldart, L. P., 1982, Exploration Seismology, Vol. 1 and 1983, Vol. 2, Cambridge
96. Sugeno, M. and T. Yasukawa, 1993, A Fuzzy-Logic-Based Approach to Qualitative Modeling, IEEE Trans. Fuzzy Syst., 1
97. Tamhane, D., Wang, L. and Wong, P.M.: "The Role of Geology in Stochastic Reservoir Modelling: The Future Trends," SPE 54307, SPE Asia Pacific Oil and Gas Conference and Exhibition, Jakarta (1999).
98. Taner, M. T., Koehler, F. and Sherrif, R. E., 1979, Complex seismic trace analysis, Geophysics, 44, 1196-1212Tamhane, D., Wong, P. M., Aminzadeh, F. 2002, Integrating Linguistic Descriptions and Digital Signals in Petroleum Reservoirs, *International Journal of Fuzzy Systems, Vol. 4, No. 1, pp 586-591*
99. Tura, A., and Aminzadeh, F., 1999, Dynamic Reservoir Characterization and Seismically Constrained Production Optimization, Extended Abstracts of Society of Exploration Geophysicist, Houston,

100. Veezhinathan, J., Wagner D., and Ehlers, J. 1991, First break picking using neural network, in Expert Systems in Exploration, Aminzadeh, F. and Simaan, M., Eds, SEG, Tulsa, 179-202.
101. Wang, L. and Wong, P.M.: "A Systematic Reservoir Study of the Lower Cretaceous in Anan Oilfield, Erlian Basin, North China," AAPG Annual Meeting (1998).
102. Wang, L., Wong, P.M. and Shibli, S.A.R.: "Integrating Geological Quantification, Neural Networks and Geostatistics for Reservoir Mapping: A Case Study in the A'nan Oilfield, China," *SPE Reservoir Evaluation and Engineering* (1999) 2(6), in press.
103. Wang, L., Wong, P.M., Kanevski, M. and Gedeon, T.D.: "Combining Neural Networks with Kriging for Stochastic Reservoir Modeling," *In Situ* (1999) 23(2), 151-169.
104. Wong, P.M., Henderson, D.H. and Brooks, L.J.: "Permeability Determination using Neural Networks in the Ravva Field, Offshore India," *SPE Reservoir Evaluation and Engineering* (1998) 1, 99-104.
105. Wong, P.M., Jang, M., Cho, S. and Gedeon, T.D.: "Multiple Permeability Predictions using an Observational Learning Algorithm," *Computers and Geosciences* (2000), in press.
106. Wong, P.M., Tamhane, D. and Wang, L.: "A Neural Network Approach to Knowledge-Based Well Interpolation: A Case Study of a Fluvial Sandstone Reservoir," *Journal of Petroleum Geology* (1997) 20, 363-372.
107. Wong, P.M., F.X. Jiang and, I.J. Taggart, 1995a, A Critical Comparison of Neural Networks and Discrimination Analysis in Lithofacies, Porosity and Permeability Prediction, Journal of Petroleum Geology, v18, 191p.
108. Wong, P.M., T.D. Gedeon, and I.J. Taggart, 1995b, An Improved Technique in Prediction: A Neural Network Approach, IEEE Transaction on Geoscience and Remote Sensing, v33, 971p.
109. Wong, P. M., Aminzadeh, F and Nikravesh, M., 2001, Soft Computing for Reservoir Characterization, in Studies in Fuzziness, Physica Verlag, Germany
110. Yao, X.: "Evolving Artificial Neural Networks," *Proceedings of IEEE* (1999) 87, 1423-1447.
111. Yashikawa, T., Furuhashi, T. and Nikravesh, M.: "Development of an Integrated Neuro-Fuzzy-Genetic Algorithm Software," 1998 BISC-SIG-ES Workshop, Lawrence Berkeley National Laboratory, University of California, Berkeley (1998).
112. Yoshioka, K, N. Shimada, and Y. Ishii, 1996, Application of neural networks and co-kriging for predicting reservoir porosity-thickness, GeoArabia, v1, No 3.
113. Zadeh L.A. and Yager R.R." Uncertainty in Knowledge Bases"' Springer-Verlag, Berlin, 1991.
114. Zadeh L.A., "The Calculus of Fuzzy if-then-Rules", AI Expert 7 (3), 1992.
115. Zadeh L.A., "The Roles of Fuzzy Logic and Soft Computing in the Concept, Design and deployment of Intelligent Systems", BT Technol Journal, 14 (4) 1996.
116. Zadeh, L. A. and Aminzadeh, F.(1995) Soft computing in integrated exploration, proceedings of IUGG/SEG symposium on AI in Geophysics, Denver, July 12.
117. Zadeh, L.A., 1965, Fuzzy sets, Information and Control, v8, 33353.
118. Zadeh, L.A., 1973, Outline of a new approach to the analysis of complex systems and decision processes: IEEE Transactions on Systems, Man, and Cybernetics, v. SMC-3, 244
119. Zadeh, L.A.: "Fuzzy Logic, Neural Networks, and Soft Computing," *Communi-*

cations of the ACM (1994) 37(3), 77-84.

120. Zadeh, L. and Kacprzyk, J. (eds.): *Computing With Words in Information/Intelligent Systems 1: Foundations*, Physica-Verlag, Germany (1999).
121. Zadeh, L. and Kacprzyk, J. (eds.): *Computing With Words in Information/Intelligent Systems 2: Applications*, Physica-Verlag, Germany (1999).
122. Zadeh, L.A.: "Fuzzy Logic = Computing with Words," *IEEE Trans. on Fuzzy Systems* (1996) 4, 103-111.
123. Bishop, C.: *Neural Networks for Pattern Recognition*, Oxford University Press, NY (1995).

124. von Altrock, C., 1995, Fuzzy Logic and NeuroFuzzy Applications Explained: Prentice Hall PTR, 350p.

An Approach to the Mathematical Theory of Perception-Based Information

Victor Korotkikh

Faculty of Informatics and Communication
Central Queensland University
Mackay, Queensland, 4740
Australia
v.korotkich@cqu.edu.au

BISC, Computer Science Division
University of California, Berkeley
California, 94720
victor16@eecs.berkeley.edu

Abstract. In the paper we approach the mathematical theory of perception-based information by using a theory that captures a new type of processes. These processes are the formations of integer relations with the integers as the ultimate building blocks. The theory of integer relations and their formations stands out remarkably because it is irreducible, i.e., the theory can not be explained in terms of deeper principles. It is suggested that the optimal functioning of a complex system may be a law of the integer formations and perception-based information plays a key role in the use of this law.

1 Introduction

Fundamental ideas by Lotfi Zadeh about perception-based information [1]-[3] open a way to explore new avenues and raise deep questions. One of the most important is the mathematical theory of perception-based information.

The laws of human information processing may be still far away from our knowledge and experience. It seems that a better position would be to understand them within an irreducible theory, i.e., a theory not allowing a deeper explanatory base. In this case there would be no further reductions possible. No one could ask to explain such a theory and its basic notions in terms of a more profound theory. These very strong requirements challenge our belief that we could find an irreducible theory to start with. For this purpose even basic notions originated from spacetime cannot be viewed irreducible because it is increasingly regarded as an approximation to reality.

In the paper we approach the mathematical theory of perception-based information by using a theory that captures a new type of processes. These processes come to life in imaginative way, i.e., they are the formations of integer relations. The formations are hierarchical and have a natural order: starting with the integers, integer relations as elements of one level form more complex integer relations as elements of the next one. This order is captured by a concept of structural complexity.

The theory of integer relations and their formations is irreducible. The existence of integer relations and their formations, based on a universal principal, are completely determined and controlled by arithmetic. In the theory integers appear with a new meaning: they are the ultimate building blocks from which the integer relations are all formed.

We consider a connection between the theory of integer formations and perception-based information. This connection is due to an optimality condition of complex systems, which suggests that the optimal functioning of a complex system may be a law of the integer formations, i.e., the structural complexity of the complex system. Perception-based information appears naturally in the context of the optimality condition as an efficient means for organizing the control of the structural complexity of a complex system.

2 Sequences and Integer Code Series

When physical systems are considered in spacetime sequences appear naturally. They realize in a discreet form our understanding that physical systems move in space as time goes in the ordered fashion. A sequence gives an approximation to the trajectory of a physical system by recording the space position in successive time instants. Sequences are used also as the main carriers of information and objects of computation.

Let I be an integer alphabet and

$$I_n = \{s = s_1...s_n, \ s_i \in I, i = 1, ..., n\}$$

be the set of all sequences of length $n \geq 2$ with symbols in I. If $I = \{-1, +1\}$ then I_n is the set of all binary sequences of length n. Let $\delta > 0$ and $\varepsilon > 0$ be respective spacings of a spacetime lattice (δ, ε) in $1+1$ dimensions. A sequence $s = s_1...s_n \in I_n$ may encode on a lattice (δ, ε) the dynamics of a physical system that has a space position $s_i\delta$ at a time instant $i\varepsilon, i = 1, ..., n$.

An integer code series (ICS) [4] plays a key role in our consideration of sequences. It expresses integrals of piecewise constant functions in a form containing the sequence and powers of integers. In addition ICS allows us to extract these powers of integers in integer relations and view them as geometrical objects that can form into each other.

Let $W_{\delta\varepsilon}([t_m, t_{m+n}])$ be a set of piecewise constant functions

$$f : [t_m, t_{m+n}] \rightarrow \Re^1$$

such that each function f belonging to the set is constant on

$$(t_{i-1}, t_i], \ i = m+1, ..., m+n$$

and equals

$$f(t_m) = s_1\delta, \quad f(t) = s_i\delta, \ t \in (t_{i-1}, t_i], \ i = m+1, ..., m+n,$$

$$t_i = i\varepsilon, \ i = m, ..., m+n,$$

where m is an integer and $s_i, \ i = 1, ..., n$ are real numbers.

A piecewise constant function of the set $W_{\delta\varepsilon}([t_m, t_{m+n}])$ can be completely characterized by a sequence of real numbers. Namely, let a function $f \in W_{\delta\varepsilon}([t_m, t_{m+n}])$ and $s_i\delta$ be its value on an interval $(t_{i-1}, t_i], \ i = m+1, ..., m+n$, then a sequence $s = s_1...s_n$ is called a code of the function, denoted $c(f)$, as it completely determines the function in the set. This connection represents sequences in terms of piecewise constant functions (see Figure 1).

Let $f^{[k]}$ denote the kth $k = 1, 2, ...$ integral of a function

$$f \in W_{\delta\varepsilon}([t_m, t_{m+n}]),$$

i.e., a family of functions whose kth derivative equals f and $f^{[0]} = f$. The notation $f^{[k]}$ further stands for a function of the family implying that values $f^{[i]}(t_m), \ i = 1, ..., k$ specify the function. The integer code series gives the kth $k = 1, 2, ...$ integral of a function $f \in W_{\delta\varepsilon}([t_m, t_{m+n}])$ in terms of the code $c(f)$ and importantly in *powers of integers* [4].

Integer Code Series (Korotkikh). *Let $f \in W_{\delta\varepsilon}([t_m, t_{m+n}])$ be a piecewise constant function with the code $c(f) = s_1...s_n$. Then the kth $k \geq 1$ integral $f^{[k]}$ of the function f at a point $t_{m+l+1}, l = 0, ..., n-1$ can be given by*

$$f^{[k]}(t_{m+l+1}) = \sum_{i=0}^{k-1} \alpha_{kmi}((m+l+1)^i s_1 + ... + (m+1)^i s_{l+1})\delta\varepsilon^k$$

$$+ \sum_{i=1}^{k} \beta_{k,l+1,i} f^{[i]}(t_m)\varepsilon^{k-i}, \qquad (1)$$

where coefficients

$$\alpha_{kmi} = \binom{k}{i}((-1)^{k-i-1}(m+1)^{k-i} + (-1)^{k-i}m^{k-i})/k!,$$

$$\beta_{k,l+1,i} = \frac{(l+1)^{k-i}}{(k-i)!}, \quad i = 1, ..., k$$

and $\binom{k}{i}, i = 0, ..., k-1$ is the binomial coefficient.

3 Structural Numbers of Sequences and their Geometrical Interpretation

It is useful to consider a sequence $s = s_1 ... s_n \in I_n$ on a lattice (δ, ε) in terms of its

Structural Numbers $\vartheta_1(s), ..., \vartheta_k(s), ...$

$$s = s_1 ... s_n \Longrightarrow \vartheta_1(s), ..., \vartheta_k(s), ...$$

defined by the formula [4]

$$\vartheta_k(s) = \sum_{i=0}^{k-1} \alpha_{kmi}((m+n)^i s_1 + (m+n-1)^i s_2 + ... + (m+1)^i s_n)\delta\varepsilon^k, \quad (2)$$

where $k = 1, 2, ...$.

The meaning of the structural numbers is as follows. Let a mapping $\rho_{m\delta\varepsilon}$ take a sequence $s = s_1 ... s_n, s_i \in \Re^1, i = 1, ..., n$ to a function $f \in W_{\delta\varepsilon}[t_m, t_{m+n}]$, denoted $f = \rho_{m\delta\varepsilon}(s)$, such that $c(f) = s$ (see Figure 1) and whose kth integral satisfies $f^{[k]}(t_m) = 0,\ k = 1, 2, ...$. It follows, by comparing (1) and (2), that the kth structural number $\vartheta_k(s), k = 1, 2, ...$ of a sequence $s \in I_n$ on a lattice (δ, ε) is the value of the kth integral of a function

$$f = \rho_{m\delta\varepsilon}(s) \in W_{\delta\varepsilon}[t_m, t_{m+n}]$$

at the point t_{m+n}

$$\vartheta_k(s) = f^{[k]}(t_{m+n}), \quad k = 1, 2, ... \quad (3)$$

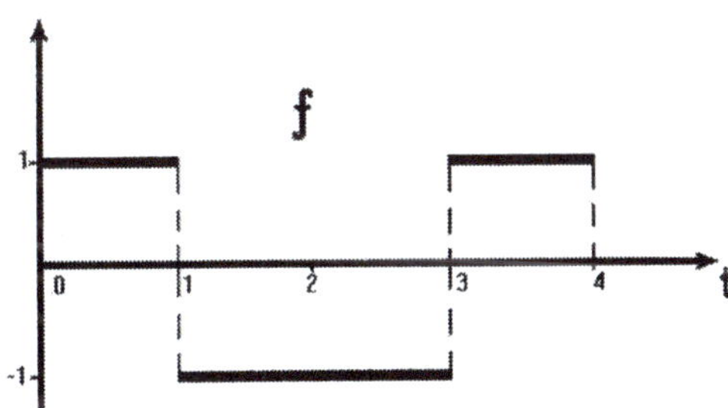

Fig. 1. Graph of a function $f = \rho_{011}(s),\ s = c(f) = s_1 ... s_4 = +1 - 1 - 1 + 1,\ \delta = 1,\ \varepsilon = 1, m = 0,\ n = 4,\ t_{m+n} = 4.$

To illustrate let us consider a function f such that (see Figure 1)

$$f = \rho_{011}(s),\ s = c(f) = s_1 ... s_4 = +1 - 1 - 1 + 1$$

and whose first $f^{[1]}$ and second $f^{[2]}$ integrals are shown in Figure 2 and Figure 3. From the figures we see that

$$f^{[1]}(t_4) = 0, \quad f^{[2]}(t_4) = 0$$

and by using (3), conclude

$$\vartheta_1(s) = 0, \quad \vartheta_2(s) = 0.$$

Importantly, by definition structural numbers are simply definite integrals but expressed in terms of the integer code series. Indeed, it is clear that the kth $k = 1, 2, ...$ structural number of a sequence s on a lattice (δ, ε) is the definite integral of a function $f^{[k-1]}$ on the interval $[t_m, t_{m+n}]$, where $f = \rho_{m\delta\varepsilon}(s)$. From this perspective structural numbers of a sequence carry information about the sequence as a whole and can be seen as its global characteristics.

Therefore, structural numbers of sequences have the geometrical interpretation of the definite integral. For example, the third structural number of sequence $s = +1 - 1 - 1 + 1$ (see Figure 1) is given by

$$\vartheta_3(s) = \int_{t_0}^{t_4} f^{[2]}(t)dt$$

and is equal to the area of the geometrical pattern located under the graph of the function $f^{[2]}$ and above the t-axis (see Figure 3).

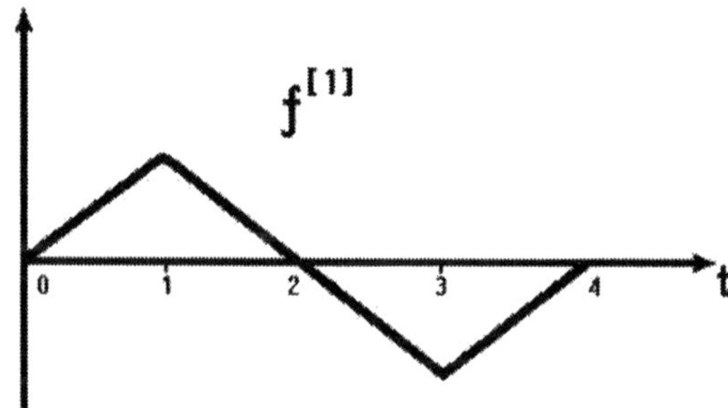

Fig. 2. First integral $f^{[1]}$ of function f in Figure 1.

The integer code series reveals a new role of the definite integral. It equates integer relations and geometrical patterns as well as their hierarchical formations. In particular, from one side ICS can express the definite integral in terms of powers of integers and arrange them in integer relations. From another side the geometrical interpretation of the definite integral links these integer relations with a geometrical pattern whose area is measured by the integral. This fact allows us to show that integer relations have a geometrical meaning and can form into each other.

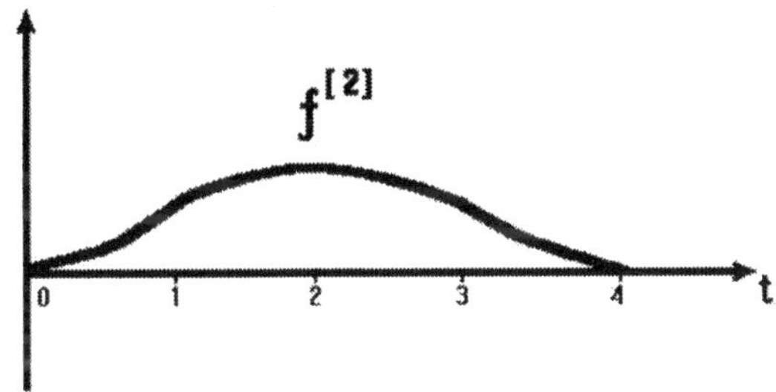

Fig. 3. Second integral $f^{[2]}$ of function f in Figure 1.

4 A Nonlocal Correlation Between Sequences

We present a nonlocal correlation between sequences [4].

The structural numbers can describe a sequence uniquely [4]. For a pair of different sequences $s, s' \in I_n$ let $C(s, s')$ denote an integer $k \geq 1$ such that the first k structural numbers of the sequences are equal

$$\vartheta_1(s) = \vartheta_1(s'), ..., \vartheta_k(s) = \vartheta_k(s'), \tag{4}$$

but the $(k+1)$th structural numbers are not

$$\vartheta_{k+1}(s) \neq \vartheta_{k+1}(s'). \tag{5}$$

If $\vartheta_1(s) \neq \vartheta_1(s')$ then $C(s, s') = 0$.

A sequence $s \in I_n$ can be uniquely specified with respect to another sequence $s' \in I_n$ by using the first $C(s, s')$ of their structural numbers.

Conditions (4) and (5) describe a nonlocal correlation between sequences $s = s_1...s_n$, $s' = s'_1...s'_n$ and $C(s, s')$ is a measure of it. In particular, the nonlocal correlation relates the components $s_i, s'_i, i = 1, ..., n$ because they are parts of sequences s, s', which as wholes must satisfy condition (4).

The following interpretation of condition (4) helps us to see why the correlation is nonlocal. Namely, condition (4) says that the graphs of functions $f^{[i]}, g^{[i]}, i = 1, ..., C(s, s')$ as trajectories associated with sequences s, s' start and finish at the same points

$$f^{[i]}(t_m) = g^{[i]}(t_m) = 0, \quad f^{[i]}(t_{m+n}) = g^{[i]}(t_{m+n}), \ i = 1, ..., C(s, s'),$$

but

$$f^{[C(s,s')+1]}(t_{m+n}) \neq g^{[C(s,s')+1]}(t_{m+n}),$$

where $f = \rho_{m\delta\varepsilon}(s)$ and $g = \rho_{m\delta\varepsilon}(s')$.

Figure 4 and Figure 5 illustrate the situation for sequences

$$s = +1 - 1 + 1 - 1 - 1 + 1 + 1 + 1, \quad s' = -1 - 1 + 1 + 1 + 1 + 1 + 1 - 1.$$

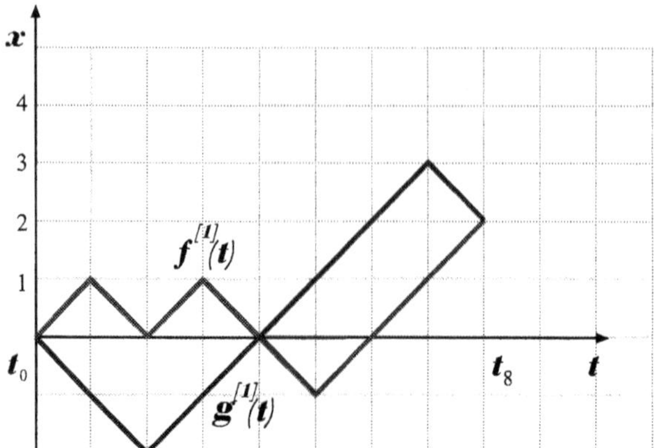

Fig. 4. First integrals $f^{[1]}$ and $g^{[1]}$ show the correlation. Starting at the same point at t_0, the integrals move differently in the interval but come together at t_8. As the move from one step to another is restricted, the integrals in the interval must be correlated to meet at the right end of the interval.

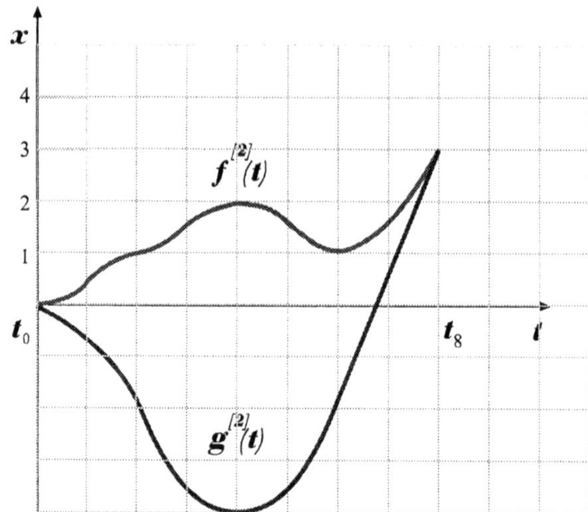

Fig. 5. Second integrals $f^{[2]}$ and $g^{[2]}$ show the correlation. Starting at the same point at t_0, the integrals move differently in the interval but come together at t_8.

In this case the correlation between the sequences s, s' results in the correlation between the first and second integrals (see Figure 4 and Figure 5)

$$f^{[1]}(t_8) = g^{[1]}(t_8), \quad f^{[2]}(t_8) = g^{[2]}(t_8)$$

of the functions $f = \rho_{011}(s), g = \rho_{011}(s')$. We can see in Figure 5 that since

$$f^{[2]}(t) > g^{[2]}(t), \quad t_0 < t < t_8,$$

then the correlation between the sequences does not provide the correlation between the third integrals of the functions

$$f^{[3]}(t_8) \neq g^{[3]}(t_8).$$

It turns out that this nonlocal correlation is connected with a new type of processes [4]. The delicacy of these processes is that they come in imaginative way: they are the formations of integer relations with the integers as the ultimate building blocks. These integer formations set up nonlocal correlations between integers. Therefore, the nonlocal correlation between two sequences is represented in terms of nonlocal correlations between integers. In the following sections we present how these processes can be revealed and where their power comes from.

5 A New Type of Processes: the Formations of Integer Relations and Integer Particles

An important step in discovering the formations of integer relations is made, when the nonlocal correlation $C(s, s')$ between sequences is considered in terms of the integer code series.

By using (3), for sequences $s = s_1...s_n$, $s' = s'_1...s'_n \in I_n$ such that $C(s, s') \geq 1$ we have

$$f^{[k]}(t_{m+n}) = \sum_{i=0}^{k-1} \alpha_{kmi}((m+n)^i s_1 + (m+n-1)^i s_2 + ... + (m+1)^i s_n)\varepsilon^k \delta,$$

$$g^{[k]}(t_{m+n}) = \sum_{i=0}^{k-1} \alpha_{kmi}((m+n)^i s'_1 + (m+n-1)^i s'_2 + ... + (m+1)^i s'_n)\varepsilon^k \delta,$$

where $f = \rho_{m\varepsilon\delta}(s)$, $g = \rho_{m\varepsilon\delta}(s')$ and $k = 1, ..., C(s, s')$.

This allows us to represent the nonlocal correlation $C(s, s')$ between sequences $s = s_1...s_n$, $s = s'_1...s'_n$ in terms of their components. Namely, it is proved that if $C(s, s') \geq 1$ then we have $C(s, s') \leq n$ and condition (4) reduces to a system of $C(s, s')$ equations

$$(m+n)^0(s_1 - s'_1) + ... + (m+1)^0(s_n - s'_n) = 0$$

$$. \quad . \quad . \quad . \quad . \quad .$$

$$(m+n)^{C(s,s')-1}(s_1 - s'_1) + ... + (m+1)^{C(s,s')-1}(s_n - s'_n) = 0 \qquad (6)$$

and condition (5) results in an inequality

$$(m+n)^{C(s,s')}(s_1 - s'_1) + ... + (m+1)^{C(s,s')}(s_n - s'_n) \neq 0. \qquad (7)$$

in integers $(s_i - s'_i)$, $i = 1, ..., n$ [4].

Notice, that if $C(s,s') = n$ then system (6) appears with the matrix

$$\begin{pmatrix} (m+n)^0 & (m+n-1)^0 & \dots & (m+1)^0 \\ (m+n)^1 & (m+n-1)^1 & \dots & (m+1)^1 \\ \cdot & \dots & & \cdot \\ (m+n)^{n-1} & (m+n-1)^{n-1} & \dots & (m+1)^{n-1} \end{pmatrix}$$

whose determinant is Vandermonde one.

Moreover, it is proposed that system (6) can be interpreted as a record of a *formation process* [4]. These processes are of a new type, i.e., they are hierarchical formations of integer relations. We illustrate this interpretation.

Notice, system (6) can be viewed as a system of integer relations expressing specific correlations between powers i, $i = 0, ..., k-1$ of integers $m+n, ..., m+1$. This is especially clear in concrete cases.

For example, consider sequences

$$s = +1-1-1+1-1+1+1-1-1+1+1-1+1-1-1+1,$$

$$s' = -1+1+1-1+1-1-1+1+1-1-1+1-1+1+1-1,$$

which are the initial segments of length 16 of the Prouhet-Thue-Morse (PTM) sequence starting with $+1$ and -1 respectively and $n = 16, m = 0$. It can be shown that $C(s,s') = 4$ and (6) becomes a system of specific integer relations

$$+16^0-15^0-14^0+13^0-12^0+11^0+10^0-9^0-8^0+7^0+6^0-5^0+4^0-3^0-2^0+1^0=0$$

$$+16^1-15^1-14^1+13^1-12^1+11^1+10^1-9^1-8^1+7^1+6^1-5^1+4^1-3^1-2^1+1^1=0$$

$$+16^2-15^2-14^2+13^2-12^2+11^2+10^2-9^2-8^2+7^2+6^2-5^2+4^2-3^2-2^2+1^2=0$$

$$+16^3-15^3-14^3+13^3-12^3+11^3+10^3-9^3-8^3+7^3+6^3-5^3+4^3-3^3-2^3+1^3=0 \quad (8)$$

whereas (7) does not follow the character in (8)

$$+16^4-15^4-14^4+13^4-12^4+11^4+10^4-9^4-8^4+7^4+6^4-5^4+4^4-3^4-2^4+1^4 \neq 0. \quad (9)$$

For clarity common factor 2 is cancelled out in (8) and (9).

System (6) opens a way to a reality of hierarchical formations of integer relations. Integer relations behave there in an interesting manner that could be rather expected from physical systems. We show and explain a picture of this reality by using (8) and (9).

There are integer relations of the similar character that are not immediately evident from system (8), but can be identified by a more careful analysis. These integer relations demonstrate connections, indicated by edges (see Figure 6). The connections between an integer relation and a number of integer relations are in place if the integer relation can be produced from the integer relations by a mathematical operation or principle.

The operation takes a number of integer relations with the same power and produces from them an integer relation in the following way. Firstly, it modifies the left part of the integer relations by increasing the power of the

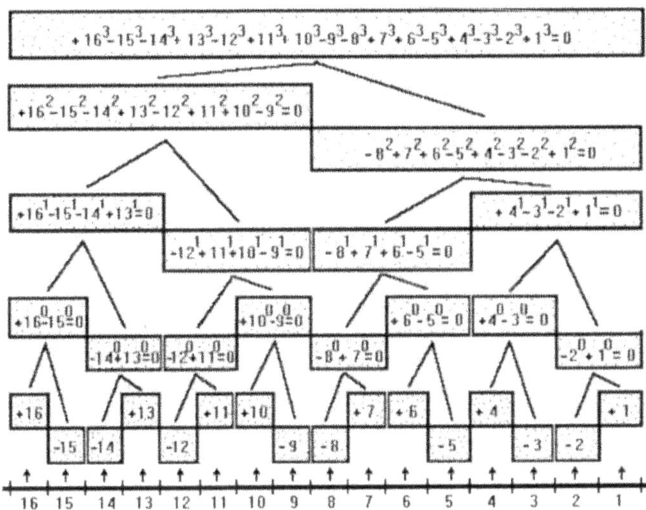

Fig. 6. Represented in this form system (8) and inequality (9) appear as a hierarchical formation of integer relations with integers 16, ..., 1 as the ultimate building blocks. In the hierarchical formation all integer relations share the same organizing principle.

integers by 1. Secondly, the operation combines the modified parts together. Surprisingly, they result in zero and give the integer relation. It may be said that the mathematical operation harmoniously integrates the integer relations in the integer relation as a new whole (see Figure 6).

An interesting picture of formations appears when the integer relations are interpreted as objects or even as some sort of integer particles that combine and form a composite particle according to the operation (see Figure 7 and Figure 8 as an illustration). In this case the operation can be seen as the organizing principle of these formations as it is shared by all the integer relations.

We are not used to see that an integer relation can be formed from another integer relations. This process is unusual for us but we are familiar with the formations of physical systems. Physical systems can interact. When certain conditions are in place they combine as parts and form a composite system as a new whole. In the composite system the parts are harmoniously integrated so that it exhibits new emergent properties not existing in the collection of parts. In this situation the whole is more than the simple sum of the parts. These phenomena of physical systems give us perceptions we try to explain by the concept of complexity. In particular, it is usually said that a composite system is more complex than the parts.

Composite systems can also interact and form even more complex systems as new wholes and so on. This results in the hierarchy of complexity levels

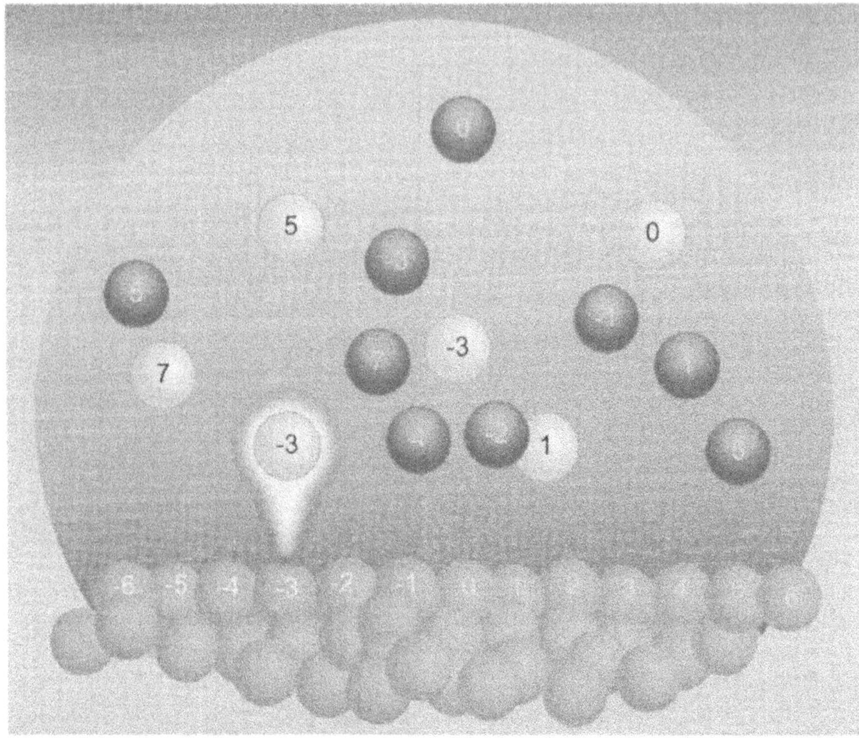

Fig. 7. In the figure integers seem like particles that can combine and form composite particles. There can be positive integer particles (shown in black) and negative integer particles (shown in white). For example, we can see a positive integer particle $+(-3)$ and a negative integer particle $-(-3)$.

we observe in physical systems. Complexity theory aims to find a universal organizing principle that governs the formations of physical systems across the complexity levels.

It comes to our attention that the formations of integer relations in some sense resemble the formations of physical systems. This analogy becomes clearer when the formations of integer relations are arranged in a hierarchical formation (as in Figure 6) with each integer relation belonging to one of four levels. We can observe that following the organizing principle elements of each level, except the top one, combine and form elements of the next level.

The two complementary levels added below the four levels complete the picture and make the hierarchical formation consistent (see Figure 6). It is worth to note that integers $16, ..., 1$ act as the ultimate building blocks of this hierarchical formation. The first complementary level, called the integers level, may be thought as a source that can generate on the second complementary level, called zero level, integers in two states, i.e., negative and positive (see Figure 7). For example, integer 3 is generated in the negative

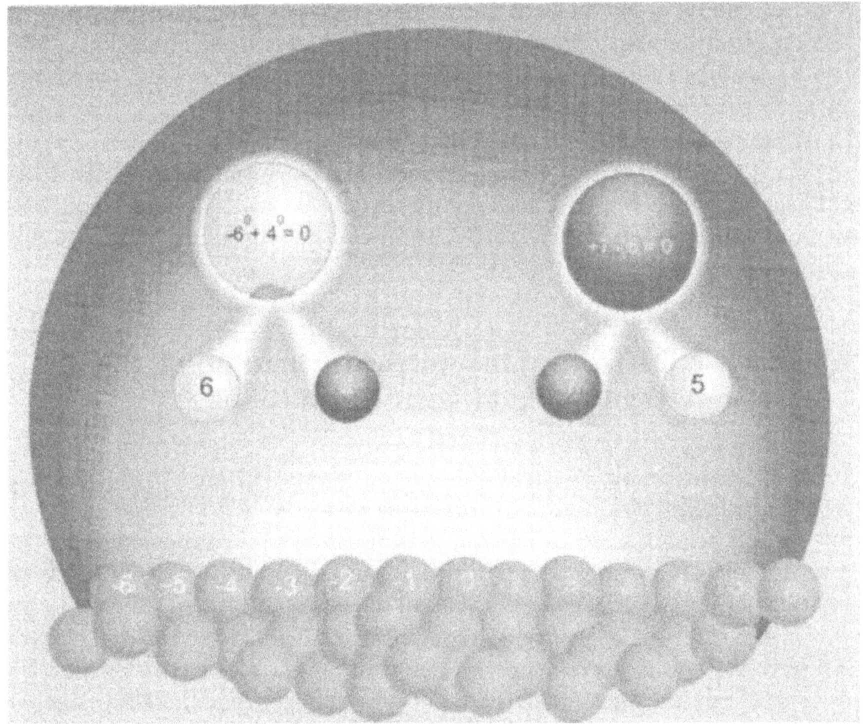

Fig. 8. The figure illustrates the formation of composite integer particles from integer particles and how they can be described. Negative integer particle 6 and positive integer particle 4 under the "interaction" produce $-(6)^0 + 4^0$, which is a composite integer particle made of them because $-(6)^0 + 4^0 = 0$.

state while integer 15 is generated in the positive state on the zero level. The elements and their states generated on the zero level of this hierarchical formation are shown in Figure 6. The elements of the zero level combine and form elements of the first level, which are the integer relations with power 0. In these formations the state of an element of the zero level is translated into the arithmetic sign in the integer relation.

When all these levels are in place, the hierarchical formation can be represented by a diagram as in Figure 6. The diagram reads that integers in certain states are generated on the zero level, which in turn form integer relations of the first level, which in turn form integer relations of the second level and so on till the fourth level. There is only one element at the top of the diagram, which can not form an element of the next level by itself because of (9). The diagram shows how each integer relation is formed level by level.

Therefore, by considering the nonlocal correlation between the sequences s, s' in terms of structural numbers, we get system (8) and inequality (9). The analysis of system (8) and inequality (9) reveals a complexity type process.

This process is a hierarchical formation of integer relations. This sets up a connection between the sequences and the hierarchical formation, which can be used to characterize the sequences s, s' and the nonlocal correlation between them in particular.

For example, we may say that the complexity of sequence s with respect to sequence s' is quite enough to generate (8), which is intuitively perceived as a complex phenomenon. At the same time it may be said that this complexity is not enough, as we observe in (9), to make this phenomenon by one unit more complex.

6 Integer Patterns: Geometrical Meaning of the Integer Relations and their Formations

In the previous section system (8) and inequality (9) are interpreted as a record of a hierarchical formation of integer relations. The concepts of integer and integer relations are an integral part of our mental equipment. They are a set of abstract symbols subject to the laws of arithmetic. Usually, integer relations appear as solutions to Diophantine equations.

However, integer relations do not appear to us rooted in reality. They do not have the power to evoke in our minds images of objects that like physical ones can form into each other. The concept of integer relation is too abstract to give us insight what we think the formation is all about. It is difficult to imagine how integer relations can be really formed from other integer relations, because it is difficult to visualize how this may actually happen.

To make a step in this direction integer relations in the previous section were imagined like integer particles. But this representation lacks any quantitative means to describe the integer relations and their formations. For example, in Figure 9 a hierarchical formation of integer relations is shown as a hierarchical formation of integer particles pictured as spheres. However, there is no connection between the integer relations and the spheres so that these geometrical objects and their transformations could be used to describe the integer relations and their formations. A question arises: is it possible to see how integer relations look like and how they form into each other?

To extend the scope of integer relations so that it would be possible to answer the question and in particular to develop a new understanding of them as geometrical objects that can form into each other a notion of integer pattern is introduced [4]. Integer patterns provide a geometrical meaning and visualize the integer relations and their hierarchical formations. We begin to present integer patterns by a notion of

Pattern of a Function. *Let $f : [a, b] \to \Re^1$ be a piecewise constant or continuous function. Let*

$$P(f, t) = \{(t, x) \in \Re^2, \min\{0, f(t)\} \leq x \leq \max\{0, f(t)\}\},$$

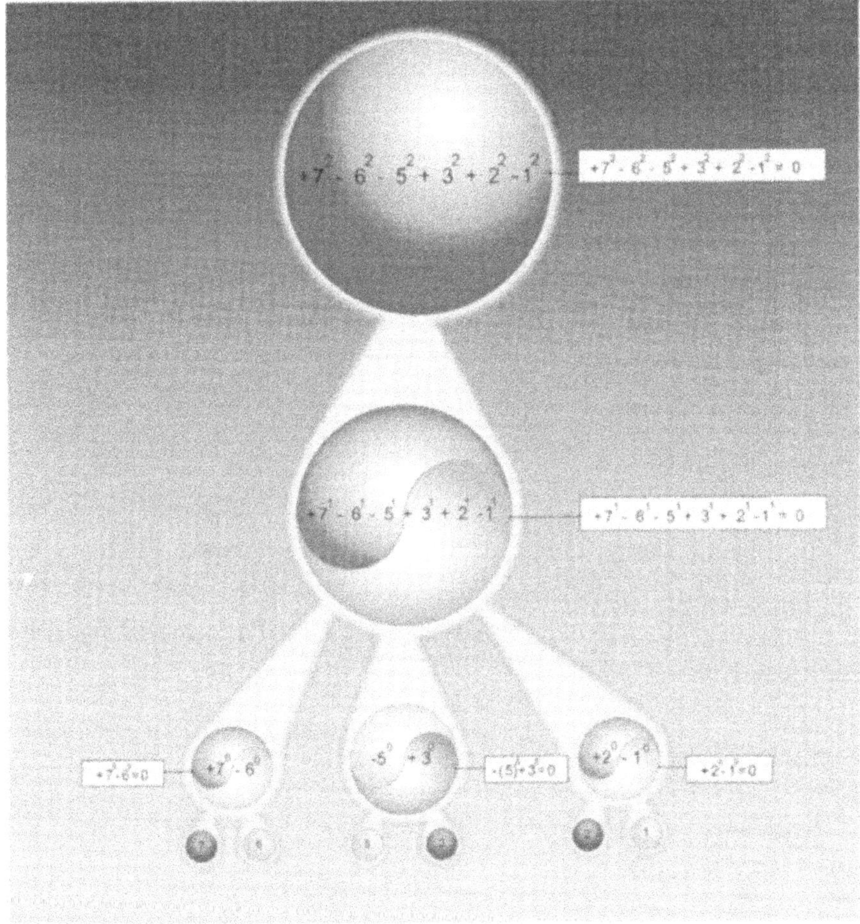

Fig. 9. A hierarchical formation of integer relations is shown as a hierarchical formation of integer particles pictured as spheres. In each integer relation of the first level, the parts are opposite but inseparable and complementary to one another to make the relation. The well-known Chinese symbol of yin and yang, which stands for the conflict and synthesis of the opposites, is used to illustrate this fact.

if function f is continuous at a point $t \in [a,b] \subset \Re^1$, and

$$P(f,t) =$$

$$\{(t,x) \in \Re^2, \min\{0, f(t-0), f(t+0)\} \leq x \leq \max\{0, f(t-0), f(t+0)\}\},$$

if function f is discontinuous at the point $t \in (a,b)$. Moreover, if function f is discontinuous at the point $t = a$ then

$$P(f,t) = \{(t,x) \in \Re^2, \min\{0, f(a+0)\} \leq x \leq \max\{0, f(a+0)\}\},$$

and if function f is discontinuous at the point $t = b$ then

$$P(f,t) = \{(t,x) \in \Re^2, \min\{0, f(b-0)\} \leq x \leq \max\{0, f(b-0)\}\}.$$

The pattern of a function f is defined by

$$P(f,[a,b]) = \{P(f,t), t \in [a,b]\}.$$

To illustrate the definition the pattern of a function f is brought out by shading in Figure 10.

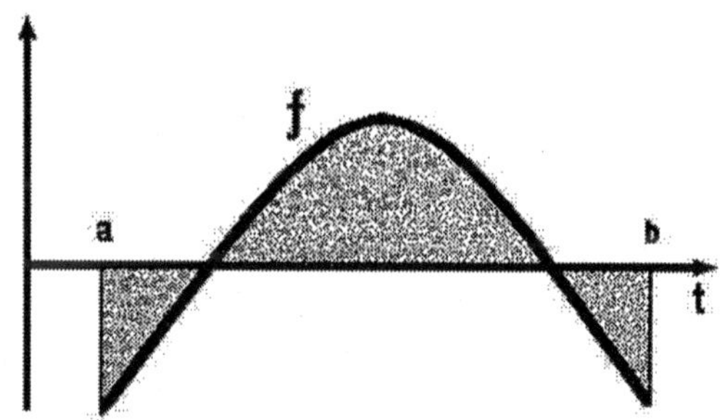

Fig. 10. The pattern $P(f,[a,b])$ (shaded) of a function f. The shaded area allows us to see the pattern as a geometrical object.

Furthermore, we describe integer patterns as patterns with a special character. For a function

$$f \in W_{\delta\varepsilon}([t_m, t_{m+n}]), \quad f = \rho_{m\delta\varepsilon}(s), \quad s = s_1 ... s_n \in I_n \tag{10}$$

let $T(f,[t_m,t_{m+n}])$ be a set of intervals $[t_{m+i}, t_{m+i+1}] \subset [t_m, t_{m+n}]$ such that

$$s_{i+1} \neq 0, \; i = 0, ..., n-1.$$

Let $T(f^{[k]}, [t_m, t_{m+n}]), \; k = 1, 2, ...$ be a set of intervals

$$[t_i, t_j] \subseteq [t_m, t_{m+n}], \quad m \leq i, \quad i+1 < j \leq m+n$$

such that

$$f^{[r]}(t_i) = 0, \quad f^{[r]}(t_j) = 0, \quad r = 1, ..., k,$$

and there exists no $t_l, l = i+1, ..., j-1$ such that $f^{[r]}(t_l) = 0$ for all $r = 1, ..., k$ at one time.

Integer Pattern. *The pattern $P(f^{[k]}, [t_i, t_j]), \; k = 0, 1, ...$ of the kth integral of a function f given by (10) is called an integer pattern if*

$$[t_i, t_j] \in T(f^{[k]}, [t_m, t_{m+n}]) \neq \emptyset.$$

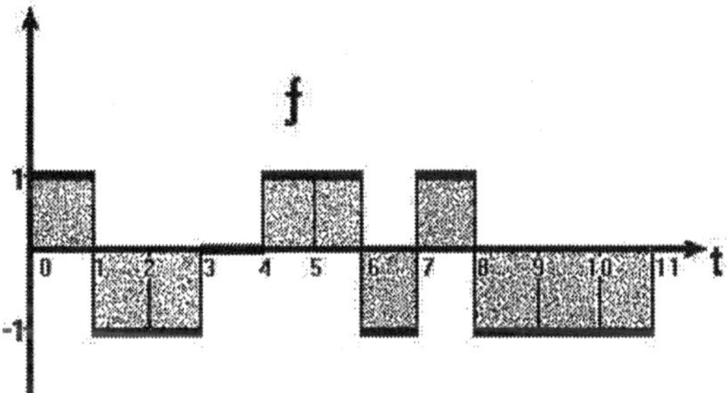

Fig. 11. Integer patterns (shaded) of a function f, where $f = \rho_{011}(s)$, $s = +1 - 1 - 1\ 0 + 1 + 1 - 1 + 1 - 1 - 1 - 1 \in I_{11}$, $I = \{-1, 0, +1\}$.

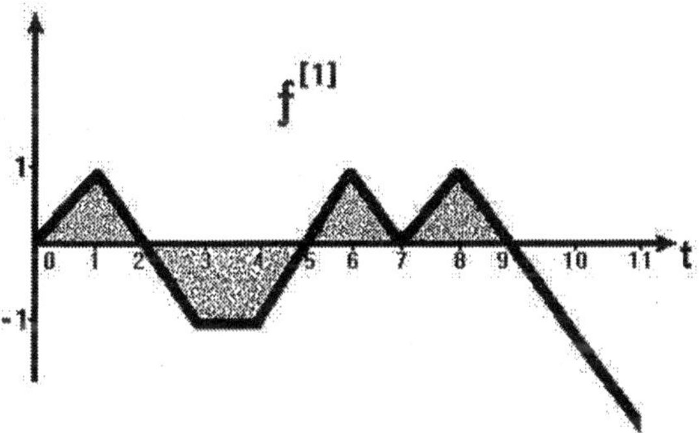

Fig. 12. Integer patterns (shaded) of the first integral $f^{[1]}$ of function f in Figure 11.

We illustrate the definition of integer pattern by examples. For instance, Figure 11 shows integer patterns

$$P(f, [t_0, t_1]), \quad P(f, [t_1, t_2]), \quad P(f, [t_2, t_3]), \quad P(f, [t_4, t_5]), \quad P(f, [t_5, t_6]),$$

$$P(f, [t_6, t_7]), \quad P(f, [t_7, t_8]), \quad P(f, [t_8, t_9]) \quad P(f, [t_9, t_{10}]), \quad P(f, [t_{10}, t_{11}])$$

of a function $f = \rho_{011}(s)$, where

$$s = +1 - 1 - 1\ 0 + 1 + 1 - 1 + 1 - 1 - 1 - 1 \in I_{11}, \quad I = \{-1, 0, +1\},$$

and Figure 12 shows integer patterns

$$P(f^{[1]}, [t_0, t_2]), \quad P(f^{[1]}, [t_2, t_5]), \quad P(f^{[1]}, [t_5, t_7]), \quad P(f^{[1]}, [t_7, t_9])$$

of its first integral $f^{[1]}$.

The integer pattern $P(f^{[2]}, [t_0, t_7])$ of the second integral $f^{[2]}$ and the integer pattern $P(f^{[3]}, [t_0, t_7])$ of the third integral $f^{[3]}$ are brought out by shading in Figure 13 and Figure 14 accordingly. Integer pattern $P(f^{[2]}, [t_0, t_7])$ consists of two patterns

$$P(f^{[2]}, [0, 3.5]), \quad P(f^{[2]}, [3.5, 7]), \tag{11}$$

which are not integer patterns because two conditions are not satisfied,

$$f^{[1]}(3.5) \neq 0$$

and 3.5 is not an integer.

These two patterns are identical in shape and like opposite to each other. They may be viewed coupled to give the integer pattern $P(f^{[2]}, [t_0, t_7])$. The integer pattern $P(f^{[2]}, [t_0, t_7])$ forms into the integer pattern $P(f^{[3]}, [t_0, t_7])$ under the integration of $f^{[2]}$. This may be pictured as if patterns (11) merged together in the formation of the integer pattern $P(f^{[3]}, [t_0, t_7])$.

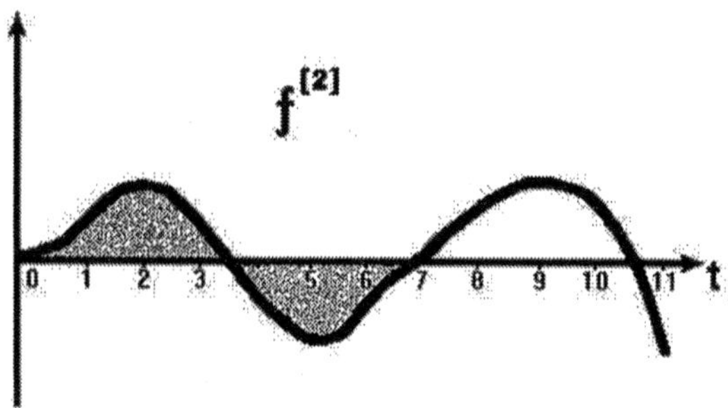

Fig. 13. The integer pattern $P(f^{[2]}, [t_0, t_7])$ of the second integral $f^{[2]}$ of function f depicted in Figure 11.

It follows from general results presented in the next section that integer pattern $P(f^{[2]}, [t_0, t_7])$ corresponds to an integer relation

$$\boxed{+7^1 - 6^1 - 5^1 + 3^1 + 2^1 - 1^1 = 0} \tag{12}$$

and integer pattern $P(f^{[3]}, [t_0, t_7])$ corresponds to an integer relation

$$\boxed{+7^2 - 6^2 - 5^2 + 3^2 + 2^2 - 1^2 = 0} \tag{13}$$

Clearly, the organizing principle forms integer relation (13) from integer relation (12). The correspondence between the integer patterns and the integer relations illustrates the geometrical interpretation of the organizing principle. Namely, the organizing principle forms (13) from (12) as the integration of $f^{[2]}$ forms $P(f^{[3]}, [t_0, t_7])$ from $P(f^{[2]}, [t_0, t_7])$.

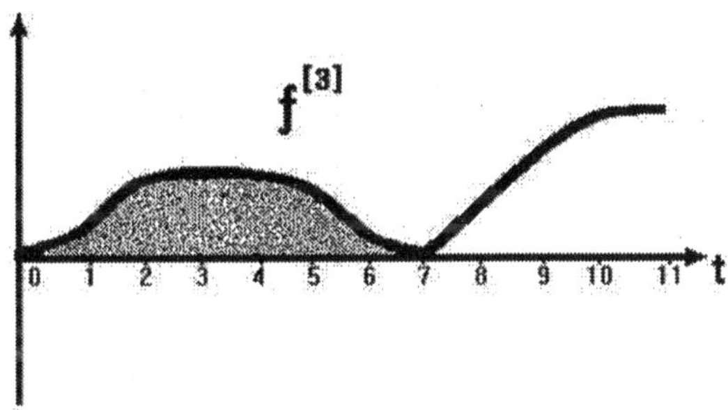

Fig. 14. The integer pattern $P(f^{[3]}, [t_0, t_7])$ of the third integral $f^{[3]}$ of function f depicted in Figure 11. The integer pattern $P(f^{[2]}, [t_0, t_7])$ in Figure 13 forms into the integer pattern $P(f^{[3]}, [t_0, t_7])$ under the integration of $f^{[2]}$.

Figure 13 and Figure 14 in comparison with Figure 9 allow us to see how integer relations (12) and (13) look like and how they form into each other. Naturally, these integer patterns propose themselves to quantitatively describe these integer relations. In particular, the areas of the integer patterns can be used to measure the integer relations.

7 Isomorphism between Formations of Integer Relations and Integer Patterns

The key idea about system (6) and inequality (7) is to interpret them as a record of a hierarchical formation of integer relations [4].

Firstly, it is shown that system (6)

$$(s_1 - s'_1)(m+n)^0 + ... + (s_n - s'_n)(m+1)^0 = 0$$

$$. \quad . \quad . \quad . \quad . \quad .$$

$$(s_1 - s'_1)(m+n)^{C(s,s')-1} + ... + (s_n - s'_n)(m+1)^{C(s,s')-1} = 0 \qquad (14)$$

can be associated with a hierarchical set $WR(s, s', n, m, I_n)$ of $C(s, s') \geq 1$ levels whose elements of level $k = 1, ..., C(s, s')$ are integer relations of the form

$$\boxed{\lambda_1 d_1^{k-1} + ... + \lambda_l d_l^{k-1} = 0} \qquad (15)$$

where $\lambda_i, d_i,\ i = 1, ..., l$ are integers such that $d_i > d_{i+1},\ i = 1, ..., l-1$ and k is the power of integers $d_i,\ i = 1, ..., l$.

Secondly, it is interpreted that elements of level $k = 2, ..., C(s, s')$ of the set $WR(s, s', n, m, I_n)$ are formed from elements of level $(k-1)$ according to an organizing principle. The principle can be described mathematically as follows. If $r \geq 1$ integer relations

$$\boxed{\lambda_{i1} d_{i1}^{k-1} + ... + \lambda_{il(i)} d_{il(i)}^{k-1} = 0} \qquad (16)$$

of level $k = 1, 2..., C(s, s') - 1$ with the ith $i = 1, ..., r$ relation containing $l(i)$ terms satisfy

$$\boxed{\sum_{i=1}^{r} \lambda_{i1} d_{i1}^{k} + ... + \lambda_{il(i)} d_{il(i)}^{k} = 0} \tag{17}$$

and the inclusion of each of the integer relations (16) is important for (17), then it is said that integer relation (17) is formed from integer relations (16).

In general, integer relations of type (16) do not form an integer relation, because integer relation (17) is more than their simple sum

$$\sum_{i=1}^{r} \lambda_{1i} d_{i1}^{k-1} + ... + \lambda_{il(i)} d_{il(i)}^{k-1} = 0. \tag{18}$$

The power of integers d_{ij}, $i = 1, ..., r$, $j = 1, ..., l(i)$ in (17) is increased by 1 compared to (18). This means that integer relations (16) must have a special property to form integer relation (17).

The interpretation of the set $WR(s, s', n, m, I_n)$ as a hierarchical formation becomes complete when two complementary levels are added below the levels of the set. The functioning of these two levels can be made consistent with the integer relations and their formations (see Figure 6).

In particular, the first complementary level, called the integers level, consists of integers. It may be thought as a source that can generate integers in any amount at the second complementary level, called the zero level, in two states, i.e., negative and positive.

For the set $WR(s, s', n, m, I_n)$ integer $(m + n - i), i = 0, ..., n - 1$ as an element is generated in an amount of $(s_i - s_i')$ in a state specified by the sign of $(s_i - s_i')$. The elements of the zero level combine and form elements of the first level of the set $WR(s, s', n, m, I_n)$. In these formations the state of an element of the zero level is translated into the arithmetic sign in the integer relation. Then the elements of the first level form the elements of the second level and so on till level $C(s, s')$. This gives that the elements of the levels, starting with the integers of the integers level, become connected into one hierarchical formation $WR(s, s', n, m, I_n)$. Remarkably, the integers act in this hierarchical formation as its ultimate building blocks.

System (14) stands as a record of the hierarchical formation. Inequality (7) in this case becomes

$$(s_1 - s_1')(m + n)^{C(s,s')} + ... + (s_n - s_n')(m + 1)^{C(s,s')} \neq 0 \tag{19}$$

and tells us that the hierarchical formation can not propagate to the next level $C(s, s') + 1$. Namely, the elements of level $C(s, s')$ can not all form elements on the next level because of (19).

Usually, the left side of integer relation

$$\lambda_1 d_1^{k-1} + ... + \lambda_l d_l^{k-1} = 0 \tag{20}$$

is a recipe for the manipulations of the symbols based on the laws of arithmetic. The integer relation (20) reads that the quantity on the left side, i.e.,

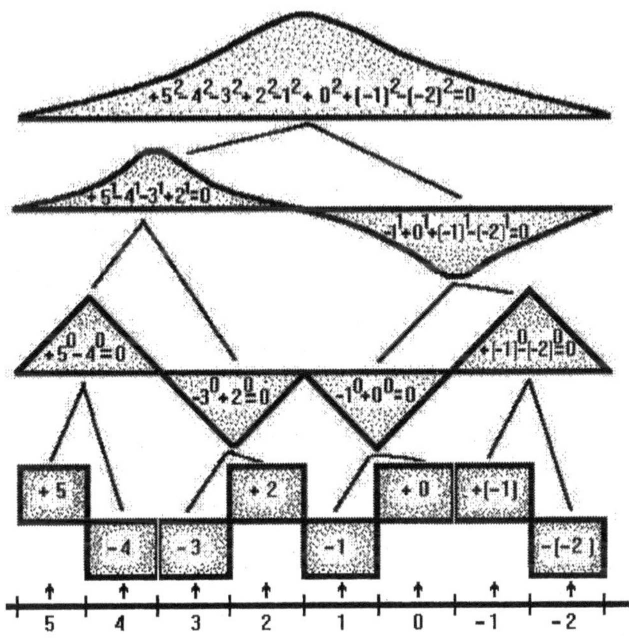

Fig. 15. A hierarchical formation of integer relations and a corresponding hierarchical formation of integer patterns are displayed together to show their unified character. The figure allows us to see how the integer relations form into each other. Note that an integer relation can be measured by the area of a corresponding integer pattern.

the result of the manipulations, is equal to the quantity on the right, i.e., zero.

A new meaning of (20) appears in the context of the integer formations. In our case integer relation

$$\boxed{\lambda_1 d_1^{k-1} + ... + \lambda_l d_l^{k-1} = 0} \tag{21}$$

is an element of level $k = 1, ..., C(s, s')$ of a hierarchical formation. It is formed from elements of the previous level and is a building block for elements of the next level. The left side of (21), called the code of the integer relation, contains information about how the element is formed. The fact that the code of an integer relation equals zero is the criterion of its existence. To show the new meaning of the integer relations as elements of the hierarchical formations they are placed in boxes like in (21).

We may ask the question: is there something happening when the formation of a new integer relation takes place? Integer relations do not appear to us rooted in reality and have no power to evoke in our minds images of objects like physical ones. As a result the notion of formation of integer relations is too abstract to give us insight what we think the formation is all about and tells us nothing about its nature. It is difficult to imagine how integer relations can be really formed from other integer relations, because it is not possible to visualize how this may actually happen. For example, the formations of elements in Figure 6 appear as a mystery because the laws involved are not clear.

Remarkably, by using the integer code series it is proved that there exists an isomorphism $\psi_{\delta\varepsilon}$ between elements of a set $WR(s, s', n, m, I_n)$, i.e., integers and integer relations, and integer patterns of integrals $f^{[k]}, k = 0, ..., C(s, s')$ of a function

$$f = \rho_{m\delta\varepsilon}(s - s') \in W_{\delta\varepsilon}([t_m, t_{m+n}]) \tag{22}$$

as well as between their hierarchical formations [4].

As a result a hierarchical formation of integer patterns $WP_{\delta\varepsilon}(s, s', n, m, I_n)$ is defined by the isomorphism

$$\psi_{\delta\varepsilon} : WR(s, s', n, m, I_n) \Longleftrightarrow WP_{\delta\varepsilon}(s, s', n, m, I_n).$$

The isomorphism $\psi_{\delta\varepsilon}$ gives a geometrical meaning to the integer relations and their formations in terms of integer patterns and their formations. In particular, the organizing principle appears in terms of function integration.

The isomorphism visualizes a hierarchical formation of integer relations

$$WR(s, s', n, m, I_n)$$

by a corresponding hierarchical formation of integer patterns

$$WP_{\delta\varepsilon}(s, s', n, m, I_n)$$

which can be represented as follows.

The integer patterns of the kth $k = 0, ..., C(s, s')$ integral $f^{[k]}$ of a function f given by (22) are viewed as elements of level k of the hierarchical formation

$$WP_{\delta\varepsilon}(s, s', n, m, I_n).$$

There is a coordinate system with the graph of the $(k+1)$th integral $f^{[k+1]}$, $k = 0, ..., C(s, s') - 1$ at level $(k+1)$ of the hierarchical formation. This coordinate system is placed at some distance above level k, which has a coordinate system with the graph of the kth integral $f^{[k]}$. There are edges between integer patterns on level k and integer patterns on level $(k+1)$. The edges show how the integration of the kth integral $f^{[k]}$ forms integer patterns of level $(k+1)$ from integer patterns of level k.

The isomorphism is illustrated in Figure 15, where a hierarchical formation of integer relations and a corresponding hierarchical formation of integer patterns are shown together. We can see that the integer relations appear as two-dimensional geometrical objects that form into each other. In particular, an integer relation has a shape of a corresponding integer pattern and can be measured by its area. The hierarchical formation of integer patterns reveals self-similarity, local and nonlocal symmetries. Figure 15 gives a static picture of the hierarchical formation and may be thought as a sequence of snapshots, one at each level of the hierarchical formation.

Results of the previous sections can be summarized by a diagram that shows a connection between sequences and hierarchical formations of integer relations and integer patterns

$$\{\,\textit{Sequences in Spacetime}\,\}$$

$$\Downarrow$$

$$\{\,\textit{Structural Numbers of Sequences (2)}\,\}$$

$$\Downarrow$$

$$\{\,\textit{Nonlocal Correlation between Sequences (4)}\,\}$$

$$\Downarrow$$

$$\left\{\begin{matrix}\textit{System of Linear Equations (6)}\\ \textit{and Inequality (7)}\end{matrix}\right\}$$

$$\Downarrow$$

$$\left\{\begin{matrix}\textit{Hierarchical Formation}\\ \textit{of Integer Relations}\end{matrix}\right\} \underset{\textit{isomorphism}}{\Longleftrightarrow} \left\{\begin{matrix}\textit{Hierarchical Formation}\\ \textit{of Integer Patterns}\end{matrix}\right\}$$

8 Formations of Integer Relations: Irreducible Theory

In the previous sections we considered a nonlocal correlation between two sequences and showed that it had the nature of a hierarchical formation of integer relations. It may be said that an integer formation sets up a nonlocal correlation between spacetime objects, i.e., sequences.

These hierarchical formations of integer relations may be considered as a reality of its own right. The reality of the integer formations, called a web of relations, can be defined as follows

$$WR(I_n) = \bigcup_{m\in\mathbf{Z}} \bigcup_{s\in I_n} \bigcup_{s'\in I_n} WR(s, s', n, m, I_n),$$

$$WR(I) = \lim_{n\to\infty} WR(I_n), \quad WR = WR(\mathbf{Z}),$$

where all the pairs of sequences $s, s' \in I_n$ are considered, $n \to \infty$ and integer alphabet I is the set of all integers $\mathbf{Z}$ [4].

The isomorphism $\psi_{\delta\varepsilon}$ defines a web of integer patterns that for a given δ and ε incorporates the hierarchical formations of integer patterns corresponding to WR into one whole

$$\psi_{\delta\varepsilon} : WR \to WP_{\delta\varepsilon}.$$

Due to the isomorphism WR and $WP_{\delta\varepsilon}$ can be viewed as one structure, i.e., called the web of relations, whose elements have a dual character. Each element of the structure is an integer relation as well as an integer pattern specified by δ and ε. The web of relations has rich properties.

Firstly, the web of relations is an irreducible theory because it is based on the integers only. This theory is about the integer relations and their formations. The existence of these integer relations and their formations, based on the organizing principle, are completely determined by the arithmetic. In the theory the integers appear with a new meaning: they are the ultimate building blocks from which the integer relations are all formed.

For example, consider two propositions about integers written as integer relations

$$+5^1 - 4^1 - 3^1 + 2^1 = 0 \qquad -1^1 + 0^1 + (-1)^1 - (-2)^1 = 0. \tag{23}$$

These propositions are true because in each of them the left side indeed equals the right side. The existence of these integer relations follows from the existence of the integers involved and the laws of arithmetic.

From our point of view integer relations (23) are elements

$$\boxed{+5^1 - 4^1 - 3^1 + 2^1 = 0}$$

$$\boxed{-1^1 + 0^1 + (-1)^1 - (-2)^1 = 0} \tag{24}$$

of the second level of the web of relations because they are formed from elements of the first level (see Figure 15), i.e., the first one from elements

$$\boxed{+5^0 - 4^0 = 0}$$

$$\boxed{-3^0 + 2^0 = 0}$$

and the second one from elements

$$\boxed{-1^0 + 0^0 = 0}$$

$$\boxed{+(-1)^0 - (-2)^0 = 0}$$

The existence of integer relations (23) gives the solid ground for the existence of elements (24).

Can elements (24) form an element of the third level? The answer to this question is "yes" and we have (see Figure 15)

$$\boxed{+5^2 - 4^2 - 3^2 + 2^2 - 1^2 + 0^2 + (-1)^2 - (-2)^2 = 0} \tag{25}$$

This formation is completely determined by the laws of arithmetic. However, in its turn two elements

$$\boxed{+5^1 - 4^1 - 3^1 + 2^1 = 0}$$

$$\boxed{-1^1 + (-1)^1 + (-2)^1 - (-4)^1 = 0} \quad (26)$$

of the second level of the web of relations can not form an element of the third level because according to the same laws

$$+5^2 - 4^2 - 3^2 + 2^2 - 1^2 + (-1)^2 + (-2)^2 - (-4)^2 \neq 0.$$

Thus, the laws of arithmetic forbid the formation of an element from the elements (26) and we can claim that this element does not exist in the web of relations.

The formation of an element in the web of relations is based on the organizing principle and is completely controlled by the laws of arithmetic. The organizing principle has a universal character as it applies equally to all elements and all levels of the web of relations.

Secondly, the integer formation has some resemblance with formations of physical systems. Many physical phenomena can be seen as hierarchical formations of more elementary phenomena that occur at different scale levels. These phenomena are characterized in terms of complexity. The hierarchical formation of integer relations has the building-block character of the elements on each of its levels. This helps to interpret the hierarchical formation as a process that combines elements of one level to form more complex elements of the next level. This interpretation allows us to see the hierarchical formations of integer relations in terms of complexity and suggests a quantitative concept of structural complexity presented in the next section.

Thirdly, the geometrical meaning of an integer relation as an integer pattern allows us to measure the elements of the web of relations. Measurable quantities of the integer pattern can be used in the quantitative characterization of the hierarchical formations. For example, a natural quantity of an element of the web of relations can be the area of the integer pattern.

The geometrical meaning makes possible to visualize the hierarchical formations. This helps us to observe in integer formations self-similarity, local and nonlocal symmetries (as in Figure 15). These notions are among the key concepts of physical systems.

9 Structural Complexity

In the previous sections we discussed an analogy between the hierarchical formation of integer relations and formations of physical systems. The analogy helps us to see an integer formation as a building-up process that combines elements of one level to form more complex elements of the next level. This suggests to interpret the integer formations in terms of complexity. In this section we continue the discussion and as a result present a concept of structural complexity [4].

Physical systems can interact. They combine as parts when certain conditions are in place and form a composite system. In the composite system the parts are harmoniously integrated. As the result, the composite system exhibits emergent properties that do not exist in the collection of the parts. To characterize this situation it is said that the whole is more than the simple sum of the parts. These phenomena of physical systems are viewed in terms of complexity. In particular, it is said that a composite system is more complex than the parts.

Composite systems can interact also and form even more complex systems and so on. This results in the hierarchy of complexity levels we observe in physical systems.

Complexity theory aims to understand the formations of physical systems across the complexity levels [5]. In this capacity it should probably have two ingredients: elements and a universal organizing principle specifying how elements of one level form composite elements of the next level. If such a complexity theory exists, it would be interesting to know the identity of the fundamental elements as they are supposed to model the formations of physical systems.

The web of relations may be useful in the search for the complexity theory. In particular, it is an irreducible theory. The web of relations has the integer relations as its elements and the organizing principle specifying how integer relations of one level form integer relations of the next level. Moreover, the hierarchical formations in some sense mirror the complexity levels of physical systems.

Indeed, there are physical systems of different complexity levels on one side and integer relations of different levels on the other. In addition, from one side physical systems, by virtue of interactions can form more complex systems, which are *qualitatively* more than the simple sum of the parts. From the other side integer relations, owing to the organizing principle, can form new integer relations, which are *quantitatively* more than their simple sum.

There is more information about some aspects of the web of relations than about corresponding parts in physical systems. Firstly, the ultimate building blocks of the integer formations are known. They are the integers. At the same time in physical reality the fundamental question about the ultimate building blocks is still debated. Strings are current candidates for this role. Secondly, the integer relations follow the same organizing principle, which applies to all levels. This principle is well-defined and has a clear geometrical interpretation. In physical systems the character of interaction can be different from one level to another and may be unknown.

The interpretation of the hierarchical formations in terms of complexity can be naturally expressed by a quantitative concept of complexity. Namely, in a hierarchical formation elements of level $k \geq 1$ can be viewed more complex by 1 than elements of level $k-1$ from which they are formed. This defines that the complexity of an element on level k equals k assuming that integers at the zero level have zero complexity.

In its turn the complexity of a hierarchical formation can be characterized by the maximum complexity of its elements. A hierarchical formation of integer relations $WR(s,s',n,m,I_n)$ starts at the zero level and successively forms all its elements of level k from all its elements of level $(k-1)$ until it reaches the highest level $C(s,s')$. The elements of level $C(s,s')$ can be viewed as the result of the hierarchical formation and have the maximum complexity.

The complexity of a hierarchical formation of integer relations

$$WR(s,s',n,m,I_n)$$

is a characteristic of a corresponding pair of sequences s,s'. We can simply say that the pair of sequences s,s' generates the hierarchical formation $WR(s,s',n,m,I_n)$. This results in a concept of complexity [4].

Structural Complexity. *The structural complexity of a sequence $s \in I_n$ with respect to a sequence $s' \in I_n$ is the complexity of the hierarchical formation of integer relations*

$$WR(s,s',n,m,I_n),$$

which makes the nonlocal correlation between these sequences s,s' and equals $C(s,s')$.

The structural complexity of a sequence $s \in I_n$ is the maximum complexity of the hierarchical formations of the sequence with respect to all other sequences from I_n and is defined by

$$C(s) = \max_{s' \in I_n} C(s,s').$$

10 Structural Complexity of a Complex System and Nonlocal Correlation Matrix

Let us consider a complex system consisting of $N \geq 2$ parts. A state of the system is described by a $N \times n$ matrix

$$S = \{s_{ij}\}_{i=1,...,N,\ j=1,...,n},$$

where a sequence $\bar{s}_i = s_{i1}...s_{in} \in I_n$ specifies part $i, i = 1,...,N$ of the system. Let $\mathbf{S}(I_n, N)$ be the set of all states of the complex system.

The structural complexity $C(S)$ of a state $S \in \mathbf{S}(I_n, N)$ of a complex system is given by hierarchical formations of integer relations

$$WR(S) = \bigcup_{s \in S} \bigcup_{s' \in S} WR(s,s',n,m,I_n),$$

which make up the nonlocal correlation between the parts of the system.

Integer relations produced by integer formations of $WR(S)$ may form together integer relations of higher levels. In particular, integer relations of a hierarchical formation and integer relations of another hierarchical formation of the same level may form integer relations of a higher level. These integer relations together with integer relations produced on this level by a different hierarchical formation may form integer relations of a more higher level.

For example, consider a state $S \in \mathbf{S}(B_4, 3)$ such that

$$S = \begin{pmatrix} +1 & -1 & -1 & +1 \\ -1 & +1 & +1 & -1 \\ +1 & -1 & +1 & -1 \end{pmatrix}$$

where

$$\bar{s} = +1 - 1 - 1 + 1, \quad \bar{s}' = -1 + 1 + 1 - 1, \quad \bar{s}'' = +1 - 1 + 1 - 1.$$

A hierarchical formation of integer relations $WR(s, s', 4, 0, B_4)$ results in an integer relation

$$\boxed{+4^1 - 3^1 - 2^1 + 1^1 = 0} \tag{27}$$

of the second level and hierarchical formations

$$WR(s', s'', 4, 0, B_4), \quad WR(s'', s, 4, 0, B_4)$$

give integer relations

$$\boxed{-4^0 + 3^0 = 0}$$

$$\boxed{+2^0 - 1^0 = 0} \tag{28}$$

of the first level accordingly. Integer relations (28) form an integer relation

$$\boxed{-4^1 + 3^1 + 2^1 - 1^1 = 0} \tag{29}$$

of the second level. In a proper representation integer relations (27) and (29) become

$$\boxed{+8^1 - 7^1 - 6^1 + 5^1 = 0}$$

$$\boxed{-4^1 + 3^1 + 2^1 - 1^1 = 0}$$

and form an integer relation

$$\boxed{+8^2 - 7^2 - 6^2 + 5^2 - 4^2 + 3^2 + 2^2 - 1^2 = 0}$$

of the third level.

We consider a $N \times N$ matrix, defined by

$$NC(S) = \{C(\bar{s}_i, \bar{s}_j)\}_{i,j=1,\ldots,N}$$

and called a nonlocal correlation matrix, to describe the structural complexity $C(S)$ of a state $S \in \mathbf{S}(I_n, N)$, where $C(\bar{s}_i, \bar{s}_j)$ is the structural complexity of a sequence $\bar{s}_i, i = 1, ..., N$ with respect to a sequence $\bar{s}_j, j = 1, ..., N$. By using the nonlocal correlation matrix $NC(S)$ the structural complexity $C(S)$ of a state S is simply viewed in terms of the maximum levels of the hierarchical formations. Elements of the matrix $NC(S)$ describe the nonlocal correlations between the pairs of the sequences of S. This explains the name of the matrix.

It may be useful to compare the nonlocal correlation matrix $NC(S)$ and the linear correlation matrix $V(S)$ of a state S. In the linear correlation matrix the elements are linear correlations coefficients, in the nonlocal correlation matrix the elements are structural complexities. The character of the linear correlation coefficient does not suggest a natural way to compare the states in terms of linear correlation. At the same time the character of structural complexity suggests us to compare the states in terms of the nonlocal correlation.

In particular, the integer formations have a natural order: starting with integers, integer relations as elements of one level form more complex integer relations as elements of the next one. This means that integer formation of level $k \geq 1$ is more complex than integer formation of level $k-1$, i.e., the elements of level k are made of the elements of level $k-1$. Consequently, an integer formation of level k produces a stronger nonlocal correlation than an integer formation of level $k-1$. The order of the integer formations is captured by the concept of structural complexity, which is nothing but the level of the integer formation.

This order of the hierarchical formations of integer relations suggests to impose a partial order on the states of the complex system and compare them in terms of structural complexity by using the nonlocal correlation matrices. This partial order on the set of states is defined as follows.

Let

$$S = \{s_{ij}\}_{i=1,...,N;\ j=1,...,n}, \quad S' = \{s'_{ij}\}_{i=1,...,N;\ j=1,...,n},$$

be two different states of a complex system with nonlocal correlation matrices

$$NC(S) = \{C(\bar{s}_i, \bar{s}_j)\}_{i,j=1,...,N}, \quad NC(S') = \{C(\bar{s}'_i, \bar{s}'_j)\}_{i,j=1,...,N}.$$

If $C(\bar{s}_i, \bar{s}_j) \leq C(\bar{s}'_i, \bar{s}'_j)$ for all $i, j = 1, ..., N$ and there exists at least one pair of elements of matrices $NC(S), NC(S')$ such that $C(\bar{s}_i, \bar{s}_j) < C(s'_i, s'_j)$ then

$$NC(S) < NC(S')$$

and it is assumed that the structural complexity of state S is less than the structural complexity of state S'

$$C(S) < C(S').$$

Thus, if

$$NC(S) < NC(S')$$

then it is suggested that

$$C(S) < C(S')$$

and vice versa.

Therefore, by using the nonlocal correlation matrix we can view and compare the states of a complex system in terms of the integer formations. In the next section we are interested to know whether the functioning of a complex system may be a law of the structural complexity.

It is important to evaluate efficiently the structural complexity $C(S)$ of a state S. For this purpose it is proposed to use a set of linear correlation matrices derived from the state S [6], [7]. The linear correlation matrix $V(S)$ is used at the first level of this approximation. Let

$$Spec(V(S)) = (\lambda_1, ..., \lambda_N)$$

be the eigenvalue spectrum of the linear correlation matrix $V(S)$ of a state $S \in \mathbf{S}(I_n, N)$ and

$$tr(V^2(S)) = \lambda_1^2 + ... + \lambda_N^2,$$

then the quadratic trace $tr(V^2(S))$ is an estimator of the structural complexity.

Namely, it is assumed that

$$N \leq tr(V^2(S)) \leq N^2 \tag{30}$$

and if

$$tr(V^2(S')) < tr(V^2(S))$$

then the structural complexity of a state S' is greater than the structural complexity of a state S

$$C(S) < C(S'), \quad NC(S) < NC(S')$$

and vice versa.

The lower and upper bounds of (30) are given by states S whose linear correlation matrix is

$$V(S) = \begin{pmatrix} 1\,1 \dots 1\,1 \\ 1\,1 \dots 1\,1 \\ . \; . \dots . \; . \\ 1\,1 \dots 1\,1 \\ 1\,1 \dots 1\,1 \end{pmatrix} \qquad Spec(V(S)) = (N, 0, ..., 0)$$

and

$$V(S) = \begin{pmatrix} 1\,0 \dots 0\,0 \\ 0\,1 \dots 0\,0 \\ . \; . \dots . \; . \\ 0\,0 \dots 1\,0 \\ 0\,0 \dots 0\,1 \end{pmatrix} \qquad Spec(V(S)) = (1, ..., 1).$$

respectively.

11 Optimality Condition of Complex Systems and Perception-Based Information

We consider a connection between the theory of integer formations and perception-based information. This connection is due to an optimality condition of complex systems suggesting that the optimal functioning of a complex system may be a law of the integer formations,i.e., the structural complexity of the complex system. Perception-based information appears naturally in the context of the optimality condition as an efficient means for organizing the control of the structural complexity of a complex system.

The optimality condition has a general character. It describes the optimal performance of a complex system for a particular problem only in terms of the structural complexities of the system and the problem. In particular, it specifies that the performance of a complex system is a convex function of its structural complexity and the optimal performance is obtained for a particular problem when the structural complexity of the system equals the structural complexity of the problem.

The optimality condition suggests how to control the structural complexity of a complex system in improving its performance. Namely, it says that if the structural complexity of a complex system is less than the structural complexity of a problem, then the first one must be increased until it equals the second one. If the structural complexity of the system is greater than the structural complexity of the problem then the first one must be decreased until the same condition is in place.

The optimality condition is put forward based on computational experiments presented in this section [8]. The experiments test how the performance of optimization method as a complex system may depend on its structural complexity

Let **P** be a class of optimization problems and $M(v), 0 \leq v \leq 1$ be an optimization method that for each value of a parameter v solves a problem $P \in \mathbf{P}$ of the class. The optimization method $M(v)$ can be seen as a complex system, which in the solution of a problem P is characterized by a state $S(P,v)$. Let $F(P,v)$ be the value of a functional F that describes the performance of the optimization method $M(v)$ for a problem P.

Suppose, the structural complexity $C(S(P,v))$ of the optimization method $M(v)$ can be controlled by the parameter v, $0 \leq v \leq 1$. In particular, a problem $P \in \mathbf{P}$ specifies a sequence $v_{Pi}, i = 1, ..., q(P)$

$$0 \leq v_{P1} < v_{P2} < ... < v_{P,q(P)-1} < v_{P,q(P)} \leq 1$$

increasing the structural complexity of the optimization method

$$C(S(P, v_{P1})) < C(S(P, v_{P2})) < ... < C(S(P, v_{P,q(P)-1})) < C(S(P, v_{P,q(P)}))$$

$$C(S(P, v_{Pi})) \leq C(S(P, v)) \leq C(S(P, v_{P,i+1})),$$

$$v_{Pi} \leq v \leq v_{P,i+1}, \quad i = 1, ..., q(P) - 1$$

as it solves the problem P for each of these values of the parameter v and demonstrates a sequence of performances

$$F(P, v_{P1}), F(P, v_{P2}), ..., F(P, v_{P,q(P)-1}), F(P, v_{P,q(P)}).$$

Thus, it is assumed that the structural complexity of the optimization method can be increased by increasing the parameter v, $0 \leq v \leq 1$.

We are interested to know whether the performance $F(P, v)$ of the optimization method may be a special function of the structural complexity and this property is common to all problems of the class.

Computational experiments made for a benchmark class of travelling salesman problems (TSP) [9] give evidence to suggest that the performance of an optimization method is a convex function of the structural complexity [8]. Namely, for each problem tested there is one point where the performance of the method peaks as its structural complexity increases on an interval.

The method used in the experiments admits the following description. In a group of N interacting agents, each agent solves the TSP problem, i.e., to find the shortest closed path between $n + 1$ cities going through each city exactly once. The agents start in the same city and each of them first chooses a next city at random.

Then at each city an agent chooses between two strategies, i.e., to pick up a next city purely at random or travel to the next city, which is currently closest to the agent. In other words, at each city an agent uses a random strategy, labelled $+1$ or the greedy strategy, labelled -1 in order to get to a next city. Thus, an agent can be described by a binary sequence of length n as the agent step by step chooses between these two strategies to visit the cities.

Each of the agents uses a fuzzy if-then rule to make a decision about a next strategy. This rule allows us to control the structural complexity of the agents. This ability is realized by using *perception-based information.* The rule has the following description.

1. If your last strategy is *successful*, continue with the same strategy.

2. If your last strategy is *unsuccessful*, consult a PTM generator which strategy to use next.

Successful and unsuccessful strategies by their meaning are perception-based information. It is used to control the structural complexity of agents in the following way.

After each step distances travelled by the agents so far become known to everyone. An agent takes this information into account to evaluate whether his last strategy is successful or unsuccessful. If according to the information an agent finds out that the distance he travelled so far deviates from the shortest one to date by not more than a value of a parameter v then his last strategy is successful and unsuccessful otherwise.

The parameter v is defined as follows. Let D_{ij} be the distance travelled by agent i, $i = 1, ..., N$ after j steps $j = 1, ..., n$ and

$$D_j^- = \min_{i=1,...,N} D_{ij}, \quad D_j^+ = \max_{i=1,...,N} D_{ij}.$$

All distances D_{ij} travelled by the agents after j steps belong to the interval $[D_j^-, D_j^+]$. The parameter v specifies a point

$$D_j(v) = D_j^+ - v(D_j^+ - D_j^-), \quad 0 \leq v \leq 1 + \gamma, \quad \gamma > 0$$

dividing the interval into two parts and determines what the agents view as a successful strategy. If the distance travelled by an agent after j steps $j = 1, ..., n$ belongs to the interval $[D_j^-, D_j(v)]$, called successful, then the agent's last strategy is successful. If the distance travelled by an agent belongs to the interval $(D_j(v), D_j^+]$, called unsuccessful, then the agent's last strategy is unsuccessful.

The optimization method for the limiting values of the parameter v gives the following states.

1. When $v = 0$, then $D_j(v) = D_j^+$. This means that the successful interval $[D_j^-, D_j(v)]$ coincides with the whole interval $[D_j^-, D_j^+]$ and irrespective of the distances travelled by the agents, an agent's strategy at each step is always successful. Therefore, an agent starting with a random strategy and following the fuzzy rule always uses the random strategy. The binary sequence of each agent consists completely of +1's and the agents for this value of the parameter v are described by a state

$$S(0) = \begin{pmatrix} +1 & +1 & ... & +1 & +1 \\ +1 & +1 & ... & +1 & +1 \\ . & . & ... & . & . \\ +1 & +1 & ... & +1 & +1 \\ +1 & +1 & ... & +1 & +1 \end{pmatrix}$$

2. In the opposite limit, when $v = 1 + \gamma$, then $D_j(v) < D_j^-$. This means that the unsuccessful interval $(D_j(v), D_j^+]$ covers the whole interval $[D_j^-, D_j^+]$ and irrespective of the distances travelled by the agents, an agent's strategy at each step is always unsuccessful. Therefore, following the fuzzy rule an agent asks PTM generator at each step which strategy to choose next. The binary sequence of an agent is the initial segment of length n of the PTM sequence starting with +1 and the agents for this value of the parameter v are described by a state

$$S(1) = \begin{pmatrix} +1 & -1 & ... & -1 & +1 \\ +1 & -1 & ... & -1 & +1 \\ . & . & ... & . & . \\ +1 & -1 & ... & -1 & +1 \\ +1 & -1 & ... & -1 & +1 \end{pmatrix}$$

The fuzzy rule allows us to control the structural complexity of the optimization method. As the parameter v is increased from 0 to $1 + \gamma$, the successful interval gets smaller and smaller, thus making the agents to ask the PTM generator more and more often. It is supposed that this may increase the structural complexity of the optimization method in a manner required for the experiments.

Let $D_{in}(P, v)$ be the distance travelled by agent i, $i = 1, ..., N$ for a problem P when the value of the parameter is v. The performance of the optimization method for a problem $P \in \mathbf{P}$ is characterized by the average distance travelled by the agents

$$D(P, v) = \frac{\sum_{i=1}^{N} D_{in}(P, v)}{N},$$

which is sought to be minimized.

Extensive computational experiments, by using the benchmark class of TSP problems [9], are made to investigate how the performance of the optimization method changes as its structural complexity increases [8]. In the experiments the interval $[0, 1+\gamma]$ of the parameter v is subdivided by points

$$v_i = i\triangle v, \quad i = 0, 1, ..., l(\triangle v),$$

with resolution $\triangle v$, where $l(\triangle v) = \lfloor \frac{1+\varepsilon}{\triangle v} \rfloor + 1$. For all problems of the class

$$\triangle v = 0.1,\ 0.05,\ 0.01,\ 0.005,\ 0.001$$

and

$$N = 100, 200, ..., 10000.$$

The optimization method is applied to solve each TSP problem P of the class successively for values v_i of the parameter v and its performance $D(P, v_i), i = 1, ..., l(\triangle v)$ is recorded for these values. As a result a function $D_{\triangle v, N}(P, v)$ of the discrete parameter v is computed as an approximation to the function $D(P, v)$.

As the resolution $\triangle v$ gets smaller and the number N of the agents increases, it is observed that the function $D_{\triangle v, N}(P, v)$ converges to a function $D(P, v)$. Importantly, this process reveals that $D(P, v)$ is a *convex function* of the parameter v with a global minimum at a point $v^*(P)$.

Figure 16 and Figure 17 illustrate the result of the experiments for problems pr76 and kroD100 when the number of agents is $N = 10000$.

It is found computationally that the structural complexity $C(S(P, v))$ of the optimization method for each problem P is a convex function of the parameter v with a global maximum at $v_c(P)$. The structural complexity of a state $S(P, v)$ is evaluated by the quadratic trace $tr(V^2(S(v)))$ of the linear correlation matrix $V(S(P, v))$. For each problem P it is checked that $v^*(P) < v_c(P)$ to conclude that the performance of the optimization problem is a convex function of the structural complexity.

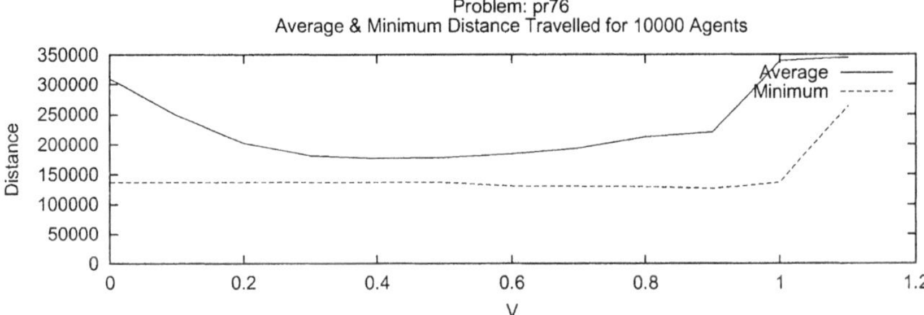

Fig. 16. The figure shows the main result of the experiments for a problem pr76 when $N = 10000$. The performance of the optimization method as a function of the perception-based parameter v converges to a smooth convex function with one global minimum.

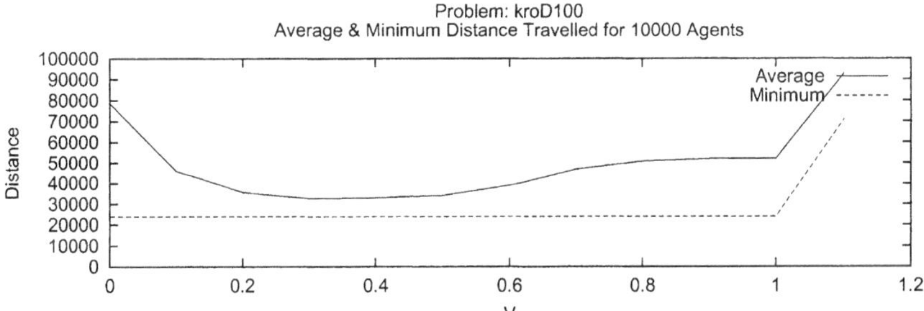

Fig. 17. The figure shows the main result of the experiments for a problem kroD100 when $N = 10000$. The performance of the optimization method as a function of the perception-based parameter v converges to a smooth convex function with one global minimum.

Moreover, the experiments give us facts to suggest that for each problem P tested, structural complexity $C(S(P, v^*(P)))$, i.e., where the performance of the optimization method peaks for the problem, is a characteristic of the problem itself. For this problem the performance of the optimization method increases as long as its structural complexity increases up to $C(S(P, v^*(P)))$ and decreases as long as the structural complexity continues to increase after. Structural complexity $C(S(P, v^*(P)))$ can be defined as the structural complexity of the problem P.

This allows us to propose a general optimality condition: the optimization method shows its best performance for a TSP problem when its structural complexity equals the structural complexity of the problem. Moreover, the optimality condition is a guide to control the structural complexity of the optimization method in improving its performance. Namely, if the structural complexity of the optimization method is less than the structural complexity of a problem then it must be increased until they become equal. If the structural complexity of the optimization method is greater than the structural complexity of the problem then it must be decreased until the same condition is in place.

12 Conclusions

In the paper we approached the mathematical theory of perception-based information by using a theory of hierarchical formations of integer relations. We considered a connection between the theory of integer formations and perception-based information. This connection is due to an optimality condition of complex systems suggesting that the optimal functioning of a complex system may be a law of the integer formations. Perception-based information appears naturally in the context of the optimality condition as an efficient means for organizing the control of the structural complexity of a complex system.

The integer relations are true propositions. The formations of the integer relations set up a structure of these propositions. In this structure each proposition is given by how it is formed or inferenced and is a building block in the formation of other propositions. The structure has natural ability to integrate information. For example, two pieces of information represented in the structure as propositions may turn to be components of a process forming new propositions.

There is an isomorphism between integer relations and their formations from one side and two-dimensional geometrical patterns and their formations from another. Consequently, an integer relation or proposition of the structure can be viewed as a geometrical pattern. This geometrical pattern results from a universal transformation applied to geometrical patterns of propositions from which the proposition is inferenced. The area of the geometrical pattern can measure the proposition.

The isomorphism opens a way to view propositions and their inference in geometrical terms. This allows us to consider geometrical transformations that make propositions and their inference imprecise but satisfy certain properties. In this context a perception may be defined as a class of geometrical patterns with the geometrical pattern of a corresponding proposition at a center of the class.

The geometrical transformations may be seen as operations in processing of the perceptions and suggest an approach to model human information processing.

Acknowledgements. I especially grateful to Lotfi Zadeh for fruitful discussions. I would like to thank Darryl Bond for implementing a software system and technical assistance in conducting the experiments. This work was supported in part by the CQU Merit Grant IN2329. The support made by the BISC and Chevron-Texaco is very much acknowledged.

References

1. L. A. Zadeh, *From Computing with Numbers to Computing with Words - From Manipulation of Measurements to Manipulation of Perceptions*, IEEE Trans. On Circ. and Syst. - I: Fund. Theory and Appl., Vol. 45, No. 1, Jan. 1999, pp. 105-119.
2. L. A. Zadeh, *A New Direction in AI: Towards a Computational Theory of Perceptions*, AI Magazine, Vol. 22, No. 1, Spring 2001, pp. 73-84 .
3. L. A. Zadeh, *Towards a Perception-based Theory of Probabilistic Reasoning with imprecise Probabilities*, Journal of Probabilistic Planning and Inference, Vol. 105, 2002, pp. 233-264.
4. V. Korotkich, *A Mathematical Structure for Emergent Computation*, Kluwer Academic Publishers, Dordrecht/Boston/London, 1999.
5. J.H. Holland, *Emergence: From Chaos to Order*, Perseus Books, Massachusetts, 1998.
6. G. Korotkikh, *A Computational Approach in Dealing with Uncertainty of Financial Markets*, PhD Thesis, Central Queensland University, 2002.
7. G. Korotkikh and V. Korotkikh, *On the Role of Nonlocal Correlations in Optimization*, Optimization and Industry: New Frontiers, P. Pardalos and V. Korotkikh (Eds.), Dordrecht/Boston/London, 2003, pp. 181-219.
8. V. Korotkich, *On Self-Organization of Cooperative Systems and Fuzzy Logic*, Fuzzy Logic: A Framework for the New Millennium, V. Dimitrov and V. Korotkich (Eds.), Physica-Verlag, Heidelberg/New York, 2002, pp. 147-167.
9. G. Reinelt, *TSPLIB 1.2*, (Online), Available from: URL ftp://ftp.wiwi.uni-frankfurt.de/pub/TSPLIB 1.2, 2002.

Fuzzy Neural Networks Based on Fuzzy Logic Algebras Valued Relations

Roberto Tagliaferri, Angelo Ciaramella
Department of Mathematics and Computer Science, University of Salerno, and INFM unit of Salerno, 84081 Baronissi (SA), Italy

Antonio Di Nola
Department of Mathematics and Computer Science, University of Salerno, 84081 Baronissi (SA), Italy

Radim Bělohlávek
Institute Res. and Appl. Fuzzy Modeling, University of Ostrava, Bráfova 7, 701 03 Ostrava, Czech Republic

Abstract

In this paper a method to build a fuzzy neural network based on fuzzy relations with truth values in a suitable algebraic structure is proposed. The properties of the network are analyzed in detail proving some interesting theorems on the stability of parameter variations. Furthermore the architecture of the fuzzy network is illustrated including both the fuzzification and defuzzification modules.

Keywords - neural networks, fuzzy systems, fuzzy relation, fuzzy neural networks

1 Introduction

Since Zadeh firstly proposed the concept of "Fuzzy Sets" in 1965 [22], fuzzy set theory and its applications have been developed quickly and widely [17], [18]. Mainly in fields as control and artificial intelligence, it has been proved that fuzzy logic is a powerful mathematical tool for dealing with modeling and control aspects of complex processes, which transparently expresses the conflicting character of the precision of the model and the degree of its generality, i.e. the principle of incompatibility [19], [23]. On the other hand, neural networks has become in the last ten years a hot research topic in all the fields of pattern

recognition and control [5]. Among the most important features of neural networks we recognize first of all the massively parallel architecture and dynamics which is inspired by the structure of human brain, the adaptation capabilities, and the fault tolerance [1].
Neural networks and fuzzy methods are closely related with two levels of human cognition. Neural networks concern the microlevel (the level of brain) while fuzzy methods concern the macrolevel (the level of mental phenomena). The microlevel and macrolevel are strongly connected, the microlevel implementing the macrolevel. The study of an analogous combination of the neural networks and fuzzy methods is therefore a promising research task. Not surprisingly, neural and fuzzy models can be combined in many ways. The resulting models can be built up of separate neural and fuzzy modules performing relatively independent tasks. The neuro-fuzzy models can be based on neural models supplied by some features of fuzzy models (esp. processing of indeterminate information), or conversely, based on fuzzy models supplied by features of neural networks (esp. adaptability).
An overview of such models, called fuzzy neural networks, can be found e.g. in [9, 13, 16, 20, 21]. A particular type of fuzzy neural networks are those following the structure of "IF-THEN" fuzzy reasoning rules [15, 16, 20]. In this paper, we design a fuzzy neural network based on a fuzzy relational "IF-THEN" reasoning scheme [20]. In our approach we try to elicit a general approach to the design of fuzzy neural networks of this type: the parameters of the net (weights, etc.) are elements of a suitable structure of truth values (e.g. residuated lattices, MV-algebras, etc.).
The importance and originality of the proposed neural network is mainly in modeling the data extracting fuzzy relations that play an important role in describing dynamical characteristics also of complex systems. We analyze some features of our network and design a fuzzy logic system by using the fuzzy neural net as its main part.
The paper is organized as follows: in the next section some basic concepts about fuzzy relation are described, and their role in fuzzy reasoning are also discussed. The architecture of the fuzzy neural network (FUZNN) is illustrated and its formal properties are analyzed in Section 3.

2 Fuzzy Relations

The concept of fuzzy relation has acquired an importance in fuzzy set theory, and in the research fields of control, modeling and knowledge engineering [10, 11, 16, 21]. In fact, if we recall how the first fuzzy controller proposed by Mamdani successfully run, then we can easily understand the important part taken in the field of control by fuzzy relation, even if the reasoning results are obtained by the so called "simplified fuzzy reasoning method" for the sake of speed up and memory save [2] and not by the composition operation of fuzzy relation. Hence, it is worthwhile to recall the mechanism of the first fuzzy controller. Firstly, we give a definition of a fuzzy relation.

Fuzzy sets are usually defined as mappings (generalized characteristic functions) of a universal set U into the real interval $[0,1]$ (as defined in [22]) which plays the role of the set of the truth degrees. However, if further algebraic manipulation should be performed, then the set of truth values should be equipped with an algebraic structure, natural from the logical point of view. A general structure satisfying these requirements has been proposed in [12] (see e.g. [14] for its properties and its role in fuzzy logic and fuzzy set theory), namely the structure of a complete residuated lattice.
By definition, a complete residuated lattice is an algebra $\mathbf{L} = \langle L, \wedge, \vee, \otimes, \rightarrow, 0, 1\rangle$ where

(i) $\langle L, \wedge, \vee, 0, 1\rangle$ is a complete lattice with the least element 0 and the greatest element 1 (i.e. infima ($\bigwedge$) and suprema ($\bigvee$) of subsets of L exist),

(ii) $\langle L, \otimes, 1\rangle$ is a commutative monoid, i.e. $\otimes$ is associative ($x \otimes (y \otimes z) = (x \otimes y) \otimes z$), commutative ($x \otimes y = y \otimes x$), and the identity $x \otimes 1 = x$ holds,

(iii) $\otimes$ and $\rightarrow$ satisfy the adjointness property, i.e. $x \leq y \rightarrow z$ iff $x \otimes y \leq z$ holds.

The abstraction from $[0,1]$ to complete residuated lattices enables us to formulate the properties (of fuzzy models) for a broad class of useful stuctures of truth values. If it is desirable (e.g. stronger properties of $\mathbf{L}$ are needed) one may use several special types of residuated lattices, e.g. MV-algebras.
In applications of such models, one selects the particular structure $\mathbf{L}$ having all the general properties of the model at disposal. The most applied complete residuated lattices are those with $L = [0,1]$ with the following structures:

- Łukasiewicz ($a \otimes b = \max(a+b-1, 0)$, $a \rightarrow b = \min(1-a+b, 1)$),
- Gödel (where $a \otimes b = \min(a,b)$, $a \rightarrow b = 1$ if $a \leq b$ and $= b$),
- product ($a \otimes b = a \cdot b$, $a \rightarrow b = 1$ if $a \leq b$ and $= b/a$).

In this perspective, the definitions of a fuzzy set and of a fuzzy relation are as follows [12]:

Definition 1 *Let U be a set of objects, and be called the universe of discourse, and $\mathbf{L}$ be a complete residuated lattice. A fuzzy set (or $\mathbf{L}$-fuzzy set) in U is a mapping $A : U \rightarrow L$.*
A fuzzy relation R between the universes U and V is a fuzzy set in the direct product $U \times V$, i.e. $R : U \times V \rightarrow L$.

Note that sometimes one distinguishes between the fuzzy set and its characteristic function. $A(u)$ is called the membership degree of the element u in A. Putting $L = [0,1]$ and $\otimes = \min$ we get the notion of a fuzzy set introduced by Zadeh [22].

If there is no danger of confusion we leave **L** unspecified and assume that it is an arbitrary complete residuated lattice.
A fuzzy relation R between finite universes can be represented in an obvious way by a matrix $R = [r_{ij}]_{m\times n}$, $i = 1, \ldots, m$, $j = 1, \ldots, n$, called fuzzy relation matrix.

Definition 2 *Let R and S be fuzzy relations between U and V, V and W, respectively. The composition of R and S is a fuzzy relation $R \circ S$ between U and W defined by*

$$R \circ S(u,w) = \bigvee_{v\in V} [R(u,v) \otimes S(v,w)]. \tag{1}$$

Some properties of the composition of fuzzy relations are given in [11] and are omitted here. Now for simplicity, we assume that a fuzzy system has one input and one output and the fuzzy rules are the following:

$$\begin{aligned}
&\text{Rule 1}: \text{ if } x \text{ is } A_1 \text{ then } y \text{ is } B_1\\
&\text{Rule 2}: \text{ if } x \text{ is } A_2 \text{ then } y \text{ is } B_2\\
&\ldots\\
&\text{Rule n}: \text{ if } x \text{ is } A_n \text{ then } y \text{ is } B_n
\end{aligned} \tag{2}$$

where A_i and B_i, $i = 1, \ldots, n$, are fuzzy sets in $U \subseteq R$ and $V \subseteq R$, respectively, and x and y are linguistic variables.

The procedure for constructing a Mamdani-type fuzzy controller is usually divided into two steps:

(**Step 1**) Find a fuzzy relation implicated by Equation 2:

$$R = \bigcup_{i=1}^{n} R_i \tag{3}$$

i.e.

$$R(x,y) = \bigvee_{i=1}^{n} R_i(x,y) \tag{4}$$

where $R_i(x,y) = A_i(x,y) \wedge B_i(x,y)$, $i = 1, \ldots, n$, and $\forall x \in U$ and $y \in V$.

(**Step 2**) Given an input A, find the fuzzy system output by using the composition of the fuzzy relation:

$$B = A \circ R \tag{5}$$

where

$$B(y) = \bigvee_{x\in U} [A(x) \otimes R(x,y)]. \tag{6}$$

The systems point of view on Equation 5 is depicted in Figure 1. These equations tell us that fuzzy relation R plays an important role in describing the dynamic characteristics of a system, as the transfer function in control theory does. Although we discussed here a very simple case of fuzzy system, the conclusions remain true also for complex fuzzy systems.

3 The architecture of the Fuzzy Relation Based Neural Network

3.1 Introduction

Since a fuzzy relation expresses the dynamic features of a system described by a fuzzy model, it is natural to design a fuzzy neural network (FUZNN) based on the fuzzy relation. Let U and V be two universes of discourse of dimensions m and n, respectively, so that we can depict the scheme of Equation 6 in the form of the neural network structure shown in Figure 2.
It should be pointed out that the weights of the net between the two layers as well as the input and output values are truth values, i.e. elements of some complete residuated lattice (the structure of truth values).

3.2 Designing a fuzzy neural network

What Equation 6 and Figure 2 express is only a simple case of a fuzzy system [20]. Here, we begin to show how to design a FUZNN for a more complex fuzzy system. Let us assume that a fuzzy system with multi-input and one-output consists of the following fuzzy rules:

$$\begin{array}{c}\text{If } x_1 \text{ is } A_{11} \text{ and } x_2 \text{ is } A_{12} \text{ and } \ldots \text{ and } x_m \text{ is } A_{1m} \text{ then } y \text{ is } B_1\\ \text{else}\\ \\ \ldots\\ \\ \text{else}\\ \text{If } x_1 \text{ is } A_{n1} \text{ and } x_2 \text{ is } A_{n2} \text{ and } \ldots \text{ and } x_m \text{ is } A_{nm} \text{ then } y \text{ is } B_n\end{array} \tag{7}$$

where A_{ij} and B_i, $i = 1, \ldots, n$ and $j = 1, \ldots, m$, are fuzzy sets in $U \subset R$ and $V \subset R$, respectively, and x_j and y are linguistic variables.
We can decompose a complex fuzzy model expressed by Equation 7 into one consisting of several simple models like those defined by Equation 2 ([2]) in the following way:

$$\begin{array}{c}\left\{\begin{array}{l}\text{If } x_1 \text{ is } A_{11} \text{ then } y \text{ is } B_1\\ \text{and}\\ \ldots\\ \text{and}\\ \text{If } x_m \text{ is } A_{1m} \text{ then } y \text{ is } B_1\end{array}\right. \qquad R_1\\ \text{else}\\ \\ \ldots\\ \\ \text{else}\end{array} \tag{8}$$

$$\left\{\begin{array}{l}\text{If } x_1 \text{ is } A_{n1} \text{ then } y \text{ is } B_n \\ \text{and} \\ \dots \\ \text{and} \\ \text{If } x_m \text{ is } A_{nm} \text{ then } y \text{ is } B_n\end{array}\right. \qquad R_n$$

Recall that if A is a fuzzy set in U_i, $1 \leq i \leq n$, then the cylindric extension of A into $U_1 \times \cdots U_n$ is a fuzzy set A^* in $U_1 \times \cdots U_n$ defined by $A^*(x_1, \dots, x_n) = A(x_i)$. The following theorem gives a description of the output of the system of Equation 7 in the spirit of Equation 8.

Theorem 1 *Given the inputs $A_1, \dots, A_n$ to the system described by Equation 7, then for the corresponding output B it holds that*

$$B(y) = \left(\bigcap_{j=1}^{m} A_j^* \right) \circ R \tag{9}$$

where

$$R = \bigcup_{i=1}^{n} \left(\bigcap_{j=1}^{m} R_{ij}^* \right), \tag{10}$$

A_j^*, R_{ij}^* are the cylindric extensions of A_j, R_{ij} to U^m, $U^m \times V$, respectively, and $R_{ij}(x, y) = A_{ij}(x) \wedge B_i(y)$ is a fuzzy relation between U and V, $i = 1, \dots, n$ and $j = 1, \dots, m$.

Proof. The proof easily follows from the fact that for the fuzzy sets $A_1, \dots, A_m$ and for the fuzzy set A in U^m defined by $A(x_1, \dots, x_m) = A_1(x_1) \wedge \dots \wedge A_m(x_m)$ it holds $A = \bigcap_{j=1}^{m} A_j^*$ (and similarly for R_{ij}).

For the sake of simplicity we can derive the following approximation of Equation 9:

$$B(y) = \bigcap_{j=1}^{m} (A_j^* \circ R) \tag{11}$$

which is equivalent to

$$B(y) = \bigcap_{j=1}^{m} (A_j \circ R_j) \tag{12}$$

where R_j is a fuzzy relation between U and V defined by $R_j(x_j, y) = \bigvee_{x_1, \dots, x_m} R(x_1, \dots, x_m, y)$, i.e. R_j is the projection of R to the j-th component U and to Y.

It should be pointed out that the difference between the approximation expressed by Equation 12 and the formula of Equation 9 will not be important in the following because R_j in Equation 12 will be determined by the learning process.
Therefore, if we use the Neural Net of Figure 2 as a unit, then it is easy for us to build the FUZNN shown in Figure 3, which can describe the dynamical features of the fuzzy model of Equation 7.

3.3 The fuzzy neural network and its properties

In this section we illustrate the FUZNN architecture which realizes the fuzzy Relation of Equation 12. In our case we suppose to have m input variables $(x_1, \ldots, x_m)$ fuzzified into p input levels and one output which is composed of k discretized levels.
The idea is to construct the FUZNN in blocks (the TRAN-NN's of Figure 3) which are related to the $(A_j \circ R_j)$ part of Equation 12.
In what follows, we treat R_j as independent (i.e. not as projections of some R) which is simpler and more general. The blocks are composed by several TRAN-NN units whose description is better shown in Figures 4 and 5. Each TRAN-NN unit realizes the fuzzy relation for a single input variable discretized into p values $(x_{j1}, \ldots, x_{jp})$ and a single discretized output y_s (see Figure 5). On the other hand, the Min Operation Array of Figure 3 is used to realize the min operation of Equation 12 and its description is more detailed in Figure 6.

Two important questions arise at this point (note that both of them are usually left unanswered in most papers on fuzzy neural networks). First, what is the sensitivity of the input-output behaviour of our network, i.e. do similar inputs lead to similar outputs? Second, do small changes in parameters (i.e. the relations R_j) lead to small changes in the behaviour? It has been shown in [4] that on the level of abstraction we deal with (i.e. parameters and signals are truth values of a residuated lattice), such questions may be naturally formulated.

Recall first that a similarity relation in a universe X is a binary fuzzy relation E between X and X satisfying for every $x, y, z \in X$ the following properties:

$E(x,x) = 1$ (reflexivity), $E(x,y) = E(y,x)$ (symmetry), and $E(x,y) \otimes E(y,z) \leq E(x,z)$.

We first show how to measure the similarity of fuzzy sets. For the following proposition, see e.g. [3]:

Proposition 2 *The fuzzy relation E defined by*

$$E(A_1, A_2) = \bigwedge_{x \in X} (A_1(x) \leftrightarrow A_2(x)) \tag{13}$$

where A_1, A_2 are fuzzy sets in X is a similarity relation in the universe of all fuzzy sets in a given universe X.

Here, $\leftrightarrow$ is a biresiduum operation defined by $x \leftrightarrow y = (x \rightarrow y) \wedge (y \rightarrow x)$. Note that the degree of similarity is indeed introduced in a natural way. Namely,

interpreting (13) on the linguistic level, two fuzzy sets are similar iff they contain the same elements.
If there is no danger of confusion we will not make any reference to the set X and write just E. The following theorem gives an answer to the first question: the similarity of two output fuzzy sets is at least as high as the similarity of the corresponding input fuzzy sets:

Theorem 3 *Let $A_1, A'_1, \ldots, A_m, A'_m$ be input fuzzy sets to our FUZNN (i.e. the network realizing Equation 12). Then for the corresponding output fuzzy sets B, B' it holds*

$$E(A_1, A'_1) \wedge \cdots \wedge E(A_m, A'_m) \leq E(B, B').$$

Proof. We show

$$\bigwedge_{j=1}^{m} E(A_j, A'_j) \leq E(B, B') = \bigwedge_{y \in Y} (B(y) \to B'(y)) \wedge \bigwedge_{y \in Y} (B'(y) \to B(y))$$

by checking that the left side of the inequality is less or equal than both $\bigwedge_{y \in Y} B(y) \to B'(y)$ and $\bigwedge_{y \in Y} B'(y) \to B(y)$. Due to symmetry, we show only

$$\bigwedge_{j=1}^{m} E(A_j, A'_j) \leq \bigwedge_{y \in Y} B(y) \to B'(y)$$

which holds iff for each $y \in Y$ we have

$$\bigwedge_{j=1}^{m} E(A_j, A'_j) \leq B(y) \to B'(y)$$

which is, by adjunction, equivalent to

$$B(y) \otimes \bigwedge_{j=1}^{m} E(A_j, A'_j) \leq B'(y).$$

We now reason as follows

$$\begin{aligned} B(y) \otimes \bigwedge_{j=1}^{m} E(A_j, A'_j) &= (\bigwedge_{j=1}^{m} \bigvee_{x \in U} A_j(x) \otimes R_j(x,y)) \otimes \bigwedge_{j=1}^{m} E(A_j, A'_j) \\ &\leq \bigwedge_{j=1}^{m} ((\bigvee_{x \in U} A_j(x) \otimes R_j(x,y)) \otimes E(A_j, A'_j)) \\ &= \bigwedge_{j=1}^{m} \bigvee_{x \in U} (A_j(x) \otimes E(A_j, A'_j) \otimes R_j(x,y)) \\ &\leq \bigwedge_{j=1}^{m} \bigvee_{x \in U} (A_j(x) \otimes (A_j(x) \to A'_j(x)) \otimes R_j(x,y)) \end{aligned}$$

$$\leq \bigwedge_{j=1}^{m} \bigvee_{x \in U} (A'_j(x) \otimes R_j(x,y))$$

$$= (\bigcap_{j=1}^{m} A'_j \circ R_j)(y) = B'(y).$$

The assertion is proved.

The next theorem answers the second question. Given two FUZNN's given by R_j and R'_j $(j = 1, \ldots, m)$ then the following is satisfied: if B and B' are the outputs of the corresponding FUZNN's to the same input $A_1, \ldots, A_m$, then the degree of similarity between the outputs is at least as high as the degree of similarity between the parameters (weights) of the FUZNN's. More precisely:

Theorem 4 *Let $A_1, \ldots, A_m$ be input fuzzy sets to two FUZNN's given by the fuzzy relations R_j and R'_j $(j = 1, \ldots, m)$, respectively. Then for the corresponding output fuzzy sets B, B' it holds*

$$E(R_1, R'_1) \wedge \cdots \wedge E(R_m, R'_m) \leq E(B, B').$$

Proof. The proof is analogous to the proof of Theorem 3 and thus omitted.

Our network is therefore stable, small changes of parameters (of either the inputs or the net itself) do not lead to big changes in output.

In real world applications, the inputs and outputs of complex systems are crisp, real numbers, then we can change our FUZNN to include the fuzzification and defuzzification modules as shown in Figure 7 and figure 8.

In this way, a new simple approach to build a FUZNN based on the fuzzy relation has been proposed and shown from Figure 3 to Figure 8, in detail.

Its structure is simple and flexible. We have analyzed the input-output behaviour as well as its sensitivity to the weight changes.

3.4 The activation functions

In this section we detailed how the FUZNN works, illustrating the input, the output and the activation functions of the network units, in the case of m inputs which are discretized into p input levels by a fuzzifier and one output which is obtained by the defuzzification of k discretized levels.
For our illustration, we take for the structure $\mathbf{L}$ of truth values the Łukasiewicz MV-algebra on [0,1] (i.e. $a \otimes b = \max(a + b - 1, 0)$, $a \to b = \min(1 - a + b, 1)$). Let us suppose to consider the $j - th$ TRAN-NN of Figure 3. Its inputs are the p membership values $(A_j(x_{j1}), \ldots, A_j(x_{jp}))$ of the $j - th$ input universe of discourse. Its outputs are the k membership values $(B_j(y_1), \ldots, B_j(y_k))$ of the discretized output universe of the discourse. The output of the first unit

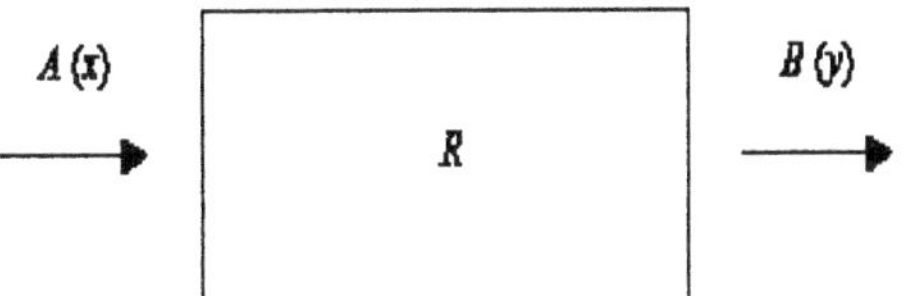

Figure 1: A simple diagram to illustrate the fuzzy relational Equation 5

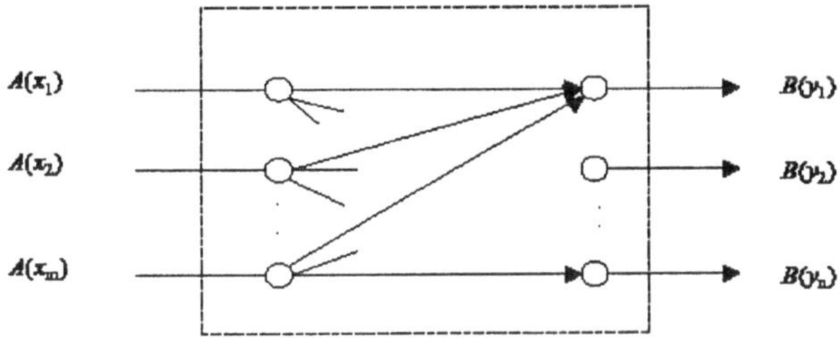

Figure 2: The Fuzzy Relational Neural Net of equation 6 for m inputs and n outputs

of B_j, $B_j(y_1)$, is obtained by the max-$\otimes$ composition by using the Lukasiewicz operations:

$$B_j(y_v) = Sup\left((A_j(x_{j1}) + w_{v1}^{(j)} - 1 \bigvee 0), \ldots, (A_j(x_{jp}) + w_{vp}^{(j)} - 1 \bigvee 0)\right) \quad (14)$$

where $w_{vs}^{(j)} = R_j(x_{js}, y_v)$. The $B(y)$ of Figure 6 is the vector composed by the k min gates $(B^1, \ldots, B^k)$ which give in output the membership vector $(B(y_1), \ldots, B(y_k))$ obtained by applying minimum operation to the outputs of the TRAN-NN's. For the generic $s-th$ B^s, we have:

$$B^s(y_s) = B(y_s) = min\,(B_1(y_s), \ldots, B_m(y_s)) \quad (15)$$

The output of the FUZNN is obtained, as shown in Figure 7, by defuzzyfying the B^s units. The simplest way is to have a weighted sum of them as follows:

$$y = \sum_{i=1}^{k} w_i B^i \quad (16)$$

4 Concluding remarks

In this paper we designed a fuzzy neural network based on a fuzzy relation and algebraic structures of fuzzy logic, and proposed some interesting theorems on

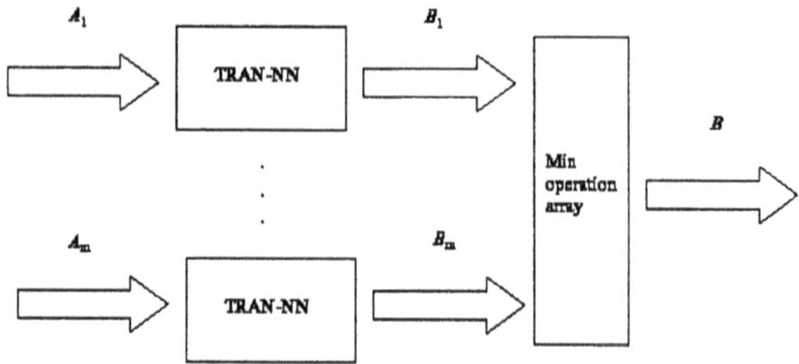

Figure 3: A block diagram of FUZNN: $(A_1, \ldots, A_n)$ are the m inputs, B is the output

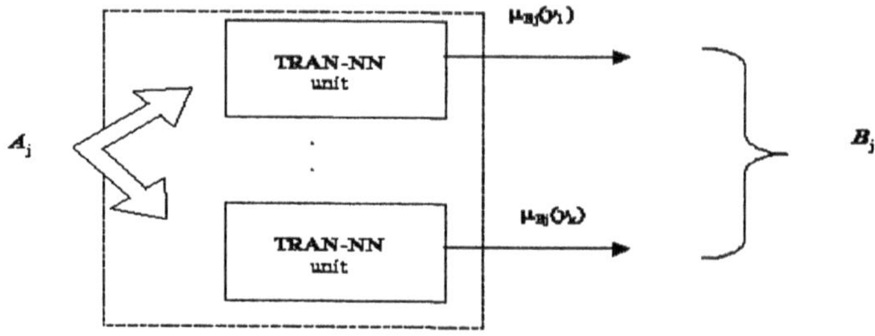

Figure 4: A block diagram illustrating the decomposition of the j-th TRAN-NN into TRAN-NN units. In input we have the p membership values of the j-th universe of discourse and in output we have the k-th membership values of the j-th discretized output universe of discourse.

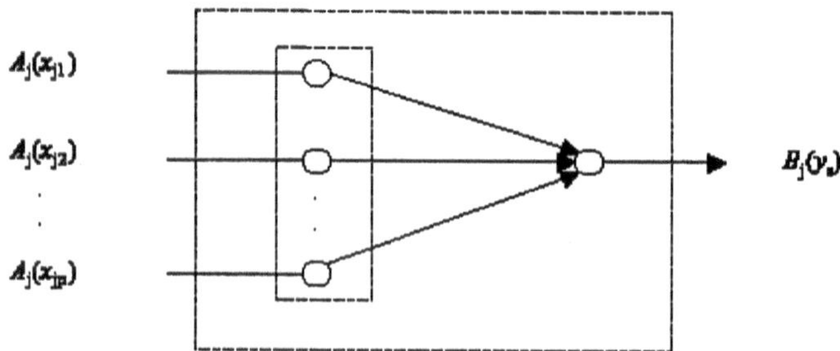

Figure 5: A TRAN-NN unit in TRAN-NN. In input we have the p membership values of the j-th universe of discourse and in output we have the membership value of the s-th element of the j-th discretized output universe of discourse.

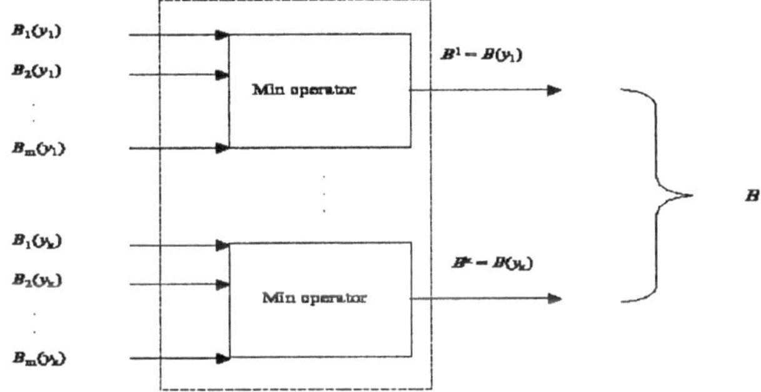

Figure 6: Internal architecture of the Min operation array of Figure 3, with k discrete levels on the universe Y.

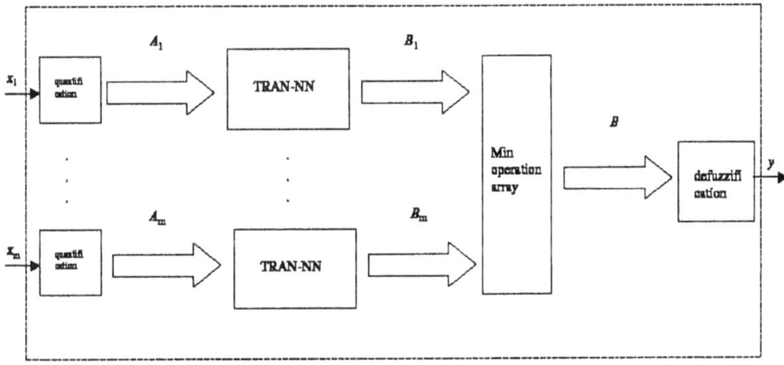

Figure 7: Block diagram of FUZNN with crisp inputs and outputs. $(x_1, \ldots, x_m)$ are the m inputs and y is the output.

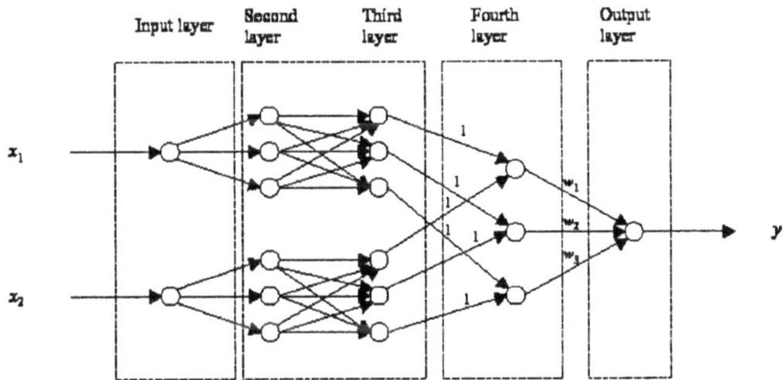

Figure 8: An example of a FUZNN with two inputs, one output, the input and the output universes of discourse discretized into three levels.

the stability of its parameter variations. The main result were obtained:

1. a particular simple architecture for "IF-THEN" reasoning to prepare it to different learning algorithms [7];

2. the usefulness of an appropriate logico-algebraic approach: in fact, such a network is, at a reasonable level, general (a particular network-instance is derived from our network by substitution of a desired structure of truth values) and has relevant easy to formulate input-output properties.

Some extensions of the model are under study, and several experiments are running also to experimentally validate the model [8].

5 Acknowledgments

This work has been partially supported by IIASS "E. R. Caianiello" (Tagliaferri) and by the project VS96037 of the MSMT CR (Bělohlávek).

References

[1] Arbib M. (ed.), The Handbook of Brain Theory and Neural Networks. Cambridge, Massachusetts, London, MIT Press, 1995.

[2] Baldwin J. F., Pilsworth B. W., A Model of Fuzzy Reasoning through Multi-valued Logic and Set Theory, Intern. J. Man-Machine Studies, 11, 351-380, 1979.

[3] Bělohlávek R., Networks Processing Indeterminacy. PhD thesis, Ostrava, 1998 (available on request).

[4] Bělohlávek R., Fuzzy logical bidirectional associative memory (submitted, 1999).

[5] Bishop C. M, Neural Networks for Pattern Recognition, New York: Oxford University Press, 1995.

[6] Buckley J. J., Hayashi Y., Fuzzy Neural Nets and Applications, Fuzzy Sets and Artificial Intelligence, 1, 11–41, 1992.

[7] Ciaramella A., Tagliaferri R., Pedrycz W., Fuzzy Relations Neural Network, Proceedings of the 10th IEEE International Conference on Fuzzy Systems, December 2001, paper P287

[8] Ciaramella A., Soft Computing Methodologies for Data Analysis, PhD Thesis, University of Salerno, 2003.

[9] Buckley J. J., Hayashi Y., Fuzzy Neural Nets and Applications, Fuzzy Sets and Artificial Intelligence, 1, 11–41, 1992.

[10] Di Nola A., Pedrycz W., Sessa S., Coping with Uncertainty for Knowledge and Inference Mechanism, Kybernetes, 15, 243-249, 1986.

[11] Di Nola A., Sessa S., Pedrycz W., Sanchez E., Fuzzy Relation Equations and Their Applications to Knowledge Engineering, Kluwer Academic Publisher, 1989.

[12] Goguen J. A., L-fuzzy sets, J. Math. Anal. Appl., 18, 145–174, 1967.

[13] Gupta M. M., Rao D. H., On the principles of fuzzy neural networks, Fuzzy Sets and Systems, 61, 1–18, 1994.

[14] Höhle U., On the fundamentals of fuzzy set theory, J. Math. Anal. Appl., 201, 786–826, 1996.

[15] Lin C. T. and Lee C. S. G., Neural Network Based Fuzzy Logic Control and Decision System, IEEE Trans. on Computer, 40, 12, 1320-1336, 1991.

[16] Lin C. T. and Lee C. S. G., Neural Fuzzy Systems: a Neuro-fuzzy Synergism to Intelligent Systems, Upper Saddle River-NJ: Prentice Hall, 1996.

[17] Ma J. L., (ed.), A Bibliography of Fuzzy Systems, Xueyuan Press, Beijing, 1989.

[18] Mizumoto M., Fuzzy Sets and Their Operators II, Information and Control, 50, 166-174, 1981.

[19] Pedrycz W., Fuzzy Control and Fuzzy Systems, New York: John Wiley, 1989.

[20] Pedrycz W., Fuzzy neural networks and neurocomputations, Fuzzy Sets and Systems, 56, 1–28, 1993.

[21] Wang L. X., Adaptive Fuzzy System and Control, Englewood Cliffs-NJ: Prentice Hall, 1993.

[22] Zadeh L. A., Fuzzy Sets, Information and Control, 8, 338-353, 1965

[23] Zadeh L. A., The Concept of a Linguistic Variable and Application to Approximate Reasoning, Information Sciences, 8, 199-249, 1975.

Simulating continuous dynamical systems under conditions of uncertainty: the probability and the possibility approaches

Gianluca Bontempi

Département d'Informatique, Université Libre de Bruxelles, Belgium
email: gbonte@ulb.ac.be

Abstract. Probability and possibility theory deal with different types of uncertainty. The paper discusses the role of these formalisms in the simulation of continuous dynamical systems where parameters and/or initial conditions are uncertain.

We derive analytically the evolution law of possibility distributions in continuous dynamical system and we compare it with consolidated results in probability theory. The analysis illustrates the limits of conventional Monte Carlo techniques for the simulation of dynamical systems where the uncertainty is expressed in a non probabilistic form. We show that the interaction existing between the probability distribution and the dynamics of the system may produce, in some cases, an inaccurate outcome of the Monte Carlo simulation, making then advisable the adoption of an appropriate possibilistic technique.

Also, we propose a new algorithm for the numerical simulation of a differential system, where the uncertainty of parameters and/or initial conditions is represented by fuzzy distributions. Finally, we discuss the simulation of a dynamical system where the evolution of the possibilistic uncertainty is simulated by using both probabilistic and possibilistic techniques.

1 Introduction

The notion of continuous dynamical system is a basic concept in system theory: it is the formalization of a natural phenomenon in a set of real variables, defined as the *state*, and a set of deterministic differential equations, defined as the *model*. The resolution of the system of differential equations provides the time evolution of the state variables, defined as the *behavior* of the system.

The differential model of a dynamical system contains elements and operators that constitute a formal description of the phenomenon under examination. However, due to errors in the measurements or lack of precise knowledge of the process, these elements may not be uniquely defined. The idea that there can be a lack of determinism in a deterministic set of equations appears at first to be a contradiction in terms. What enters here is the important distinction between mathematical precision and imprecise observations. The determinism we speak about in dynamical models is the property of the mathematical uniqueness of the solution, namely, given a precise initial condition of the system, there is precisely one evolution of the system satisfying both the equations of the model and the initial conditions. However, this mathematical determinism does not guarantee a completely certain evolution, because it is not always possible to know initial condition and parameters of the model with an infinite accuracy.

Uncertainty modeling has long been recognized by system theory and artificial intelligence as a topic worthy of investigation. In system theory uncertainty is classically treated in probabilistic form by the theory of stochastic processes. Research in artificial intelligence has enriched the spectrum of available techniques to deal with uncertainty by proposing new non probabilistic formalisms, as the theory of possibility (based on the theory of fuzzy sets) and the theory of evidence [22]. A long-standing debate took place in literature on the relationship between probability and possibility as ways to model uncertainty [1]. The purpose of this paper is not to add a further contribution to this debate. We do not intend to answer to the question about what is the best formalism to represent uncertainty. Instead, we will give for granted that there exist configurations where the scientist assumes that a probabilistic representation of the uncertainty is preferable and situations when he deems that a possibilistic framework is a better alternative. Our goal is to stress that these two distinct formalisms have different properties and that consequently require distinct resolution methods. In particular we will focus on the simulation of continuous dynamical systems where initial conditions and/or parameters are described by uncertainty distributions.

A common practice in the simulation of dynamical systems consists in adopting probabilistic techniques, like Monte Carlo, to deal with any form of uncertainty, including possibilistic distributions. This paper aims to warn against the dangers of the habit in situations where the uncertainty is better described in terms of a possibility distribution rather than a probabilistic one.

According to the Liouville equation, the main feature of probabilistic simulation techniques is the interaction coupling the evolution of the probability distribution to the dynamics of the system. On the contrary, we show that the possibility formalism is characterized by a decoupling of the laws modeling the dynamics of the system and the ones describing the evolution the uncertainty distribution. This means that the use of a probabilistic method to simulate a possibilistic model may induce the attribution of probabilities not significantly different from zero to trajectories, which represent instead behaviors of the system deserved to be taken into consideration in a possibilistic framework. As a consequence, the adoption of specific possibilistic simulation techniques is recommended for qualitative models if we wish to have a description of the behavior consistent with the uncertainty affecting the model.

These considerations are relevant in practical situations where the aim of the simulation is to reveal critical behaviors of a system described in qualitative terms. Think, for instance, to the qualitative simulation of a nuclear power plant for which we aim to detect dangerous behaviors resulting from possible faults. In these situations the adoption of a probabilistic method could withdraw from sight hazardous evolutions of the system characterized by low probability but with a possibility value significantly larger than zero.

In this paper we will represent possibility distributions through the formalism of fuzzy sets. The notion of fuzzy differential equation and fuzzy integral is first introduced by Sugeno [27]. A second fuzzy integral is developed by Dubois and Prade [5–7] based on the notion of a fuzzy set mapping and the Lebesgue integral. An existence and an uniqueness theorem for the solution of a fuzzy differential equation is provided by Kaleva [12]. Fuzzy differential systems are related to the theory of stochastic diffusion processes by Leland [16]. He defines the notion of fuzzy Brownian motion and derives a partial differential equation which describes the evolution of fuzzy distributions analogous to the Kolmogorov forward equation for irregular stochastic differential equations. Unfortunately, these equations are not computationally tractable once the dimension of the system grows up. For that reason, we restrict to consider regular systems where initial conditions and/or parameters are uncertain.

The theoretical contribution of the paper is a comparison of the analytical laws underlying (i) the evolution of probability distributions in regular stochastic differential equations and (ii) the evolution of possibility distributions in deterministic dynamical system with fuzzy initial conditions and/or parameters.

The second contribution is an improvement of the fuzzy simulation technique *Qua.Si. (Qualitative Simulator)* proposed by Bonarini and Bontempi in [2,3] to simulate the evolution of possibility distributions in differential continuous systems with fuzzy initial conditions and/or parameters. The basic idea is to reduce the simulation problem to a constrained multivariate optimization task, for which well known resolution techniques are available in literature.

The experimental session compares the *Qua.Si.* technique to the Monte Carlo method in the simulation of a Lotka-Volterra dynamical system, where uncertainty is described by possibilistic distributions. The experiment shows that for some imprecise configurations the Monte Carlo technique returns an incomplete description of the dynamics, missing possible evolutions of the system. The experimental results provide arguments supporting the adoption of specific possibilistic method to deal with the simulation of qualitative models.

The organization of this paper is as follows. We first review some fundamentals of dynamical systems in Section 2. Section 3 introduces probability measures and possibility measures as special types of fuzzy measures [28]. Section 4 discusses how a probabilistic distribution evolves in a regular stochastic differential equation. Section 5 reviews some fundamentals of the theory of fuzzy sets and numbers [30] while Section 6 presents fuzzy differential equations. Section 6.1 introduces the *Qua.Si.* method for numerical resolution of fuzzy differential systems: here the problem of simulating a fuzzy dynamics is decomposed in a set of subproblems where intervals are propagated in the state space. We will show how each of these subproblems may be reduced to a problem of constrained optimization. An analytical and computational comparison between probabilistic and possibilistic methods in dynamical systems is provided in Section 7 and 8. Finally, in Section 9 we compare the simulation outputs of a dynamical system affected by possibilistic uncertainty, adopting the Monte Carlo and the *Qua.Si.* technique, respectively.

2 Dynamical systems and ordinary differential equations

Consider the autonomous and continuous nth order dynamical system

$$\dot{y} = F(y), \qquad y \in \Re^n, \tag{1}$$

where $y = [y_1, \ldots, y_n]^T$ is the $[n \times 1]$ state vector, t is the variable denoting time, $F = [F_1, \ldots, F_n]^T$ and F_i is defined and continuous in the strip $-\infty \leq y_i \leq \infty$ for $i = 1, \ldots, n$. Let the initial condition be $y(0) = y_0$ with $y_0 \in \Re^n$. We denote by

$$y(t, y_0) \equiv [y_1(t, y_0), \ldots, y_n(t, y0)]^T \tag{2}$$

the *solution* at time t of the system of differential equations (1) with initial condition y_0.

To ensure that only one solution passes through each point, we will assume that F satisfy the Lipschitz condition [26] for all $t \in [t_0, T]$. As a consequence, the general solution $y(t, y_0)$ of (1) with $y_0 \in \Re^n$ at a fixed time $t < T$ can be viewed as a continuous mapping in the phase space $Y^t : \Re^n \to \Re^n$, which carries the initial point y_0 to the point $y(t, y_0)$. The problem of finding a solution to (1), satisfying a known initial condition y_0, is called *initial value problem.* The solution $y(t, y_0)$ may be expressed in analytical or numerical form. Several numerical methods to solve ordinary differential equations are available in literature [10,15].

3 Fuzzy measures of uncertainty

In the previous section we considered continuous dynamical systems where all parameters and initial conditions are known in a precise way. We now introduce some measures to represent uncertainty in a deterministic dynamical system.

Sugeno [28] provided a general framework to the uncertainty measures by introducing the notion of *fuzzy measure.*

A fuzzy measure $g(\cdot)$ on a continuous set Y is a function

$$g : 2^Y \to [0,1]$$

which assigns to each subset of Y a number in the unit interval $[0,1]$. In order to qualify $g(\cdot)$ as a fuzzy measure, the function $g(\cdot)$ must satisfy certain properties, as the axioms of fuzzy measures defined in [14,29]. Two special types of fuzzy measures are the probability measures and possibility measures.

Let $P : 2^Y \to [0,1]$ be a *probability measure* P. By definition

$$P(A) = \int_A p(y)dy \tag{3}$$

is the probability measure of A for all $A \in 2^Y$. The function $p : Y \to \Re^+$ is the *probability density function* and satisfies the property

$$\int_Y p(y)dy = 1$$

Let us now denote a *possibility measure* by $\Pi : 2^Y \to [0,1]$. By definition the possibility measure of A for all $A \in 2^Y$ is

$$\Pi(A) = \sup_{y \in A} \pi(y) \tag{4}$$

where $\pi : Y \to [0,1]$ is the *possibility distribution function.*

Definitions (3) and (4) reveal a substantial difference between the notion of probability and possibility: while probability satisfies the property of additivity, this is not true for the possibility measure, which is sub additive. Then, given two generic sets $A, B \in 2^Y$, we have

$$P(A \cup B) = P(A) + P(B) - P(A \cap B)$$

while

$$\Pi(A \cup B) = \sup(\Pi(A), \Pi(B))$$

As we will show in the rest of the paper, the above properties induce different evolution laws for probability and possibility distributions in deterministic dynamical systems.

4 Stochastic differential equations (SDE)

The introduction of random elements into an ODE leads to the *stochastic differential equations* (SDE). The definition of a SDE depends on what we mean by derivatives of a stochastic process. A crucial point is the notion of *regularity* [25] of random functions occurring in the differential equation.

If a very irregular random process occurs (as white noise processes) we have the so-called *Ito stochastic differential equations.* In these equations the uncertainty of variables and/or parameters is represented by adding to their deterministic value the effect of a random process of the white noise type. Consider a time varying parameter $U(t)$ subject to some random effects: according to the Ito approach its uncertainty is modeled by $U(t) = D(t)+n(t)$ where the probability distribution of the *noise* term $n(t)$ is known and $D(t)$ is assumed to be non random.

When the random processes occurring in a differential equation are sufficiently regular the equations are called *regular stochastic differential equations*. In this case an uncertain constant parameter U is represented by a random variable with a known probability distribution. The theory of the Ito stochastic differential equations differs significantly from the theory of regular stochastic equations [9].

In this paper we will limit ourselves to consider the properties of the probability formalism when this is used to model incomplete knowledge about constant parameters and/or initial conditions in dynamical systems: as a consequence, the Ito representation, typically used to model dynamical systems subjected to rapidly time-varying random excitation, will be out of the scope of the paper.

Let us now consider the regular differential equation

$$\dot{y} = F(y), \qquad y \in \Re^n, \quad y(0) = y_0, \tag{5}$$

where the initial condition y_0 is the realization of a continuous random variable with known density $p(y_0)$.

The solution $y(t, y_0)$ of the stochastic initial value problem is still a regular stochastic process, which is completely characterized by a probability density $p(y(t, y_0))$.

4.1 Regular SDE and the Liouville equation

The solution of a regular stochastic initial-value problem is provided by the Liouville equation [25,9]. In order to present the Liouville equation, we first introduce the concept of integral invariant. Let $\Omega(t)$ represent an arbitrary volume at time t in the phase space of (5), and let $\Omega(t + dt)$ be the set of points

$$\Omega(t + dt) = \{y(t + dt, y(t)) : y(t) \in \Omega(t), \dot{y} = F(y)\} \tag{6}$$

that propagate according to (5) and such that $y(t)$ is contained in $\Omega(t)$.

Let us consider the integral of some function $f(y, t) : \Re^n \times \Re \to \Re$ over this moving domain $\Omega(t)$,

$$I(t) = \int \dots \int_{\Omega(t)} f(y, t) dy_1 \dots dy_n \tag{7}$$

The quantity $f(y, t)$ is called an *integral invariant* of (5) if $I(t)$ is constant in time.

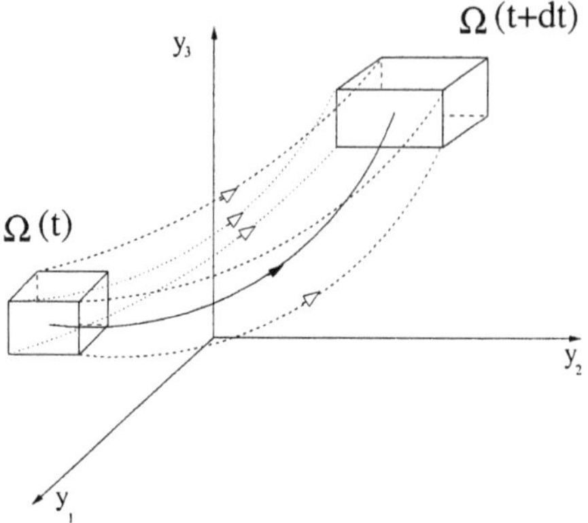

Fig. 1. Evolution of a volume $\Omega(t)$ in the phase space of a 3rd order dynamical system.

One basic result of stochastic theory is the Liouville theorem, which states that each function $f(y,t)$ whose integral with respect to the spatial variable is invariant during the evolution of the system, satisfies the following equation

$$\frac{\partial f(y,t)}{\partial t} + \sum_{i=1}^{n} \frac{\partial}{\partial y_i} [f(y,t)F_i(y)] = 0 \tag{8}$$

which is commonly known as the *Liouville equation.*

It has been demonstrated that the probability density $p(y(t,y_0),t)$ is an integral invariant [25], that is,

$$P(\Omega(t)) = \int_{\Omega(t)} p(y,t)dy$$

is constant in time. Hence, the property of integral invariance implies that the Liouville equation (8) holds for the probability density:

$$\frac{\partial p(y,t)}{\partial t} + \sum_{i=1}^{n} \frac{\partial}{\partial y_i} [p(y,t)F_i(y)] = 0 \tag{9}$$

The Liouville equation is a first order partial differential equation and can be solved in general by the Lagrange method. It may be also transformed to the following ordinary differential equation holding along the trajectories of the system [20]:

$$\frac{dp(y,t)}{dt} = -p(y,t)\text{div}F(y) \tag{10}$$

where div is the *divergence* of the system and is defined as

$$\text{div}F(y) = \sum_{i=1}^{n} \frac{\partial F_i}{\partial y_i} \tag{11}$$

The probability density of a point in the phase space may be intuitively interpreted as the density of trajectories of the dynamical system at that point. When the divergence (11) is null in the whole phase space, the dynamical system is *conservative*: this means that the volume of a region $\Omega(t)$, as well as the density of trajectories, remains constant during the evolution of the dynamical system (Fig. 2a). When the divergence is greater or smaller than zero, the system is said to be *dissipative*: if the divergence is greater than zero then the system is divergent, the volume of a region $\Omega(t)$ increases and the density decreases (Fig. 2b). The opposite happens if the system is convergent.

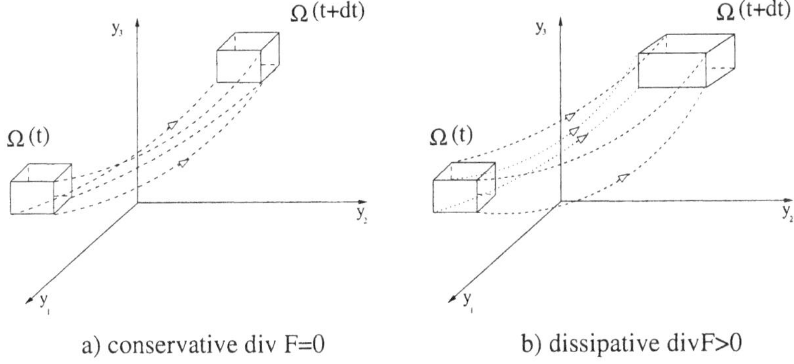

Fig. 2. Conservative and dissipative dynamics.

Hence, the evolution of the probability measure in a dynamical system is linked to the dynamics $F(\cdot)$ through Equation (10) and satisfies the following two properties.

Property 1. The probability density $p(y(t, y_0), t)$ along a trajectory $y(t, y_0)$ of a dynamical system is not constant (except for $\text{div}F = 0$).

Property 2. The probability of finding trajectories inside the time-variant volume $\Omega(t)$ is constant during the evolution of the dynamical system.

It is worth remembering that the Liouville equation is a special case (i.e. the diffusion term is null) of the Fokker-Planck-Kolmogorov equation [9] that represents the evolution of probability distribution for the Ito stochastic differential equation. Therefore the regular SDE may be seen as a particular case of the more general Ito SDE.

4.2 Numerical solution of regular SDE

By solution of a regular SDE (5) we mean a stochastic process satisfying the equation together with its probabilistic properties. Generally, as in the non random case, analytical solutions are not available. Most often, we can only obtain approximate solutions by numerical schemes.

The Liouville equation can be used for the numerical computation of the evolution of the probability density. However, the numerical resolution of a partial differential equation implies a huge computational cost.

A common numerical resolution method is Monte Carlo [23]. This is a statistical method for simulation that makes use of sequences of random numbers. It may be applied when the system is modeled by probability density functions. Thus, it is currently used to simulate the evolution of a probability density in ordinary differential equations [21].

Monte Carlo simulation of a regular SDE is composed of the repetition of three steps:

1. a random sampling of the probability density of the initial condition distribution,
2. a series of deterministic model simulations ("trials") each using as initial condition the sampled values,
3. the statistical description of the evolution of the probability density by means of the sample simulation.

The set of simulated trajectories represents then a probabilistic sample of the solution of the stochastic differential equation.

5 Fuzzy numbers and possibility distributions

The concept of fuzzy number is an extension of the notion of real number: it encodes approximate but non probabilistic quantitative knowledge [13,24]. For instance, let us consider a system (e.g. a steam generator of a nuclear power plant) where it is not easy or convenient to measure a certain variable (e.g. its internal pressure); suppose, however, that we have a qualitative, imprecise knowledge of the value of the variable (e.g. the value of the variable is *around 5 bar*). This knowledge can be easily coded in the form of a possibilistic distribution, which may be represented by a fuzzy number.

Zadeh introduced both the concept of fuzzy set [30] and the concept of possibility measure [31]. Also, Zadeh's *possibilistic principle* [31] postulates

$$\mu_A(y) = \pi_A(y) \qquad \forall y \in Y \tag{12}$$

where $\mu_A(y)$ is the membership function of a fuzzy set A and $\pi_A(y)$ is the possibility distribution associated with A. We rest on this principle to justify the use of fuzzy membership functions to represent possibilistic information.

5.1 Fuzzy functions

A fuzzy function can be understood in several ways according to where fuzziness occurs [4]. By considering deterministic models, we deal only with ordinary functions that just carry the fuzziness of their arguments without generating extra fuzziness themselves. In such cases, we can operate the so-called fuzzy extension of crisp functions by adopting the *extension principle* introduced by Zadeh [30]. This principle provides a general method for extending standard mathematical concepts in order to deal with fuzzy quantities [4].

Let $\Phi : Y_1 \times Y_2 \times \cdots \times Y_n \rightarrow Z$ be a deterministic mapping such that $z = \Phi(y_1, y_2, \ldots, y_n)$. The extension principle enables us to induce from n input fuzzy sets η_{Y_i} on Y_i an output fuzzy set μ_Z on Z through Φ given by

$$\mu_Z(t) = \sup_{t=\Phi(s_1,s_2,\ldots,s_n)} \min\left(\eta_{Y_1}(s_1), \eta_{Y_2}(s_2), \ldots, \eta_{Y_n}(s_n)\right) \tag{13}$$

or

$$\mu_Z(t) = 0 \quad \text{if} \quad \Phi^{-1}(t) = \emptyset \tag{14}$$

where $\Phi^{-1}(t)$ is the inverse image of t.

In the next section, we use the extension principle to define the notion of fuzzy differential equation.

6 Fuzzy differential equations (FDE)

Let us consider a differential equation

$$\dot{y} = F(y), \quad y \in \Re^n, \qquad y(0) = y_0, \tag{15}$$

where $\mu_0(y_0)$ is the degree of membership of the initial condition to the fuzzy distribution μ_0 defined on the n-dimensional domain Y.

We interpret the notation (15) as a fuzzy extension, according to the principle (13), of the notion of ordinary differential equation. Hence, we define a *fuzzy differential equation* (FDE) as a deterministic differential equation where some coefficients or initial conditions are uncertain and defined as fuzzy numbers: its solution is then the time evolution of a fuzzy region of uncertainty which corresponds to the possibility distribution in the phase space.

We derive now three theorems which define the properties of the solution of a FDE.

Theorem 1. *Let $\mu(\cdot, t^*, \mu_0)$ denote the fuzzy state of the fuzzy differential equation (15) at time t^*. Then its value is given by*

$$\mu(y^*, t^*, \mu_0) = \mu_0(y_0) \tag{16}$$

such that

$$y^* = y(t^*, y_0) \tag{17}$$

where $y = y(t, y_0)$ is the solution of the differential equation

$$\dot{y} = F(y), \quad y \in \Re^n, \quad y(0) = y_0.$$

Proof: The fuzzy value $\mu(y^*, t^*, \mu_0)$ of the state at time t^*, according to the extension principle (13), is

$$\mu(y^*, t^*, \mu_0) = \sup_{y^*=y(t^*,y_0)} \mu_0(y_0) \tag{18}$$

Since $F(\cdot)$ satisfies the Lipschitz condition, there is only one trajectory $y(t, y_0)$ passing through the point y^* at time t^*: it follows that $\mu(y^*, t^*, \mu_0) = \mu_0(y_0)$ where $y^* = y(t^*, y_0)$.

Theorem 2. *In a fuzzy differential equation the value of the possibility distribution is constant along the trajectories of the system.*

Proof: According to Theorem 1, $\mu(y(t, y_0), t, \mu_0) = \mu_0(y_0)$ along a trajectory $y(t, y_0)$. Then, according to the Zadeh's possibilistic principle (12)

$$\pi(y(t, y_0)) = \mu(y(t, y_0), t, \mu_0) = \mu_0(y_0) = \text{constant} \tag{19}$$

Theorem 3. *Let $\Omega(t)$ represent an arbitrary volume at time t in the phase space of the fuzzy differential system (15) and $\Pi(\Omega(t))$ be the possibility measure of this set. Then, $\Pi(\Omega(t))$ is time invariant.*

Proof: According to Theorem 2 the value of the possibility distribution $\pi(y(t, y_0))$ is time invariant along a trajectory $y(t, y_0)$. Since

$$\Pi(\Omega(t)) = \sup_{y(t,y_0)\in\Omega(t)} \pi(y(t, y_0))$$

according to definition (4), it follows that $\Pi(\Omega(t))$ is time invariant.
In order to interpret intuitively these results, let us consider a solution (trajectory) of the system passing through the point $y_1 = y(t_1, y_0)$ at time t_1 and through the point $y_2 = y(t_2, y_0)$ at time t_2. According to Theorem 2, the possibility that a trajectory passes through y_1 at t_1 is equal to the possibility that the same trajectory passes through y_2 at time t_2. Moreover, if we consider a region $\Omega(t)$ that evolves according to the dynamics of the system, according to Theorem 3, the possibility of finding a trajectory in $\Omega(t_1)$ at time t_1 equals the possibility of finding a trajectory in $\Omega(t_2)$ at time t_2.

6.1 Numerical solution of a fuzzy differential equation

Using the extension principle to compute the fuzzy value of a fuzzy mapping is not feasible in practice since this procedure requires an infinite number of computations.

The original approach we propose considers instead fuzzy calculus as an extension of interval mathematics by making use of the α-cut notion [4]. According to [19], under very general conditions, a fuzzy computation may be decomposed into several interval computations, each one having as arguments the α-cut intervals of the fuzzy arguments. Each interval computation returns an α-cut of the fuzzy output. Then, the following relation holds for the mapping (13)

$$\mu_\alpha = \varphi(\eta_1, \eta_2, \ldots, \eta_n)_\alpha = \varphi(\eta_{1\alpha}, \eta_{2\alpha}, \ldots, \eta_{n\alpha}) \tag{20}$$

where μ_α and $\eta_{i\alpha}$ denote the α-cuts of the fuzzy numbers μ_Z and η_{Y_i}, respectively. It follows that every fuzzy computation, also differential, may then be decomposed into the repetition of more interval computations. For that reason, hereafter we will focus on interval computational methods.

Interval differential equations (IDE) Let us consider a differential equation

$$\dot{y} = F(y), \quad y \in \Re^n, \quad y(0) \in \mu_{0\alpha}, \tag{21}$$

where $\mu_{0\alpha}$, $0 \leq \alpha \leq 1$, is the α-cut of the fuzzy initial condition μ_0 in the FDE (15)

This interval differential equation (IDE) is the α-cut of the corresponding fuzzy differential equation (15).

Let us define as interval solution $\mu_\alpha(t)$ of (21) the set of solutions of the ordinary differential equation whose initial conditions belong to $\mu_{0\alpha}$.

Numerical solution of IDE The numerical solution of the system (21) is apparently an easy task. We might intuitively guess that the simple extension of ODE numerical methods (e.g. Euler, Runge-Kutta) in terms of interval mathematics [17,18] could suffice. In fact, in [2] we show that a straightforward use of interval mathematics for the resolution of fuzzy (or interval) differential equation may yield incorrect results. This is due to the fact that the fuzzy (and interval) formalism is unable to represent the interaction that the differential equation establishes between variables: as a consequence, spurious values are added to the solution and the evolution of the system may reach region where no numerical solution exists [2,11].

Figure 3 illustrates the problem. The figure plots the evolution of the initial condition (i.e. the rectangle ABCD) in a oscillatory 2nd order system: if we represent the uncertain state of the dynamical system at time t' using the interval notation, we add to the solution (rotated rectangle A'B'C'D') a set of spurious trajectories (black regions) which do not correspond to any evolution of points belonging to ABCD.

In [2,3] we proposed an approach for the resolution of interval differential equations which avoids this problem. Let us consider the set of initial conditions of the interval differential equation (21). They form an hyper cube of dimensionality n which we define as the *region of uncertainty* of the system at time $t = 0$. The numerical solution of the IDE requires the simulation of the set of trajectories having as initial condition a point inside the region of uncertainty at time $t = 0$.

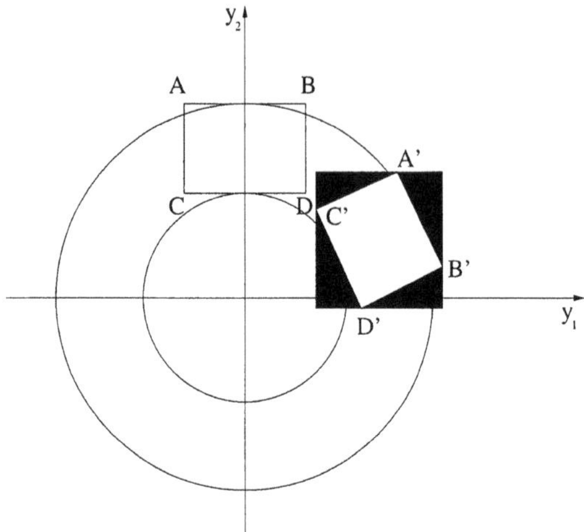

Fig. 3. Introduction of spurious trajectories in the simulation of an oscillating system.

We proved that, under general conditions of continuity and differentiability, it is sufficient to compute the evolution of the external surface of the uncertainty region [2]. We established the following theorem.

Theorem 4. *An ordinary autonomous differential equation maps the external surface of its region of uncertainty at time t into the external surface of its region of uncertainty at time $t + dt$.*

The validity of the theorem leads to the conclusion that it is sufficient to track the trajectories of the points belonging to the external surface of the region of uncertainty to simulate how the region of uncertainty evolves. However, the quantity of trajectories to be simulated is still infinite, although the order of magnitude is lower. The main problem remains how to sample, in the most convenient way, the external surface of the region of uncertainty during its time evolution.

In this paper we extend the approach presented in [2] by reducing the sampling problem to an optimization problem. The basic idea is that finding the numerical solution of an IDE is equivalent to having, at each time step t^*, the maximum and the minimum value of the state variables y_i, $i = 1, \ldots, n$. These values are obtained by maximizing and minimizing at time t^* the functions $y_i(t^*, y_0)$. Hence, we may use a constrained multivariate optimization algorithm to find the extrema of the functions $y_i(t^*, y_0)$, $i = 1, ..., n$, where t^* is fixed and the argument y_0 is constrained to range over the external surface of the initial region of uncertainty.

Note that the performance of a multivariate optimization algorithm may be improved by providing it with the partial derivatives (gradient) of the function value with respect to the arguments: in our case we should provide the derivatives at time t^* of y_i with respect to $y_0 \in Y_{0\alpha}$. To compute these derivatives, we proposed in [2] the use of the *connection matrix* described in [17].

Let $y(t, y_0)$ be the n-dimensional, real vector solution of the numeric initial-value problem. The connection matrix for the solution $y(t, y_0)$ is defined as the $[n \times n]$ matrix $C(t, y_0)$ where the generic $[i, j]$ element is

$$C_{ij}(t, y_0) = \left. \frac{\partial y_i(t, \hat{y})}{\partial \hat{y}_j} \right|_{\hat{y}=y_0} \tag{22}$$

Moore [17] demonstrated that along the trajectories of a dynamical system

$$\frac{\partial C(t, y_0)}{\partial t} = J(t, y_0) C(t, y_0) \tag{23}$$

where $C(0, y_0) = I$ with I identity matrix and $J(t, y_0)$ is the Jacobian matrix of the vector function F evaluated at $y(t, y_0)$. The $[i, j]$ element of the matrix $J(t, y_0)$ is

$$J_{ij}(t, y_0) = \left. \frac{\partial F_i(y)}{\partial y_j} \right|_{y=y(t, y_0)} \tag{24}$$

It is evident that the elements of the matrix C give the sensitivity of the solutions $y_i(t, y_0)$ to small changes of the initial values $(y_0)_i$.

By coupling the IDE system with the set of n^2 equations coming from the connection matrix differential system (23) we obtain an augmented differential system which gives at time t^* the values of the variables y_i and the derivatives (gradient) of the y_i with respect to the initial condition y_0.

The core of our algorithm is then made of the following steps:

1. the optimization routine takes a starting point y_0 in the external surface of the initial uncertainty region;
2. this point is passed to the routine of ODE numerical integration which uses y_0 as initial condition and returns both the value $y(t^*, y_0)$ and the gradient values;
3. if the optimization routine considers $y(t^*, y_0)$ as a maximum (minimum) the algorithm stops; otherwise, the value $y(t^*, y0)$ and the gradient are used to sample a new point y_0 and the execution returns to point 2.

Let us consider, for instance, a second order IDE (fig. 6) where the external surface of the region of uncertainty at time $t = 0$ is the square ABCD. Suppose that the square is transformed by the differential equation into the region A'B'C'D' at time t^*. For each of the $2^2 = 4$ portions of the external surface of ABCD (e.g., the segment AB), the optimization algorithm searches the maximum and the minimum of the correspondent transformed part (e.g., the curve A'B'). The greatest of the maxima and the smallest of the minima of each external portion are taken as the extremal values of $y_i(t^*)$.

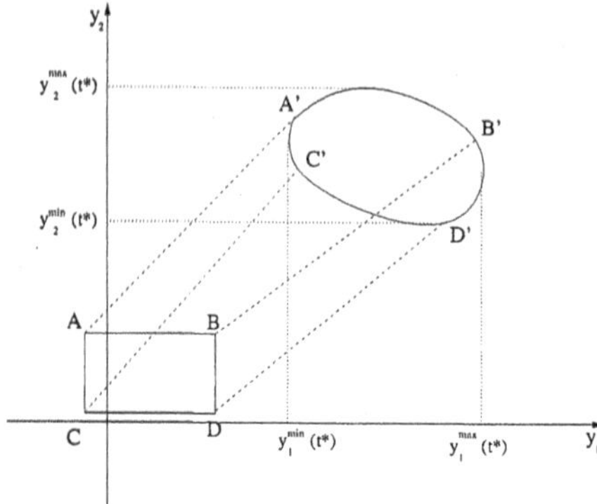

Fig. 4. Evolution of the external surface of the region of uncertainty for a 2-nd order system.

In the case of a fuzzy differential equation, the procedure is to be repeated for each α-cut that decomposes the fuzzy initial value. The number of α-cuts to be considered depends on the desired degree of accuracy. On this matter, it is important to remark that the region of uncertainty corresponding to an α-cut of level α_1 is always contained into the region of uncertainty corresponding to an α-cut of level α_2, if $\alpha_1 > \alpha_2$ [2].

The resulting algorithm, called *Qua.Si. (Qualitative Simulator)*[1], has been implemented in Matlab© language, by integrating a symbolic calculus program which computes the Jacobian of the original differential system, an algorithm of constrained optimization and a numerical solver of ordinary differential equations.

Remarks: The proposed method goes one step further with respect to the non interacting algorithm proposed in [2]: the heuristic criterion to choose the trajectories is now replaced by an optimization algorithm whose task is to choose the points to be simulated in order to reconstruct the region of uncertainty at a given t^*. This does not guarantee that the global maximum and minimum will be found but at least makes the computational complexity of each step comparable to that of well established algorithms. However, as far as complexity is concerned, two drawbacks of the present approach have to be mentioned: (i) the complexity is an exponential function of the order n of the dynamical system (2^n is the number of components of the external surface of the initial region of uncertainty) (ii) the fuzzy solution at time t^* is not computed by updating the fuzzy solution at the previous instant $(t^* - \Delta t)$: in fact, there is no guarantee that the trajectories belonging to the extremes of the region of uncertainty at time $(t^* - \Delta t)$ be the same trajectories to be considered at time t^*. Then, at each time t^* the trajectories must be simulated

[1] available at `http://iridia.ulb.ac.be/~gbonte/software/Quasi/Quasi.html`

starting from the instant $t = 0$. This implies an heavy computational cost for large t^*.

7 Probability vs. possibility in differential equations

The relation between possibility and probability theory has long been debated in literature [1]. In Sections 4 and 6 we showed that probability density and possibility distribution along the trajectories evolve according to different laws: while the probability density varies according to the Liouville equation (10), possibility remains constant (Theorem 2). Let us now discuss more in detail the relation between these two formalisms.

Let $y(t, y_0)$ be a point in the phase space of the dynamical system. To make the notation easier, in the following we will denote it by $y(t)$. The probability that the state belongs to a infinitesimal volume $\Omega(t)$ around $y(t)$ is $p(y(t), t)\Omega(t)$. Given the property of integral invariance

$$p(y(t), t)\Omega(t) = p(y(t + dt), t + dt)\Omega(t + dt)$$

the quantity $p(y(t), t)\Omega(t)$ remains constant along a trajectory. That is, the following relation holds

$$\frac{d}{dt}\left(p(y(t), t)\Omega(t)\right) = 0 \tag{25}$$

This implies

$$\Omega(t)\frac{d}{dt}p(y(t), t) + p(y(t), t)\frac{d\Omega}{dt} = 0 \tag{26}$$

and from the Liouville equation (10) we obtain

$$-p(y(t), t)\mathrm{div}F(y(t))\Omega(t) + p(y(t), t)\frac{d\Omega}{dt} = 0 \tag{27}$$

Hence, along the trajectories where $p(y(t), t)$ is different from zero we have also

$$\frac{d\Omega(t)}{dt} = \Omega(t)\mathrm{div}F(y(t)) \tag{28}$$

From the above formula and Eq. (19) we have

$$\frac{d}{dt}\left(\pi(y(t), t)\Omega(t)\right) = \Omega(t)\frac{d\pi\left(y(t), t\right)}{dt} + \pi\left(y(t), t\right)\frac{d\Omega}{dt} = \\ = \pi\left(y(t), t\right)\Omega(t)\mathrm{div}F(y(t)) \tag{29}$$

Consequently, the evolutions in time of probability and possibility are summarized by the following equations

$$\frac{d}{dt}p(y(t),t) = -p(y(t),t)\mathrm{div}F(y(t)) \tag{30}$$

$$\frac{d}{dt}\left(p(y(t),t)\Omega(t)\right) = 0 \tag{31}$$

$$\frac{d}{dt}\pi(y(t),t) = 0 \tag{32}$$

$$\frac{d}{dt}\left(\pi(y(t),t)\Omega(t)\right) = \pi(y(t),t)\mathrm{div}F(y(t)) \tag{33}$$

The first consideration we may deduce from the above equations concerns what Zadeh calls the *possibility/probability consistency principle* [31]. He defines a degree of *consistency* γ of the probability density $p(\cdot)$ with the possibility distribution $\pi(\cdot)$. We extend Zadeh's definition to a continuous domain Y by defining the quantity

$$\gamma(t) = \int_Y \pi(y(t),t)p(y(t),t)dy \tag{34}$$

Note that the closer is γ to 1, the higher is the consistency. The following theorem establishes that full probability/possibility consistency is preserved during the system's evolution:

Theorem 5. *Consider a dynamic system where the uncertainty of the initial condition is represented both in probabilistic and in possibilistic form. If the initial probability density is consistent with the initial possibility distribution ($\gamma = 1$) then consistency is preserved during the system's evolution.*

Proof: From Eq. (34) and $\gamma = 1$ it follows that $p(y) > 0$ implies $\pi(y) = 1$. Hence, the differential laws of evolution (30) and (32) show that the consistency is conserved during the system evolution.

Our second consideration refers to the definition of a parameter relating possibility and probability density evolution along a system trajectory.

Theorem 6. *Let us define $\psi(y(t),t)$ such that*

$$\psi(y(t),t) = \frac{p(y(t),t)}{\pi(y(t),t)} \tag{35}$$

The time evolution of $\psi(\cdot,\cdot)$ is governed by the following differential law

$$\frac{d}{dt}\psi(y(t),t) = -\psi(y(t),t)\,divF(y(t)) \tag{36}$$

where $\psi(0) = \frac{p(y(0),0)}{\pi(y(0),0)}$

Proof: From the time invariance of $\pi(y(t),t)$ (formula (32)) and the formula (30) we have

$$\frac{d}{dt}\psi(y(t),t) = \frac{\frac{d}{dt}p(y(t),t)}{\pi(y(t),t)} = \\ = -\frac{p(y(t),t)\mathrm{div}F(y(t))}{\pi(y(t),t)} = -\psi(y(t),t)\mathrm{div}F(y(t)) \quad (37)$$

Once we know the initial distribution $\frac{p(y(0),0)}{\pi(y(0),0)}$, the differential relation (36) allows the derivation of a probability density value $p(y(t),t)$ from the related possibility distribution $\pi(y(t),t)$ and vice versa.

In summary, Theorem 5 and 6 prove that not only a property of consistency between probability and possibility is satisfied in dynamical systems but also that it is possible to define a parameter ψ, relating probability and possibility, whose time evolution depends on the dynamics F.

8 Monte Carlo and qualitative simulation: the potential danger

The results presented in the previous section show that the numerical resolution of fuzzy differential equations demands specific non probabilistic methods.

As a matter of fact, this requirement is not always satisfied in real applications. A common habit among simulation practitioners is to adopt probabilistic formalisms to address also non probabilistic uncertainty with the implicit (strong) assumption that the solution will not be significantly affected. Even the early works on qualitative simulations proposed Monte Carlo techniques as a possible way of performing fuzzy simulation of dynamical systems [8], specifically in conditions of normalized possibility distributions.

However, the results presented in the previous section show that the adoption of the Monte Carlo approach is not correct each time we are dealing with possibility distributions, even normalized.

Let us consider the evolution laws of probability and possibility expressed by Eq. (31) and (33), respectively . In the probabilistic case, the homeomorphism between $\Omega(t)$ and $\Omega(t+dt)$ and the relation (31) imply that the probability of being at time t in $\Omega(t)$ equals the probability of being at time $t+dt$ in $\Omega(t+dt)$. As a consequence, a random sampling of $\Omega(t)$ induces a random sampling in $\Omega(t+dt)$. This property justifies the adoption of Monte Carlo in the probabilistic setting.

On the contrary, if the uncertainty is expressed by a possibility normalized distribution, the law (33) states that a random sampling of the initial condition does no imply a random sampling of the possibilistic solution. Therefore, the use of the Monte Carlo approach to evaluate the evolution of a normalized fuzzy distribution in a deterministic system is not justified.

Moreover, the probability along a trajectory changes according a differential dynamics as described in Eq. (10), while Theorem 2 states that the possibility along a trajectory is constant. In particular, Eq. (10) implies that, if for a time interval sufficiently long the divergence along a specific trajectory is positive, the probability density along that trajectory might take a value which is not significantly different from zero, even if its value at $t = 0$ was greater than zero.

These considerations should sound as a reminder for engineers doing simulations under conditions of uncertainty: the choice of the simulation tool depends on the type of uncertainty we are dealing with.

Let us imagine a model designer who represents his imprecise knowledge about a parameter as a uniform possibility distribution but who uses a Monte Carlo routine for studying the evolution of the system. What he generally expects is to generate a uniform sampling in the space of possible trajectories and have consequently a qualitative description of the potential operating conditions. Equation (10) says this might not be true in reality. The evolution of the probability density is strictly related to the dynamics of the system (in particular to its divergence), and we cannot exclude a priori that the Monte Carlo simulation will not hide to the designer some relevant trajectories he would expect to see.

This lack of observability could have catastrophic effects if, for instance, the aim of the simulation is to check whether the uncertain system will pass through some critical operating conditions: think to the simulation of an industrial plant under conditions of uncertainty, where the probability of a dangerous event takes a probability near to zero, but unfortunately a non zero possibility.

9 Experiments

Our experimental session considers the Lotka-Volterra dynamical system, described by the state equations:

$$\begin{cases} \dot{y}_1 & = k_1 a y_1 - k_2 y_1 y_2 \\ \dot{y}_2 & = k_2 y_1 y_2 - k_3 y_2 \end{cases} \tag{38}$$

where $a = 1$, $k_1 = 1$, $k_2 = 0.01$ and $k_3 = 1$. Assume that our imprecise knowledge over the initial conditions is modeled by two fuzzy rectangular sets:

$$\mu_Y(y_1) = \begin{cases} 1 & \text{if } 1 \le y_1 \le 50 \\ 0 & \text{otherwise} \end{cases} \tag{39}$$

and

$$\mu_Y(y_2) = \begin{cases} 1 & \text{if } 2 \le y_2 \le 60 \\ 0 & \text{otherwise} \end{cases} \tag{40}$$

respectively.

First we use the *Qua.Si.* algorithm described in Section 6.1 to simulate the qualitative behavior of the system (38) over the time interval $t = [0, 8]$. Fig. 5 reports the temporal evolution of the bounds of the lower α-cut of the two fuzzy state variables.

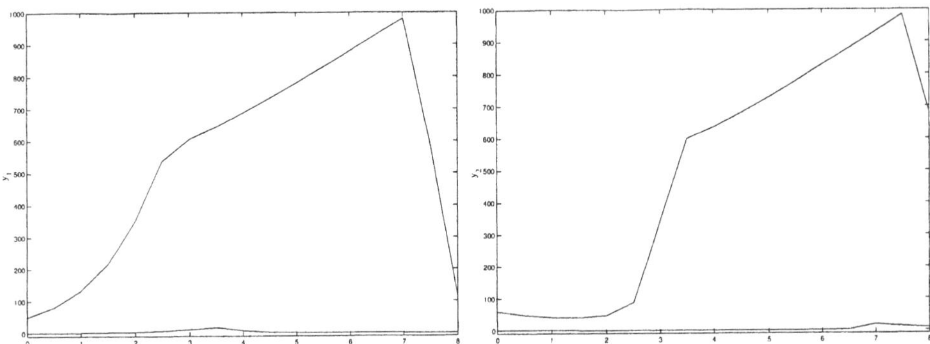

Fig. 5. Fuzzy qualitative simulation of the Lotka-Volterra system.

Then, we employ the Monte Carlo technique to simulate the same system where the initial qualitative conditions are interpreted as uniform probability distributions. The output of $N = 1000$ trials is reported in Fig. 6.

A zoom of the superposition of the upper parts of Fig. 5 and Fig. 6 is reported in Fig. 7. What we remark here is that, even after an high number of simulation, the Monte Carlo does not return some boundary trajectories. Consider for example the time instant $t = 7$. According to the fuzzy simulation, the maximum value that the variable y_2 can attain is $y_2(7) = 931.86$, corresponding to the initial condition $y_1(0) = 1.60, y2(0) = 2$. Instead, according to the Monte Carlo simulation, the extreme value is largely lower, that is 696.89. This is a direct consequence of the interaction between dynamics and probability as described in Eq. (10). In Fig. 8 we report the evolution of the probability $p(y(t, y_0))$ of the trajectory starting in $y_1(0) = 1.60, y_2(0) = 2$, as returned by (10): we see that the probability at time $t = 7$ is not significantly different from zero. This implies that the probability of discovering the upper bound of the value of y_2 at time $t = 7$ is practically null.

The experimental results confirm that the interplay between the probability distribution and the dynamics could hide trajectories (relevant from a possibilistic point of view) in the output of the probabilistic simulation. This could have hazardous consequences if the purpose of the simulation is to have a qualitative description of the uncertain behavior.

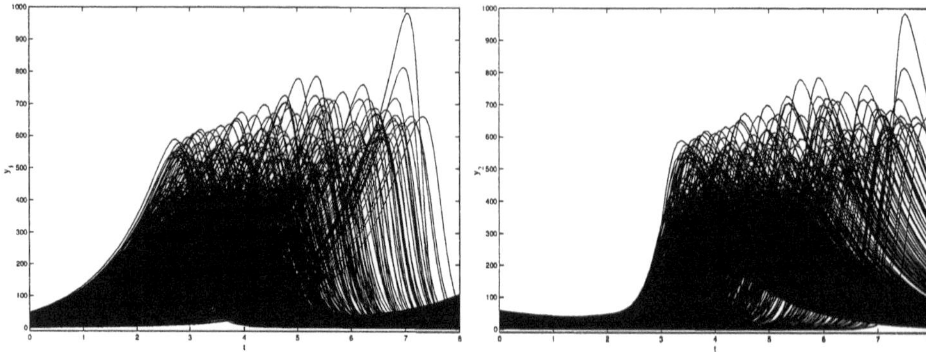

Fig. 6. Monte Carlo simulation of the Lotka-Volterra system.

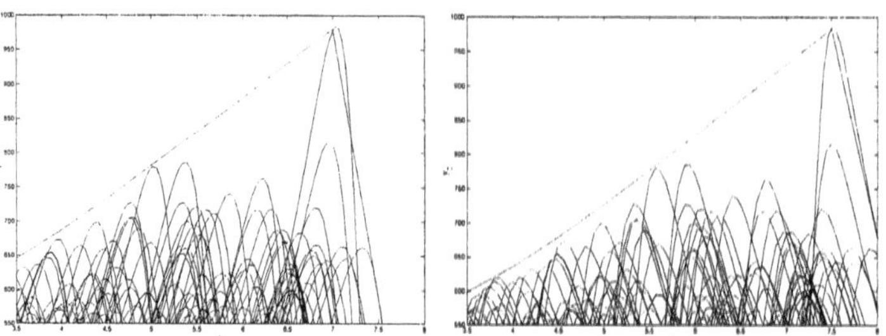

Fig. 7. Monte Carlo and fuzzy simulation (superposition and zoom).

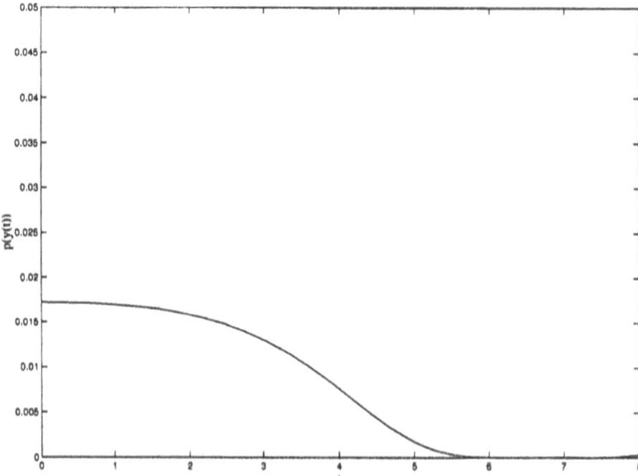

Fig. 8. Evolution of the probability $p(y(t))$ of the Lotka-Volterra trajectory which is at the boundary at time t=7.

10 Conclusions

This paper compared probability and possibility as two formalisms to represent uncertainty in continuous dynamical systems. Probability and possibility are measures on the dynamical phase space which describe two different kinds of uncertainty: the objective uncertainty that originates from random measurements, and the subjective uncertainty that stems from imprecise human knowledge. Using basic results of system theory and fuzzy measures, we showed that different, but related, laws describe the evolution of probability and possibility distributions in dynamical systems. We have made this relation precise by the introduction of a parameter, whose evolution is characterized by a differential equation related to the system dynamics.

The different analytical properties between probability and possibility suggest the adoption of specific computational methods to simulate the evolution of possibilistic and probabilistic dynamical systems. Although this remark could seem straightforward, in literature probabilistic methods have often been used to deal with possibilistic formulations. The paper criticized this attitude on formal and experimental basis by stressing the risk of producing inadvertently inaccurate simulations, which omit interesting trajectories of the system.

The experimental session presented an example where Monte Carlo returns an inadequate description of the qualitative behavior of a dynamical system, where the uncertainty is expressed by a possibility distribution. We showed that this is due to the kind of interaction existing between the dynamics of the system and the evolution law of probability.

Some fields of engineering, like fuzzy control and identification, are interested in a non stochastic representation of uncertainty in dynamical systems. We believe that a comparative analysis of possibilistic methods with related stochastic methods may be useful to better understand the differences and the relative merits of both the approaches. We hope that the study proposed in this paper, may help in this effort.

References

1. J. C. Bezdek. Editorial: Fuzziness vs. probability - again (!?). *IEEE Transactions on Fuzzy systems*, 2(1):1–3, 1994.
2. A. Bonarini and G. Bontempi. A qualitative simulation approach for fuzzy dynamical models. *ACM Transactions on Modeling and Computer Simulation (TOMACS)*, 4(4), 1994.
3. G. Bontempi. Qua.si. iii: A software tool for simulation of fuzzy dynamical systems. In A.Javor, A. Lehmann, and I. Molnar, editors, *Modeling and Simulation ESM 96 (Proceedings European Simulation Multiconference 1996)*, pages 615–619, Ghent, Belgium, 1996. SCS International.
4. D. Dubois and H. Prade. *Fuzzy Sets and Systems.* Academic Press, New York, 1980.

5. D. Dubois and H. Prade. Towards fuzzy differential calculus, part 1: integration of fuzzy mappings. *Fuzzy Sets and Systems*, 8:1–17, 1982.
6. D. Dubois and H. Prade. Towards fuzzy differential calculus, part 2: integration on fuzzy intervals. *Fuzzy Sets and Systems*, 8:105–116, 1982.
7. D. Dubois and H. Prade. Towards fuzzy differential calculus, part 3: differentiation. *Fuzzy Sets and Systems*, 8:223–233, 1982.
8. P. A. Fishwick. Fuzzy simulation: Specifying and identifying qualitative models. *International Journal of General Systems*, 19:295–316, 1991.
9. C. W. Gardiner. *Handbook of stochastic methods*. Springer-Verlag, Berlin, 1985.
10. E. Hairer, S. P. Norsett, and G. Wanner. *Solving ordinary differential equations.* Springer, 1987.
11. E. Hüllermeier. Numerical simulation methods for uncertain dynamics. Technical Report tr-ri-95-170, University of Paderborn, 1995.
12. O. Kaleva. Fuzzy differential equations. *Fuzzy Sets and Systems*, 24:301–317, 1987.
13. A. Kaufmann and M. Gupta. *Introduction to fuzzy arithmetic: theory and applications.* Van Nostran and Reinhold, New York, 1985.
14. G. J. Klir and T. A. Folger. *Fuzzy sets, uncertainty, and information.* Prentice-Hall, New York, 1988.
15. D. Lambert and J. D. Lambert. *Numerical Methods for Ordinary Differential Systems : The Initial Value Problem.* John Wiley and Son, 1991.
16. R. P. Leland. Fuzzy differential systems and malliavin calculus. *Fuzzy Sets and Systems*, 70:59–73, 1995.
17. R. E. Moore. *Interval Analysis.* Prentice-Hall, Englewood Cliffs, 1966.
18. A. Neumaier. *Interval Methods for Systems of Equations.* Cambridge University Press, 1990.
19. H. T. Nguyen. A note on the extension principle for fuzzy sets. *Journal of Mathematical Analysis Applications*, 64(2):369–380, 1978.
20. G. Nicolis and I. Prigogine. *Self Organisation in Nonequilibrium Systems.* Wiley, New York, 1977.
21. R. Rubinstein. *Simulation and the Monte Carlo method.* Wiley, 1981.
22. G. Shafer. *A mathematical theory of evidence.* Princeton University Press, Princeton, 1976.
23. Y.A. Shreider. *Monte Carlo Methods.* Pergamon Press, 1966.
24. P. Smets. Imperfect information: imprecision, uncertainty. Technical Report 93-3, Iridia -Université Libre de Bruxelles, Bruxelles, 1993.
25. K. Sobczyk. *Stochastic Differential Equations with applications to physics and engineering.* Kluwer Academic Publishers, Dordrecht, 1990.
26. J. Stoer and R. Bulirsch. *Introduction to Numerical Analysis.* Springer-Verlag, 2nd edition, 1993.
27. M. Sugeno. *Theory of fuzzy integrals and its applications.* PhD thesis, Tokyo Institute of Technology, 1974.
28. M. Sugeno. *Fuzzy Automata and Decision Processes*, chapter Fuzzy measures and fuzzy integrals: a survey., pages 89–102. North Holland, Amsterdam, m. m. gupta and g. n. saridis and b. r. gaines edition, 1977.
29. Z. Wang and G. J. Klir. *Fuzzy Measure Theory.* Plenum Press, New York, 1992.
30. L. A. Zadeh. Fuzzy sets. *Information and Control*, 8:338–353, 1965.
31. L. A. Zadeh. Fuzzy sets as a basis for a theory of possibility. *Fuzzy Sets and Systems*, 1(1):3–28, 1978.

Resolution of Min–Max Fuzzy Relational Equations

Ketty Peeva

Sofia 1000, P.O. Box 384, Faculty of Applied Mathematics, Technical University of Sofia, Bulgaria, e-mail: kgp@vmei.acad.bg

Abstract. The paper provides methodology and polynomial time algorithm for inverse problem resolution for *min–max* fuzzy relational equations. An exact and universal method is presented for solving *min–max* fuzzy linear systems of equations, completed with polynomial time algorithm. The main results are concerned with establishing the consistency of the system, computing complete solution set if the system is consistent or marking the contradictory equations if the system is inconsistent. A method and algorithm for solving *min–max* fuzzy relational equations are proposed.

1 Introduction

Santos (1968) first discussed the min–max composition of fuzzy relations for *min–max* fuzzy machines.

Sanchez (1972, 1976) first studied the inverse problem resolution for fuzzy relational equations in case of max–min or min–max compositions. De Baets (2000), Klir et al. (1995) and Peeva (2002) summarize the most important results in this subject.

This paper proposes methodology, unified method and algorithm for inverse problem resolution for *min–max* fuzzy relational equations, based on the author's results for solving *min–max* fuzzy linear systems, cf. Peeva (2001, 2002).

Basic notions are given in Section 2. Section 3 deals with *min–max* fuzzy linear systems of equations. A systematic way is presented of simplifying a *min–max* fuzzy linear system of equations so that the complete solution set can easily be found. Rather than work with the actual system, a symbolic augmented matrix, whose elements capture all the properties of the equations, is used. This universal and exact method for solving *min–max* fuzzy linear system of equations provides the inverse problem resolution for *min–max* fuzzy relational equations (Section 4). Polynomial time algorithms for solving *min–max* fuzzy linear systems of equations (Section 3) and *min–max* fuzzy relational equations (Section 4) are presented.

2 Basic notions

We recall some definitions and results and introduce the underlying notions.

We follow Gratzer (1978) for lattice theory and Klir at al. (1995, 1997) for fuzzy set theory.

The ordinary symbol $\leq$ is used for the partial order relation on a partially ordered set (poset) P. By a *greatest element* of a poset P we mean an element $b \in P$ such that $x \leq b$ for all $x \in P$; the *least element* of P being defined dually. The (unique) least and greatest elements of P, when they exist, are called *universal bounds* of P and are denoted by 0 and 1 respectively. A *lattice* is a poset L any two of whose elements x and y have a *greatest lower bound* denoted by $x \wedge y$, and a *least upper bound* denoted by $x \vee y$. A lattice is *complete* when each of its subsets X has a least upper bound denoted by $\sup X$ or $\vee X$ and a greatest lower bound denoted by $\inf X$ or $\wedge X$. A *Brouwerian lattice* is a lattice L in which for any given elements a and b the set of all $x \in L$ such that $a \wedge x \leq b$ contains a greatest element, denoted $a \alpha b$. In a *dually Brouwerian lattice* for any given elements a and b the set of all $x \in L$ such that $a \vee x \geq b$ contains a least element, denoted $a \varepsilon b$.

A totally ordered set L with universal bounds 0 and 1 and with operations join $\vee$ and meet $\wedge$ is called a *bounded chain*, written $\mathsf{L} = (L, \vee, \wedge, 0, 1)$ or simply L. Obviously L is a complete Brouwerian and a complete dually Brouwerian lattice, if we define

$$a \alpha b = \begin{cases} 1 & \text{if } a \leq b \\ b & a > b \end{cases}, \qquad a \varepsilon b = \begin{cases} b & \text{if } a < b \\ 0 & a \geq b \end{cases}.$$

Remark 1. In what follows we work with the bounded chain $\mathsf{L} = (L, \vee, \wedge, 0, 1)$ as described above, i.e. we suppose L is also a complete Brouwerian and a complete dually Brouwerian lattice.

In particular I is used for the bounded chain $\mathsf{I} = ([0, 1], \max, \min, 0, 1)$.

The matrix $A = (a_{ij})_{m \times p}$, with elements $a_{ij} \in L$ for each i, $1 \leq i \leq m$, and for each j, $1 \leq j \leq n$, is called a *matrix over* L.

The matrix $A^t = (a_{ji})_{n \times m}$ is called the *transpose* of $A = (a_{ij})_{m \times n}$.

The matrices $A = (a_{ij})_{m \times p}$ and $B = (b_{ij})_{p \times n}$ are called *conformable*, cf. Schwartzman (1994), if the number of columns in $A = (a_{ij})_{m \times p}$ equals the number of rows in $B = (b_{ij})_{p \times n}$.

We recall some kinds of matrix multiplication, cf. for instance Klir (1997).

Definition 1. Let $A = (a_{ij})_{m \times p}$ and $B = (b_{ij})_{p \times n}$ be finite conformable matrices over L. The matrix $C = (c_{ij})_{m \times n}$,

i) $C = A \bullet B$ is called *standard* ($\vee - \wedge$ or *max–min*) *product* of A and B if

$$c_{ij} = \bigvee_{k=1}^{p} (a_{ik} \wedge b_{kj}) \text{ for each } i,\ 1 \le i \le m \text{ and for each } j,\ 1 \le j \le n.$$

ii) $C = A \circ B$ is called *costandard* ($\wedge - \vee$ or *min–max*) *product* of A and B if

$$c_{ij} = \bigwedge_{k=1}^{p} (a_{ik} \vee b_{kj}) \text{ for each } i,\ 1 \le i \le m \text{ and for each } j,\ 1 \le j \le n.$$

iii) $C = A \alpha B$ is called α*-product* of A and B if

$$c_{ij} = \bigwedge_{k=1}^{p} (a_{ik} \alpha b_{kj}) \text{ for each } i,\ 1 \le i \le m \text{ and for each } j,\ 1 \le j \le n.$$

iv) $C = A \varepsilon B$ is called ε*-product* of A and B if

$$c_{ij} = \bigvee_{k=1}^{p} (a_{ik} \varepsilon b_{kj}) \text{ for each } i,\ 1 \le i \le m \text{ and for each } j,\ 1 \le j \le n.$$

3 Min–max fuzzy linear systems of equations

We study *min–max* fuzzy linear systems of equations of the form

$$\left| \begin{array}{lllll} (a_{11} \vee x_1) & \wedge \Lambda \wedge & (a_{1n} \vee x_n) & = & b_1 \\ \Lambda & \Lambda & \Lambda & \Lambda & \Lambda \\ (a_{m1} \vee x_1) & \wedge \Lambda \wedge & (a_{mn} \vee x_n) & = & b_m \end{array} \right. \Leftrightarrow \quad A \circ X = B \qquad (1)$$

where $A = (a_{ij})_{m \times n}$, $X = (x_j)_{n \times 1}$ and $B = (b_i)_{m \times 1}$ are finite matrices over L of coefficients, unknowns and second member of the system, respectively.

The goal is to investigate solving *min–max* fuzzy linear system from theoretical and computational points of view. Polynomial time algorithm for solving (1) is proposed, answering the following questions:

i) Is the system consistent?
ii) If it is inconsistent – what are the contradictory equations?
iii) If it is consistent – what is its complete solution set?

In order to make the exposition clear we define different kinds of solutions for $A \circ X = B$ and give a classification according to the solutions, following Peeva (2001, 2002).

Definition 2. Let the system $A \circ X = B$ be given.

i) $X^0 = (x_j^0)_{n\times 1}$ is called a *point solution* of $A \circ X = B$ if $A \circ X^0 = B$ holds. The set of all point solutions of $A \circ X = B$ is denoted by X^0.

ii) If $\mathrm{X}^0 \neq \varnothing$ then $A \circ X = B$ is called *consistent*, otherwise it is *inconsistent*.

iii) A point solution $X_{low}^0 \in \mathrm{X}^0$ is called a *lower point solution* of $A \circ X = B$ if for any $X^0 \in \mathrm{X}^0$ the relation $X^0 \leq X_{low}^0$ implies $X^0 = X_{low}^0$, where $\leq$ denotes the partial order induced in X^0 by the order of L. Dually, a point solution $X_{up}^0 \in \mathrm{X}^0$ is called an *upper point solution* of $A \circ X = B$ if for any $X^0 \in \mathrm{X}^0$ the relation $X_{up}^0 \leq X^0$ implies $X^0 = X_{up}^0$.

iv) An interval $X_j \subseteq L$ is called *feasible* for the j-th component of the solution, if $x_j \in X_j$ implies $a_{ij} \vee x_j \geq b_i$ for each i, $1 \leq i \leq m$. An n-tuple $(X_1, \ldots, X_n)$ of intervals $X_j \subseteq L$, $1 \leq j \leq n$, is called an *interval solution* of the system $A \circ X = B$ if any vector $X^0 = (x_j^0)_{n\times 1}$ with $x_j^0 \in X_j$, $1 \leq j \leq n$, belongs to X^0. The interval solution, that is maximal with respect to this feasibility property, is called a *maximal interval solution.*

Two *min–max* fuzzy linear systems are called *equivalent*, if each solution of the first one is a solution of the second one and vice versa.

Any interchange of equations in the system $A \circ X = B$ gives an equivalent system.

We assign to $A \circ X = B$ a new system $A^* \circ X = B$ with a coefficient matrix $A^* = (a_{ij}^*)$ carried out from A with respect to B according to the expression:

$$a_{ij}^* = \begin{cases} 0 \text{ if } a_{ij} < b_i, \\ b_i \text{ if } a_{ij} = b_i, \\ 1 \text{ if } a_{ij} > b_i. \end{cases} \tag{2}$$

We call $a_{ij}^* = 0$ to be of S-type (from smaller), $a_{ij}^* = b_i$ to be of E-type (from equal) and $a_{ij}^* = 1$ to be of G-type (from greater).

$(A^* : B)$ with elements a_{ij}^*, determined by (2), is called *augmented matrix* of the system $A^* \circ X = B$. The system $A^* \circ X = B$ is called *associated* to $A \circ X = B$.

Any *min–max* fuzzy linear system is equivalent with its associated.

Remark 2. In the following we suppose that all conventions given below are valid for the system $A \circ X = B$:

i) $b_1 \le b_2 \le \text{K} \le b_m$ is satisfied for the second member B of the system;

ii) the associated system $A^* \circ X = B$ is obtained;

iii) for any column $j, 1 \le j \le n$, we denote by k the greatest number of the row with S-type coefficient in it and by r the smallest number of the row with E-type coefficient in it.

Lemma 1. *Let the system $A \circ X = B$ in n unknowns and m equations be given.*

i) *The time complexity function for obtaining its equivalent system with $b_1 \le b_2 \le \text{K} \le b_m$ is $O(m^2)$, i.e., polynomial in the number m of the equations in the system.*

ii) *The time complexity function for computing $(A^* : B)$ is $O(mn)$.*

We give a unified and exact method for solving the system (1), resulting in:

1. A necessary and sufficient condition (Corollary 1) for consistency of the system, similar to the general test for consistency of a system of linear equations in Linear Algebra, see MacLane and Birkhoff (1979).
2. Analytical expressions for the smallest (Corollary 2) and the upper (expression (4)) solutions, leading to fast computational method to determine complete solution set of (1).
3. An effective procedure (Algorithm 1) for inverse problem resolution for (1).

The next theorem is fundamental. It contains the main results for solving (1).

Theorem 1 (Peeva (2001)). *Let the system $A \circ X = B$ be given.*

i) *If the j-th column of A^* contains S-type coefficient, $a^*_{kj} = 0$, then:*

   a) *$X_j = [b_k, 1]$ is the feasible interval for the j-th component;*

   b) *$x_j = b_k$ implies $a_{ij} \vee x_j = b_i$ for:*

   - *$i = k$,*
   - *for each $i < k$ with $a^*_{ij} \le b_i = b_k$;*
   - *for each $i > k$ with $a^*_{ij} = b_i$.*

ii) *If the j-th column of A^* does not contain any S-type coefficient, but it contains E-type coefficient $a^*_{rj} = b_r$, then*

   a) *$X_j = L$ is the feasible interval for the j-th component;*

   b) *$x_j \in [0, b_r]$ implies $a_{ij} \vee x_j = b_i$ for each $i \ge r$ with $a^*_{ij} = b_i$.*

iii) *If the j-th column of A^* contains neither S-type nor E-type coefficient, then the feasible interval is $X_j = L$ but $a_{ij} \vee x_j > b_i$ holds for any $x_j \in L$.*

We denote by S and E all S and E coefficients, respectively selected to satisfy the i-th equation, $1 \le i \le m$, by the term $a_{ij} \vee x_j = b_i$ according to Theorem 1 ib), iib). Those S and E coefficients are called *selected*.

From Theorem 1 we obtain

Corollary 1 (Peeva (2001)). *Let the system $A \circ X = B$ in n unknowns and m equations be given.*

i) *It is consistent iff there exists at least one selected coefficient $a^*_{ij} \in \{ \mathsf{S}, \mathsf{E} \}$ for each i, $1 \le i \le m$.*

ii) *It is inconsistent, if there exists an equation with no selected coefficient. Then this equation is in contradiction with the other equations.*

iii) *The time complexity function for establishing the consistency of the system is $O(mn)$.*

We introduce a vector $IND = IND_{m \times 1}$ in order to indicate the consistency of the system. The natural number in the i-th component of IND shows the number of different ways to satisfy i-th equation of the system. If there exists at least one zero component in IND, the system is inconsistent.

Corollary 2. *Let the system $A \circ X = B$ be consistent. Then:*

i) *It has a unique least solution $X_{least} = (x_j)_{nx1}$, where $x_j = b_k$ if the j-th column of A^* contains* S type *coefficient and $x_j = 0$ otherwise.*

ii) *The time complexity function for computing X_{least} is $O(mn)$.*

iii) *If $IND(i) = 1$ for each $i = 1, K, m$, then the system has unique upper solution.*

Remark 3. Computing X_{least} by Corollary 2 i) is equivalent to its computing by the formula $X_{least} = A^t \varepsilon B$, proved by Sanchez (1976) for *min-max* fuzzy relational equations.

In order to compute the upper solutions of (1), we introduce a *help matrix* $H = (h_{ij})_{m \times n}$ with elements

$$h_{ij} = \begin{cases} b_i & \text{if} \quad a^*_{ij} = \mathsf{S} \quad \text{or} \quad a^*_{ij} = \mathsf{E} \\ 1 & \text{otherwise} \end{cases}.$$

For any i, $1 \le i \le m$, the elements $h_{ij} = b_i \ne 1$ in H mark different ways to satisfy the i-th equation of the system. If $h_{ij} = b_i \ne 1$ in the j-th column of H, we write $\left\langle \frac{b_i}{j} \right\rangle$. In this case $h_{ij} \vee x_j = b_i \vee x_j$ gives a way to satisfy the i-th equation of the system for any $x_j \in [0, b_i]$ and $x_j = b_i$ is the upper bound for the j-th component. Let $|H_i|$ denotes the number of the elements $h_{ij} = b_i \neq 1$ in the i-th row of H. For a consistent system $\prod_{i=1}^{m} |H_i|$ determines the upper bound of the number of its upper solutions.

Then we determine the upper solutions from H. Any solution of an equation of the system must be in co-ordination with the other equations. For this reason we consider the equations together, i.e., the concatenation W of all ways from H:

$$W = \prod_{i \in I} \left(\sum_j \left\langle \frac{b_i}{j} \right\rangle \right) \tag{3}$$

In order to compute different ways to satisfy any equation in the system in co-ordination with the others, we are interested in the properties of this concatenation.

It is *distributive* with respect to the addition. Expanding the parentheses in (3) we obtain

$$W = \sum_{(j_1, \Lambda, j_m)} \left\langle \frac{b_{i1}}{j_1} \right\rangle \left\langle \frac{b_{i2}}{j_2} \right\rangle \Lambda \left\langle \frac{b_{im}}{j_m} \right\rangle \tag{4}$$

$\left\langle \frac{b_{i1}}{j_1} \right\rangle \left\langle \frac{b_{i2}}{j_2} \right\rangle \Lambda \left\langle \frac{b_{im}}{j_m} \right\rangle$ in (4) defines a point solution $X^0 = \left(x_j^0 \right)$ with components $x_{ji}^0 = b_i$; for the missing j we put $x_j^0 = 1$. The expression (4) includes all upper solutions of the system, determined according to the next properties:

$$\left\langle \frac{b_{i1}}{j_1} \right\rangle \left\langle \frac{b_{i2}}{j_2} \right\rangle = \begin{cases} \frac{b_{i1} \wedge b_{i2}}{j} \text{ if } j = j_1 = j_2 \\ \text{unchanged if } j_1 \neq j_2 \end{cases}. \tag{5}$$

$$\begin{aligned} &\left\langle \frac{b_{i1}}{j_1} \right\rangle \left\langle \frac{b_{i2}}{j_2} \right\rangle \Lambda \left\langle \frac{b_{im}}{j_m} \right\rangle + \left\langle \frac{b_{s1}}{j_1} \right\rangle \left\langle \frac{b_{s2}}{j_2} \right\rangle \Lambda \left\langle \frac{b_{sm}}{j_m} \right\rangle = \\ &= \begin{cases} \left\langle \frac{b_{i1}}{j_1} \right\rangle \mathrm{K} \left\langle \frac{b_{im}}{j_m} \right\rangle & \text{if } b_{it} \leq b_{st} \\ \text{unchanged} & \text{otherwise} \end{cases} \end{aligned} \tag{6}$$

For a consistent system $A \circ X = B$ the upper solutions are computable and the set of all its upper solutions is finite, cf. Peeva (2001).

Bearing in mind Theorem 1 and its Corollaries, we propose polynomial time algorithm for solving the system $A \circ X = B$. It answers the questions (i) – (iii) marked in the beginning of Section 3.

Algorithm 1 *for solving* $A \circ X = B$.

1. Enter the matrices $A_{m\times n}$, $B_{m\times 1}$.
2. Compute the matrix $(A^* : B)$ with elements a_{ij}^* according to (2).
3. Initialize the vector *IND*. For $i = 1,...,m$ and for $j = 1,\Lambda ,n$, put $h_{ij} := 1$ in $H = (h_{ij})_{m\times n}$.
4. Reserve for the least solution X_{least} an $n\times 1$ column-matrix and a dynamic $n \times \prod |H_i|$ array for $\{X_{up}\}$. For each $j = 1,\Lambda ,n$ put $X_{least}(j) := 0$ and for the upper solutions $X_{up}(j) := 1$.
5. Give the initial value $j := 0$ to the running variable j.
6. Increase the running variable, $j := j + 1$.
7. If $j > n$ go to Step 10.
8. If the j-th column in A^* does not contain any *S*-type coefficient, go to Step 9.

Otherwise take the greatest number k of the row with *S*-type coefficient in the j-th column of A^*.

Put

$X_{least}(j) := b_k$,

$IND(k) := IND(k) + 1$ and $h_{kj} := b_k$,

$IND(i) := IND(i) + 1$ for each $i<k$ with $a_{ij}^* \le b_i = b_k$,

$IND(i) := IND(i) + 1$ for each $i>k$ with $a_{ij}^* = b_i$, put $h_{ij} := b_k$.

Go to Step 6.

9. If the j-th column in A^* does not contain any *E*-type coefficient, go to Step 6.

Otherwise take the smallest number r of the row with *E*-type coefficient in the j-th column of A^*.

Put $IND(i) := IND(i) + 1$ and $h_{ij} := b_r$ for $i \ge r$ with $a_{ij} = b_i$.

Go to Step 6.

10. If $IND(i) = 0$ for some $i = 1,\Lambda ,m$, then the system is inconsistent and the i-th equation is in contradiction with the others. Go to Step 14.
11. Compute $\{X_{up}\}$ from H according to (3), (4), (5) and (6).

12. If $IND(i) = 1$ for $i = 1, \ldots, m$ the system is consistent with a unique least, upper and maximal interval solution stretched from X_{least} to X_{up}. Go to Step 14.
13. The system is consistent with unique least solution and it may have more than one upper (respectively, maximal interval) solutions, as determined in Step 11.
14. End.

Example. Using Algorithm 1, solve the system

$$\left|\begin{array}{l}(0.2 \vee x_1) \wedge (0.2 \vee x_2) \wedge (0.1 \vee x_3) \wedge (0.3 \vee x_4) \wedge (0.1 \vee x_5) = 0.3 \\ (0.6 \vee x_1) \wedge (0.6 \vee x_2) \wedge (0.2 \vee x_3) \wedge (0.2 \vee x_4) \wedge (0.8 \vee x_5) = 0.3 \\ (0.4 \vee x_1) \wedge (0.5 \vee x_2) \wedge (0.5 \vee x_3) \wedge (0.5 \vee x_4) \wedge (0.4 \vee x_5) = 0.5 \\ (0.6 \vee x_1) \wedge (0.4 \vee x_2) \wedge (0.6 \vee x_3) \wedge (0.5 \vee x_4) \wedge (0.6 \vee x_5) = 0.6 \\ (0.5 \vee x_1) \wedge (0.6 \vee x_2) \wedge (0.6 \vee x_3) \wedge (1.0 \vee x_4) \wedge (0.1 \vee x_5) = 0.6\end{array}\right..$$

The system is over the bounded chain $\mathsf{I} = ([0, 1], \max, \min, 0, 1)$.

$$(A : B) = \begin{pmatrix} 0.2 & 0.2 & 0.1 & 0.3 & 0.1 & : & 0.3 \\ 0.6 & 0.6 & 0.2 & 0.2 & 0.8 & : & 0.3 \\ 0.4 & 0.5 & 0.5 & 0.5 & 0.4 & : & 0.5 \\ 0.6 & 0.4 & 0.6 & 0.5 & 0.6 & : & 0.6 \\ 0.5 & 0.6 & 0.6 & 1 & 0.1 & : & 0.6 \end{pmatrix},$$

$$(A^* : B) = \begin{pmatrix} S & S & S & E & S & : & 0.3 \\ G & G & S & S & G & : & 0.3 \\ S & E & E & E & S & : & 0.5 \\ E & S & E & S & E & : & 0.6 \\ S & E & E & G & S & : & 0.6 \end{pmatrix}.$$

Consider the first column, we mark a^*_{51} as selected, put $IND(5) = 1$, $X_{least}(1) = 0.6$ and as corollaries of this choice a^*_{41} is also selected because it is of E-type and $b_4 = b_5 = 0.6$ and $IND(4) = 1$.

For $j = 2$ we mark a^*_{42} as selected, put $IND(4) = 1 + 1 = 2$, $X_{least}(2) = 0.6$ and as corollaries of this choice a^*_{52} is also selected and $IND(5) = 1 + 1 = 2$.

For $j = 3$ we mark a^*_{23} as selected and also a^*_{13} as selected because it is S-type and $b_1 = b_2 = 0.3$, $X_{least}(3) = 0.3$; we also mark all E-type coefficients below a^*_{23}, namely a^*_{33}, a^*_{43}, a^*_{53}, as well as we make the corresponding changes in IND: $IND(2)=1$, $IND(1)=1$, $IND(3)=1$, $IND(4)=3$, $IND(5)=3$.

For $j=4$ we mark a^*_{44} as selected, put $X_{least}(4)=0.6$, $IND(4)=4$.

Finally, for $j=5$ we select a^*_{55}, put $X_{least}(5)=0.6$, $IND(5)=4$; a^*_{45} is also selected because it is of E-type and $b_4=b_5=0.6$ and $IND(4)=5$.

Hence the augmented matrix, the help matrix and the least solution are:

$$\begin{pmatrix} S & S & \mathsf{S} & E & S & : & 0.3 \\ G & G & \mathsf{S} & S & G & : & 0.3 \\ S & E & \mathsf{E} & E & S & : & 0.5 \\ \mathsf{E} & \mathsf{S} & \mathsf{E} & \mathsf{S} & \mathsf{E} & : & 0.6 \\ \mathsf{S} & \mathsf{E} & \mathsf{E} & G & \mathsf{S} & : & 0.6 \end{pmatrix},\ H=\begin{pmatrix} 1 & 1 & 0.3 & 1 & 1 \\ 1 & 1 & 0.3 & 1 & 1 \\ 1 & 1 & 0.5 & 1 & 1 \\ 0.6 & 0.6 & 0.6 & 0.6 & 0.6 \\ 0.6 & 0.6 & 0.6 & 1 & 0.6 \end{pmatrix},\ X_{least}=\begin{pmatrix} 0.6 \\ 0.6 \\ 0.3 \\ 0.6 \\ 0.6 \end{pmatrix}.$$

There exist unique upper solution (which is obvious from the fact that a^*_{23} is selected and it selects all the coefficients in the third column of the symbolic matrix) and unique maximal interval solution:

$$X_{up}=\begin{pmatrix} 1 \\ 1 \\ 0.3 \\ 1 \\ 1 \end{pmatrix},\quad X_{\max}=\begin{pmatrix} [0.6,1] \\ [0.6,1] \\ 0.3 \\ [0.6,1] \\ [0.6,1] \end{pmatrix}.$$

Theorem 2 (Peeva (2001)). *For the system* (1):

i) *it is algorithmically decidable in polynomial time whether the system is consistent or not;*
ii) *if the system is consistent — the least point solution, the upper point solutions and the maximal interval solutions are computable;*
iii) *if the system is inconsistent the numbers of the contradictory equations are obtained.*

To solve $X \circ A = B$ the operation transposition has only to be conducted.

We shall say that the column matrix $B_{m\times 1}=(b_i)_{m\times 1}$ is a $\vee-\wedge$*-convex linear combination* (CLC) of the column matrices $A_j=(a_{ij})_{m\times n}$, $1\le j\le n$, with coefficients $x_j\in L$, $1\le j\le n$, if

$B=(A_1\wedge x_1)\vee\Lambda\vee(A_n\wedge x_n)$, I.e., $b_i=\vee_{1\le j\le n}(a_{ij}\wedge x_j)$ for each i, $1\le i\le m$.

Applying Algorithm 1, it is algorithmically decidable whether $B_{m\times 1}$ is a CLC of A_j.

4 Solving min–max fuzzy relational equations

De Baets (2000) presents solution methods for various fuzzy relational equations. His exposition is completed with literature pointers for other types of fuzzy relational equations, generalizations, problems and applications. Solving fuzzy relational equations if the composition is *max-min* is analyzed by Cuninghame-Green and Chéhlárová (1995), Klir and Bo Yuan (1995), Miyakoshi and Shimbo (1987), Pappis and Sugeno (1985), in the case of t-norms — by Pedrycz (1993), Di Nola et al. (1989), Miyakoshi and Shimbo (1985), for Browerian chain — by Stamou and Tzafestas (1999), Sessa (1989), Sanchez (1977, 1982).

Let X and Y be crisp sets. A binary *fuzzy relation* $R \subseteq X \times Y$ over [0, 1] is defined as a fuzzy subset of $X \times Y$:

$$R = \{((x, y), \mu_R(x, y)) \mid (x, y) \in X \times Y, \mu_R : X \times Y \to [0, 1]\}.$$

The fuzzy relations $R \subseteq X \times Y$ and $S \subseteq Y \times Z$ with $pr_2(X \times Y) = pr_1(Y \times Z)$ are called *composable.*

When we consider a binary fuzzy relation $R \subseteq X \times Y$ over L, shortly *L-fuzzy relation*, we write $R \in L(X \times Y)$.

According to De Baets (2000), three kinds of binary compositions for composable L-fuzzy relations are considered very often:

- *round composition*, $R\theta S(x,z) = \sup_{y \in Y} C(R(x,y), S(y,z))$,
- *subcomposition*, $R < S(x,z) = \inf_{y \in Y} I(R(x,y), S(y,z))$,
- *supercomposition*, $R > S(x,z) = \inf_{y \in Y} I(S(y,z), R(x,y))$.

Here C is conjunctor, I is implicator, as defined in De Baets (2000).

The following types of problems, subject of the fuzzy relational calculus, are problems of interest for computer implementation:

i) Computing $R\theta S$, $R > S$ or $R < S$, if the L-fuzzy relations R and S are given, is called *direct problem resolution.*

ii) *Inverse problem resolution* means solving equation of type $R\theta S = T$, $R > S = T$ or $R < S = T$ in the unknown L-fuzzy relation S (or R).

Here we consider the inverse problem resolution for *min–max* fuzzy relational equations. Sanchez (1976) gives analytical expression for the least solution.

Let $R \in L(X \times Y)$ and $Q \in L(Y \times Z)$ be L-fuzzy relations. The L-fuzzy relation $T \in L(X \times Z)$,

i) $T = R \bullet Q$ is called *standard* ($\vee - \wedge$ or $\max - \min$) *composition* of R and Q, if

$$R \bullet Q\ (x,z) = \underset{y \in Y}{\vee}\ (\wedge(R(x,y),\ Q(y,z))\ , \text{ or}$$

$$R \bullet Q\ (x,z) = \max_{y \in Y}(\min(R(x,y),\ Q(y,z)).$$

ii) $T = R \circ Q$ is called *costandard* ($\wedge - \vee$ or $\min - \max$) *composition* of R and Q, if

$$R \circ Q\,(x,z) = \bigwedge_{y \in Y} (\vee(R(x,y), Q(y,z))\,, \text{ or}$$

$$R \circ Q\,(x,z) = \min_{y \in Y}(\max(R(x,y), Q(y,z))\,.$$

iii) $T = R\alpha Q$ is called α - *composition* of R and Q, if

$$R\alpha Q\,(x,z) = \bigwedge_{y \in Y} ((R(x,y)\alpha Q(y,z))\,.$$

iv) $T = R\varepsilon Q$ is called ε - *composition* of R and Q, if

$$R\varepsilon Q\,(x,z) = \bigvee_{y \in Y} ((R(x,y)\varepsilon Q(y,z))\,.$$

Let $R \in L(X \times Y)$, $Q \in L(Y \times Z)$ and $T \in L(X \times Z)$ be L-fuzzy relations. The equation

$$R \circ Q = T \tag{7}$$

where one of the fuzzy relations on the left is unknown, and the other relation on the left and the relation T are given, is called *fuzzy relational equation* in the case of costandard composition, written by the sign $\circ$.

If the relations are over finite support, we can assign to any relation a matrix and this leads to the next notion, namely – fuzzy matrix equation.

Let X and Y be finite crisp sets. We can present the fuzzy relation $R \in L(X \times Y)$ by a matrix over L and we can describe the compositions of L - fuzzy relations by matrix operations. Often we shall denote by the same letter the fuzzy relation and its representing matrix.

Let $A = (a_{ij})_{m \times p}$, $B = (b_{ij})_{p \times n}$ and $C = (c_{ij})_{m \times n}$ be matrices over L. The equation

$$A \circ B = C$$

where one of the matrices on the left side is unknown and the other matrix and the matrix C are given, is called *fuzzy matrix equation* in case of costandard product.

Let the fuzzy relations R, Q and T be over finite support and the composition $R \circ Q$ makes sense.

Theorem 3. *There exists polynomial time algorithm for computing the least solution and establishing solvability of (7) for Q if R and T are given.*

Proof. Follows from Theorem 1 and Algorithm 1:

In order to solve (7), we represent relations by the corresponding matrices, written for convenience with the same letters:

$$R_{m\times n} \circ Q_{n\times p} = T_{m\times p} \tag{8}$$

We split $Q_{n\times p}$ and $T_{m\times p}$ by columns and apply the following equivalence:

$$R_{m\times n} \circ Q_{n\times p} = T_{m\times p} \Leftrightarrow \left| \begin{array}{c} R_{m\times n} \circ Q^{(1)} = T^{(1)} \\ \Lambda \\ R_{m\times n} \circ Q^{(p)} = T^{(p)} \end{array} \right. \tag{9}$$

For $j = 1,\dots,p$, the j-th column of Q is denoted by $Q^{(j)}$ and the j-th column of T is denoted by $T^{(j)}$. Now, instead of (8) we solve p fuzzy linear systems, having the same matrix $R_{m\times n}$, with Algorithm 1. For any equation we take the corresponding columns (with like indices) of Q and T. □

Algorithm 2 *for solving* $R_{m\times n} \circ Q_{n\times p} = T_{m\times p}$.

1. Enter the matrices that represent the relations R and T.
2. Transform the fuzzy relational equation in the form (9).
3. Solve (9) by Algorithm 1.
4. Express the solution $Q_{n\times p}$.
5. End.

Algorithm 2 gives the complete solution set when the equation has solution, or it establishes inconsistency.

Corollary 5 (Sanchez (1976)). *Let* $R \in L(X \times Y)$ *and* $T \in L(X \times Z)$ *be L-fuzzy relations. If* $R \circ Q = T$ *is solvable for* Q *, then* $R^{-1} \varepsilon T$ *is its least solution.*

Corollary 6. *Let* $A = (a_{ij})_{m\times p}$ *and* $C = (c_{ij})_{m\times n}$ *be two fuzzy matrices. If* $A \circ B = C$ *is solvable for* B *, then* $A^t \varepsilon C$ *is its least solution.*

References

Cuninghame-Green R. A. and Chéhlárová K. (1995): Residuation in fuzzy algebra and some applications. Fuzzy Sets and Systems **71**, 227-239.

De Baets B. (2000): Analytical solution methods for fuzzy relational equations. In the series: Fundamentals of Fuzzy Sets, The Handbook of Fuzzy Sets Series, D. Dubois and H. Prade (eds.), Vol. 1, Kluwer Academic Publishers, pp. 291 – 340.

Di Nola A. et al. (1989): Fuzzy Relation Equations and Their Application to Knowledge Engineering. Kluwer Academic Press, Dordrecht.

Gratzer G. (1978): General Lattice Theory. Akademie-Verlag, Berlin.

Higashi M. and Klir G. J. (1984): Resolution of finite fuzzy relation equations. Fuzzy Sets and Systems **13**: 65 - 82.

Klir G. and Bo Yuan (1995): Fuzzy Sets and Fuzzy Logic: Theory and Applications. Prentice Hall PTR, NJ.

Klir G., Clair U. H. St. and Bo Yuan (1997): Fuzzy Set Theory Foundations and Applications. Prentice Hall PRT.

MacLane S. and Birkhoff G. (1979): Algebra. Macmillan, New York.

Miyakoshi M. and Shimbo M. (1985): Solutions of composite fuzzy relational equations with triangular norms. Fuzzy Sets and Systems **16**, 53 - 63.

Miyakoshi M. and Shimbo M. (1987): Sets of solution-set-invariant matrices of simple fuzzy relation equations. Fuzzy Sets and Systems **21**, 59 – 83.

Pappis C. P. and Sugeno M. (1985): Fuzzy relational equations and the inverse problem. Fuzzy Sets and Systems **15**, 79 - 90.

Pedrycz W. (1993): s-t Fuzzy relational equations. Fuzzy Sets and Systems **59**, 189-195.

Peeva K. (1992): Fuzzy linear systems. Fuzzy Sets and Systems **49**, 339 - 355.

Peeva K. (2001): Min–max-fuzzy linear systems of equation., In 26th Summer School Applications of Mathematics in Engineering and Economics, B. Cheshankov, M. Todorov (eds.), Sozopol 2000, Heron Press, 254-259.

Peeva K. (2002): Methods and algorithms for solving fuzzy linear systems – Applications in Artificial Intelligence areas. Dr. Sci. Thesis, Technical University of Sofia, Sofia.

Sanchez E. (1972): Equations de Relations Floves. Thèse Biologie Humaine, Marseille.

Sanchez E. (1976): Resolution of composite fuzzy relation equations. Information and Control **30**, 38-48.

Sanchez E. (1977): Solutions in composite fuzzy relations equations: Application to medical diagnosis in Brouwerian logic. In Fuzzy Automata and Decision Processes, Proceedings, M. M. Gupta, G. N. Saridis, B. R. Gaines (eds.), Elsevier, North-Holland, Amsterdam.

Sanchez E. (1982): Resolution of composite fuzzy relation equations. Information and Control **36** (1), 38-48.

Santos E. S. (1968): Maximin, minimax and composite sequential mashines. J. Math. Anal. Appl. **24,** 246-259.

Sessa S. (1989): Finite fuzzy relation equations with unique solution in complete Brouwerian Lattices. Fuzzy Sets and Systems **29**, 103 – 113.

Stamou G. B. and Tzafestas S. G. (1999): Fuzzy relation equations and fuzzy inference systems: An inside approach. IEEE Trans. On Syst., Man and Cybern. - part B: Cybernetics **29**(6), 694-702.

Fuzzy Relation Equations with Words

Vilém Novák

[1] University of Ostrava
Institute for Research and Applications of Fuzzy Modeling
30. dubna 22,
701 03 Ostrava 1, Czech Republic
[2] Institute of Information and Automation Theory
Academy of Sciences of the Czech Republic
Pod vodárenskou věží 4, 186 02 Praha 8, Czech Republic

1 Introduction

In this paper, we focus on two questions. The first one is, how a systematic theory of the specific linguistic expressions widely used in the applications of fuzzy logic can be developed. These expressions form a class of the so called evaluating linguistic expressions. These are all expressions such as "small, very small, roughly medium, more or less big", etc. and fuzzy numbers (also modified using linguistic hedges). Such a theory has been proposed in this paper. We argue that it fits well the meaning of the discussed expressions.

The second question is solution of fuzzy relation equations derived from the linguistic data consisting of the above expressions. We show that the pure evaluating linguistic expressions, i.e. those not containing fuzzy number, in most cases do not allow a solution. Thus, the only reasonable way is the use of fuzzy numbers.

2 Preliminaries

We will work with a set of truth values, which, in general, is a residuated lattice, i.e. an algebra

$$\mathcal{L} = \langle L, \vee, \wedge, \otimes, \rightarrow, \mathbf{0}, \mathbf{1} \rangle \tag{1}$$

with four binary operations and two constants such that:

(i) $\langle L, \vee, \wedge, \mathbf{0}, \mathbf{1} \rangle$ is a lattice with the ordering $\leq$ defined using the operations $\vee, \wedge$ as usual, and $\mathbf{0}, \mathbf{1}$ are its least and the greatest elements, respectively;
(ii) $\langle L, \otimes, \mathbf{1} \rangle$ is a commutative monoid, that is, $\otimes$ is a commutative and associative operation with the identity $a \otimes \mathbf{1} = a$;
(iii) the operation $\rightarrow$ is a residuation operation with respect to $\otimes$, i.e.

$$a \otimes b \leq c \quad \text{iff} \quad a \leq b \rightarrow c. \tag{2}$$

More specifically, we will suppose $\mathcal{L}$ to be one of the following.

A *BL-algebra* $\mathcal{L}_{BL}$, which is a residuated lattice fulfilling, moreover, the following conditions:

(a) *prelinearity*

$$(a \rightarrow b) \vee (b \rightarrow a) = \mathbf{1},$$

(b) *divisibility*

$$a \otimes (a \rightarrow b) = a \wedge b$$

for all $a, b \in L$. More specifically, we will suppose that $L = [0, 1]$ and $\otimes$ is some continuous t-norm.

The second possibility is that $\mathcal{L}$ is a Łukasiewicz MV-algebra

$$\mathcal{L}_{Ł} = \langle L, \otimes, \oplus, \neg, \mathbf{0}, \mathbf{1} \rangle$$

where $L = [0, 1]$ and

$$\begin{aligned} a \otimes b &= 0 \vee (a + b - 1), && \text{(Łukasiewicz conjunction)} \\ a \oplus b &= 1 \wedge (a + b), && \text{(Łukasiewicz disjunction)} \\ \neg a &= 1 - a, && \text{(negation)} \end{aligned}$$

for all $a, b \in [0, 1]$. The Łukasiewicz implication is a residuation operation in $\mathcal{L}_{Ł}$ given by

$$a \rightarrow b = \neg a \oplus b = 1 \wedge (1 - a + b).$$

Let $\otimes$ be a t-norm. By $\rightarrow$, we denote the corresponding residuation operation. Furthermore, we put $z(a) = \bigwedge_{k=1}^{\infty} a^k$ for every $a \in [0, 1]$ where the power is taken with respect to $\otimes$.

A fuzzy set $A \underset{\sim}{\subset} V$ in the universe V is a function

$$A : V \longrightarrow L$$

where L is the support of one of the above mentioned algebras. By $\mathcal{F}(V)$ we denote the set of all fuzzy sets on V, i.e. $\mathcal{F}(V) = L^V$. A binary fuzzy relation in U and V is a fuzzy set

$$R \underset{\sim}{\subset} U \times V.$$

Let $A \underset{\sim}{\subset} U$, $B \underset{\sim}{\subset} V$ be fuzzy sets. Then the residuation operation between A and B is defined by

$$(A \ominus\!\!\!\!\rightarrow B)(u, v) = A(u) \rightarrow B(v), \qquad u \in U, v \in V.$$

We will also use the following symbol for a cut of a function $f : \mathbb{R} \longrightarrow \mathbb{R}$ to $[0, 1]$:

$$f^*(x) = \begin{cases} 1 & \text{if } 1 < f(x), \\ f(x) & \text{if } 0 \leq f(x) \leq 1, \\ 0 & \text{if } f(x) < 0. \end{cases}$$

The following function will also be used:

$$\delta_z(x) = \begin{cases} 1 & \text{if } x = z, \\ 0 & \text{otherwise.} \end{cases}$$

3 Evaluating Linguistic Expressions

In the applications of fuzzy logic, an important role is played by the so called *evaluating linguistic expressions*, which are linguistic expressions characterizing a position on a bounded ordered scale (usually an interval of some numbers (real, rational, natural, etc.). Examples are "small, medium, big, very small, roughly large, extremely high", etc.

3.1 Linguistic characterization

The symbol $\langle \ldots \rangle$ used below denotes a metavariable for the kind of a word given inside the angle brackets.

Definition 1. An evaluating linguistic expression is one of the following:

(i) *Simple evaluating linguistic expression* which is either of the linguistic expressions:

(a)

$$\langle \text{pure evaluating expression} \rangle := \langle \text{linguistic hedge} \rangle \langle \text{atomic evaluating linguistic expression} \rangle$$

where $\langle$linguistic hedge$\rangle$ is either *empty hedge* (i.e. no hedge is present), or an intensifying adverb with narrowing effect (*extremely, significantly, very*) or widening effect (*more or less, roughly, quite roughly, very roughly*). Furthermore,

$$\langle \text{atomic evaluating linguistic expression} \rangle$$

is any of the adjectives "small", "medium", or "big".

(b)

$$\langle \text{fuzzy number} \rangle := \langle \text{linguistic hedge} \rangle \langle \text{number} \rangle$$

where $\langle$linguistic hedge$\rangle$ is either "about" or an intensifying adverb with widening effect and $\langle$number$\rangle$ is a name of a real number (element taken from $\mathbb{R}$).

(ii) *Compound evaluating linguistic expression*
'$\mathcal{A}$ or $\mathcal{B}$' where $\mathcal{A}, \mathcal{B}$ are evaluating linguistic expressions.

Note that introduction of the empty hedge enables to explicate all pure evaluating expressions in the same way. This makes the theory of their semantics much simpler and transparent.

Example 1. Atomic evaluating linguistic expressions are *small, medium, big.* Fuzzy numbers are, e.g. *twenty five, the value z, roughly 100*, etc. Simple evaluating linguistic expressions are *very small, more or less medium, roughly big, about twenty five, approximately* x_0, etc. Compound evaluating linguistic expressions are *roughly small or medium* , etc.

The "fuzzy number" is a linguistic characterization of some number. This means that every linguistic characterization of a number is understood imprecisely. We will take the form "about x_0" as canonical. It will play a similar role as "empty hedge" in the case of fuzzy number.

It is noticeable that atomic evaluating linguistic expressions usually form pairs of antonyms, i.e. the pairs

"nominal adjective — antonym".

Of course, there are a lot of other examples, e.g. "young — old", "ugly — nice", "stupid — clever", etc. The triple of expressions

$$\langle\text{linguistic hedge}\rangle\langle\text{nominal adjective}\rangle \text{ — } \langle\text{linguistic hedge}\rangle\langle\text{middle member}\rangle \text{ — } \langle\text{linguistic hedge}\rangle\langle\text{antonym}\rangle$$

is called the *evaluating trichotomy.* If all the linguistic hedges are empty then it is the *basic evaluating trichotomy.*

Definition 2. Let $\mathcal{A}$ be an evaluating linguistic expression. Then the linguistic expression

$$\langle\text{noun}\rangle \text{ is } \mathcal{A} \tag{3}$$

is an *evaluating predication.* If $\mathcal{A}$ is a simple evaluating linguistic expression then (3) is a a simple evaluating predication.

If $\mathcal{A}$ and $\mathcal{B}$ are evaluating predications then '$\mathcal{A}$ and $\mathcal{B}$' and '$\mathcal{A}$ or $\mathcal{B}$' are compound predications.

Example 2. Evaluating predications are, e.g. "*temperature is very high*" (here "high" is taken instead of "big"), "*pressure is not small*", "*frequency is small or medium*", "*cost is not small and not big*", "*income is roughly three million*", etc. Compound evaluating predication is, e.g. "*temperature is high and pressure is very high*".

By $\mathcal{S}$, we denote the set of linguistic expressions consisting of evaluating linguistic expressions, evaluating predications and the conditional clause

$$\text{IF } \mathcal{A} \text{ THEN } \mathcal{B} \tag{4}$$

where $\mathcal{A}$, $\mathcal{B}$ are evaluating linguistic predications. This set characterizes a fragment of natural language (English), which is largely employed in fuzzy logic.

3.2 Semantics of evaluating linguistic expressions

A general model of the semantics of linguistic expressions is based on the distinction between their intension and extension in the sense introduced by R. Carnap ([2]).

By *possible world* we understand a certain context of use, time, or place to which a linguistic proposition can relate. Then we introduce the following.

Intension of a linguistic expression, of a sentence, or of a concept, can be identified with the property denoted by it. An intension may lead to different truth values in various possible worlds but it is invariant with respect to them — it does not change when the possible world is changed. The intension is assumed to keep the Frege's compositionality principle: a more complex intension is a function of simpler ones.

Extension is a class of elements determined by an intension, which fall into the meaning of a linguistic expression in a given possible world. Thus, it depends on the particular context of use and changes whenever the possible world is changed.

Expressions of natural language are names of intensions. Let us remark that a lot of convincing arguments have been given to the statement (see, e.g., [4]) that the meaning of expressions of natural language cannot be identified with their extensions (the mentioned compositionality principle is broken).

In this paper, we introduce a formalization of these concepts for the semantics of evaluating linguistic expressions in fuzzy logic. We will follow some of the fundamental ideas considered in the intensional logic (cf. [4]).

We will work with a set W of possible worlds. These can be understood as special parameters representing particular states of affairs (or contexts of use). Moreover, let a set V be given, which represents objects. In the case of evaluating expressions, V will be a set of values of some features of objects, such as temperature, pressure, height, width, etc. It is reasonable to put $V = \mathbb{R}$.

Let us stress that on the basis of this idea, a real object (for example, a controlled system, situation, human being, etc.) can be mathematically represented by a vector of values of its features. More precisely, let o be such an object and let us distinguish its features $\varphi_1, \ldots, \varphi_n$. Each feature φ_i can attain values from some set V_i. Then a given object o is represented by a vector of values

$$o = \langle v_1, \ldots, v_n \rangle \in V_1 \times \cdots \times V_n.$$

As a special case, we may take $V_i = \mathbb{R}$, $i = 1, \ldots, n$. Then a real object is represented by a vector of real numbers. This is in full accordance with the assumed practice, and also with the way, how information in the database systems is stored. For example, a person is represented by a record being, in fact, a sequence of numbers (age, height, wage, etc.).

We will formally define *intension* as a function

$$A : W \longrightarrow \mathcal{F}(V). \tag{5}$$

The *extension* in a possible world $w \in W$ is a fuzzy set

$$A(w) \underset{\sim}{\subseteq} V.$$

Note that this definition (taken from intensional logic) is in accordance with Carnap's idea that an extension must be recapturable from an intension. It is also clear that while extension changes, intension remains the same independently on the possible world.

For the semantics of the evaluating linguistic expressions we introduce the following definition.

Definition 3. An *intensional* space is given by:

(i) A set of possible worlds $W = \{\langle v_L, v_S, v_R \rangle\}$ where $v_L, v_S, v_R \in [0, \infty)$ and $v_L < v_S < v_R$.

(ii) An algebra of truth values is either a BL-algebra $\mathcal{L}_{BL}$ where $L = [0, 1]$ and $\otimes$ is a continuous t-norm, or it is a Łukasiewicz algebra $\mathcal{L}_{\text{Ł}}$.

(iii) A couple of linear functions $L, R : W \times \mathbb{R} \longrightarrow L$ defined in each possible world $w \in W$ by

$$L_w(x) = \left(\frac{v_S - x}{v_S - v_L}\right)^* \tag{6}$$

$$R_w(x) = \left(\frac{x - v_S}{v_R - v_S}\right)^* \tag{7}$$

and a function

$$M_w(x) = \neg L_w(x) \wedge \neg R_w(x) = \left(\frac{x - v_L}{v_S - v_L}\right)^* \wedge \left(\frac{v_R - x}{v_R - v_S}\right)^*. \tag{8}$$

(iv) A set of linear functions defined in each possible world $w \in W$ by

$$\mathbf{B}_w = \{B_{w,x_0} \mid$$
$$B_{w,x_0}(x) = \left(\frac{x - x_0 + h_L}{h_L}\right)^* \wedge \left(\frac{x_0 - x + h_R}{h_R}\right)^* \text{ or } B_{w,x_0}(x) = \delta_{x_0}(x),$$
$$x_0 \in [v_L, v_R], 0 < h_L < x_0 - v_L, 0 < h_R < v_r - x_0\} \tag{9}$$

(v) A class of *abstract hedges*

$$\mathbf{Hf} = \{\nu : [0, 1] \longrightarrow [0, 1] \mid \nu \in \mathbf{Hf}^{quad} \cup \mathbf{Hf}^{lin} \tag{10}$$

where

$$\mathbf{Hf}^{quad} = \{\nu^{quad}_{a,b,c} \mid a, b \in (-\infty, 1), c \in (0.5, 1], a < b < c\}$$

$$\nu^{quad}_{a,b,c}(y) = \begin{cases} 1, & c \leq y, \\ 1 - \frac{(c-y)^2}{(c-b)(c-a)}, & b \leq y < c, \\ \frac{(y-a)^2}{(b-a)(c-a)}, & a \leq y < b, \\ 0, & y < a \end{cases} \tag{11}$$

and

$$\mathbf{Hf}^{lin} = \{\nu_{a,c}^{lin} \mid a \in (-\infty, 1], c \in (0.5, 1], a < b < c\}$$

$$\nu_{a,c}^{lin}(y) = \begin{cases} 1, & c \leq y, \\ \frac{(y-a)}{(c-a)}, & a \leq y < c, \\ 0, & y < a. \end{cases} \tag{12}$$

As an alternative to (11), we may introduce a simplified class

$$\nu_{a,c}^{quad}(y) = \begin{cases} 1, & c \leq y, \\ 1 - \frac{(c-y)^2}{(c-a)^2}, & \frac{a+c}{2} \leq y < c, \\ \frac{(y-a)^2}{(c-a)^2}, & a \leq y < \frac{a+c}{2}, \\ 0, & y < a. \end{cases}$$

Note that possible worlds are in this definition identified with intervals of real numbers. Of course, we can extend this definition to intervals in arbitrary ordered set. This is unnecessary for our explanation below. To simplify the notation, if $v \in \mathbb{R}$ and $w = \langle v_L, v_S, v_R \rangle$ is a possible world then we will often write $v \in w$ instead of $v \in [v_L, v_R]$.

The model described in this section stems from the idea that the meaning of evaluating expressions is determined by some horizon, which we can see in the given possible world. The numbers v_L, v_S, v_R represent left limit, horizon, and right limit of possible values which may fall into the meaning of the expressions in concern, respectively. Following ideas of P. Vopěnka [16], the part of the world between v_L and v_S is clear, known part, and the part between v_S and v_R "dark" (unclear) part. Small values fall into clear part while large ones fall into the second one. However, we cannot be certain about precise position of the horizon v_S. Therefore, we specify our uncertainty about this using degrees of truth that "we find ourselves still inside the clear part" and, at the same time, other degrees of truth that we are "already inside dark part". This is mathematically captured by simple linear functions $L(x), R(x)$ representing truth values uniformly diminishing when moving away from v_L (we are still less and less certain that we find ourselves in the clear part) and uniformly increasing when moving away from v_S (we are still more and more certain that we find ourselves in the dark part). Finally, there is also a part which is neither the clear nor the dark one, namely $M(x)$ (cf. the definition (8)).

If in (11), the parameters $a, b \in [0, 1)$ then the corresponding abstract hedge $\nu_{a,b,c}$ is *pure*, otherwise it is *modified.* It can be interpreted as a *deformation* of the horizon necessary when we need to specify more closely a part of the world we are interested in.

To simplify notation, we will often write only $\nu \in \mathbf{Hf}$ understanding that it is, in fact, determined by the parameters a, b, c as in (10). We will say that an abstract hedge $\nu_1 \in \mathbf{Hf}$ is sharper than $\nu_2 \in \mathbf{Hf}$, $\nu_1 < \nu_2$, if $\langle a_2, b_2, c_2\rangle < \langle a_1, b_1, c_1\rangle$.

Lemma 1. *If* $\nu_1 < \nu_2$ *then* $\nu_1(y) \leq \nu_2(y)$, $y \in [0, 1]$.

Definition 4. The following are special classes of intensions:

(i) Type *Small*

$$\mathbf{Sm} = \{\mathrm{Sm}_\nu : W \longrightarrow \mathcal{F}(\mathbb{R}) \mid \mathrm{Sm}_\nu(w) = \nu(L_w(x)), \nu \in \mathbf{Hf}\}$$

(ii) Type *Medium*

$$\mathbf{Me} = \{\mathrm{Me}_\nu : W \longrightarrow \mathcal{F}(\mathbb{R}) \mid \mathrm{Me}_\nu(w) = \nu(M_w(x)), \nu \in \mathbf{Hf}\}$$

(iii) Type *Big*

$$\mathbf{Bi} = \{\mathrm{Bi}_\nu : W \longrightarrow \mathcal{F}(\mathbb{R}) \mid \mathrm{Bi}_\nu(w) = \nu(R_w(x)), \nu \in \mathbf{Hf}\}$$

(iv) Type *fuzzy number*

$$\mathbf{Fn} = \{\mathrm{Fn}_{\nu,x_0} : W \longrightarrow P(\mathcal{F}(\mathbb{R})) \mid$$
$$\mathrm{Fn}_{\nu,x_0}(w)(x) = \{\nu(B_{w,x_0}(x)) \mid B_{w,x_0} \in \mathbf{B}_w\}, \nu \in \mathbf{Hf}\}\}$$

Note that this definition also includes crisp numbers as special case of the fuzzy ones. If x_0 is such a number then $\nu(B_{w,x_0}(x)) = 1$ for all *pure* $\nu \in \mathbf{Hf}$.

We will generally denote the class of intensions of the evaluating expressions by

$$\mathbf{Ev} = \mathbf{Sm} \cup \mathbf{Me} \cup \mathbf{Bi} \cup \mathbf{Fn}\,.$$

A special class are *pure evaluating expressions*, whose intensions are

$$\mathbf{Ev}^{pure} = \mathbf{Sm} \cup \mathbf{Me} \cup \mathbf{Bi}\,.$$

If $\mathrm{Ev} \in \mathbf{Ev}$ is an intension and $w \in W$ is a possible world then $\mathrm{Ev}(w)$ is the corresponding *extension* of Ev in w.

Remark 1 (to the notation). We are working with several classes of functions, namely $\mathbf{Sm}, \mathbf{Me}, \mathbf{Bi}, \mathbf{Fn}$ which are altogether denoted by $\mathbf{Ev}$. Their members are functions from the set W of possible worlds to some set of fuzzy sets. Thus, if $\mathrm{Ev} \in \mathbf{Ev}$ is a function then $\mathrm{Ev}(w)$ is a fuzzy set in the given possible world w. Since $\mathrm{Ev}(w)$ is itself a function, then if $u \in w$ is some element (recall that the latter means $u \in [v_L, v_R]$), then its membership degree in $\mathrm{Ev}(w)$ is $\mathrm{Ev}(w)(u)$. To simplify the notation, we will better write $\mathrm{Ev}_w(u)$ in such a case. As a special case, we will write $\mathrm{Fn}_{\nu,x_0,w}$ instead of $\mathrm{Fn}_{\nu,x_0}(w)$

When writing an element of, say $\mathrm{Sm} \in \mathbf{Sm}$, then we know that Sm is determined also by some abstract hedge ν, i.e. we mean Sm_ν. However, if ν does not matter, we usually omit it from the subscript. This convention will be used also with other symbols.

Let Ev_1, Ev_2 be intensions of the same type containing the hedges ν_1, ν_2, respectively. Then Ev_1 is *sharper than* Ev_2, $\text{Ev}_1 < \text{Ev}_2$ if $\nu_1 < \nu_2$.

Lemma 2. *Let* $\text{Ev}_1, \text{Ev}_2 \in \mathbf{Ev}^{pure}$ *be of the same type and* $\text{Ev}_1 < \text{Ev}_2$. *Then*

$$\text{Ev}_{1,w} \leq \text{Ev}_{2,w}$$

for every possible world $w \in W$.

PROOF: This immediately follows from Lemma 1 since $\nu_1 < \nu_2$ by the definition. □

Lemma 3. *In Łukasiewicz algebra* $p \rightarrow \text{Ev}_w \in \mathbf{Ev}$ *and* $p \otimes \text{Ev}_w \in \mathbf{Ev}$ *holds for every* $p \in (0, 1]$.

PROOF: This follows from rather tedious proof of the fact that $\text{Ev}(w) + k \in \text{Ev}$ for $k \in [0, 1)$. □

Linguistic assignments

Definition 5. Let $\mathcal{A}$ be a a simple evaluating linguistic expression. Then its meaning is identified with its intension and, in general, it is a function

$$\text{Int}(\mathcal{A}) \in \mathbf{Ev}$$

for which ν is a *pure hedge function.* More specifically,

$$\text{Int}(\langle\text{linguistic hedge}\rangle\, \textit{small}) \in \mathbf{Sm},$$

$\text{Int}(\langle\text{linguistic hedge}\rangle\, \textit{medium} \in \mathbf{Me}$, $\text{Int}(\langle\text{linguistic hedge}\rangle\, \textit{big}) \in \mathbf{Bi}$ and

$$\text{Int}(\langle\text{linguistic hedge}\rangle\langle\text{fuzzy number}\rangle) \in \mathbf{Fn}.$$

In more details, $\text{Int}(\langle\text{linguistic hedge}\rangle x_0) = \text{Fn}_{\nu,x_0}$.

Remark 2. If $\mathcal{A}$ is an evaluating linguistic predication (3) then we put its intension *equal* to the intension of the evaluating expression inside (3).

The extension of $\mathcal{A}$ in $w \in W$ is

$$\text{Ext}_w(\mathcal{A}) = \text{Int}(\mathcal{A})(w),$$

i.e. if $w = \langle v_L, v_S, v_R\rangle$ then it is a fuzzy set $\text{Ext}_w(\mathcal{A}) = A \underset{\sim}{\subset} [v_L, v_R]$. The construction of the extension of the basic evaluating trichotomy is depicted on Fig. 1. We have marked in the figure also the distinguished points defined

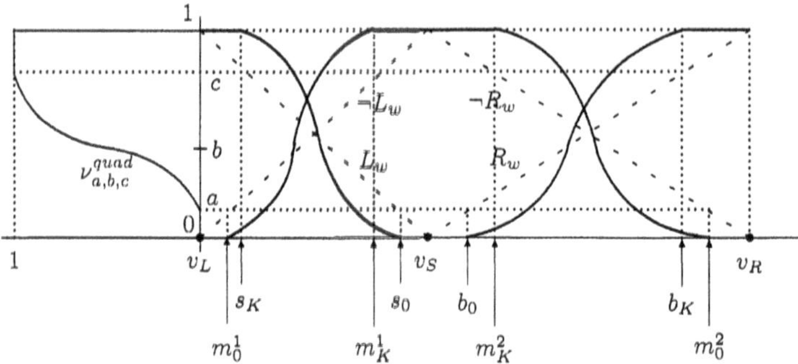

Fig. 1. Schematic picture of the construction of the extension of evaluating linguistic expressions.

in the given possible world for each intension. Let $\nu \in \mathbf{Hf}$ be given by the parameters a, b, c. Then we put:

$$
\begin{array}{lll}
s_K = L_w^{-1}(c), & s_0 = L_w^{-1}(a), & \text{(for small)} \\
m_K^1 = (\neg L_w)^{-1}(c), & m_0^1 = (\neg L_w)^{-1}(a), & \text{(for left part of medium)} \\
m_K^2 = (\neg R_w)^{-1}(c), & m_0^2 = (\neg R_w)^{-1}(a), & \text{(for right part of medium)} \\
b_K = R_w^{-1}(c), & b_0 = R_w^{-1}(a). & \text{(for big)}
\end{array}
$$

We will introduce the following terminology, which is useful for further explanation. The set $[v_L, s_K]$ is called the *kernel* of Sm_w and denoted by $\mathrm{Ker}(\mathrm{Sm}_w)$. Similarly, $\mathrm{Ker}(\mathrm{Me}_w) = [m_K^1, m_K^2]$, $\mathrm{Ker}(\mathrm{Bi}_w) = [b_K, v_R]$ are kernels of Me_w and Bi_w, respectively.

Similarly, the sets $[s_0, v_R]$, $[v_L, m_0^1] \cup [m_0^2, v_R]$ and $[v_L, b_0]$ are *zero areas* and the remaining parts are *vagueness areas.*

Lemma 4. *Let* $\mathrm{Ev}_1, \mathrm{Ev}_2$ *be intensions of different types and* $\nu_1 = \nu_{a_1,b_1,c_1} \in$ $\mathbf{Hf}$ *and* $\nu_2 = \nu_{a_2,b_2,c_2} \in \mathbf{Hf}$ *be their respective hedges. Then*

(a) $\mathrm{Ker}(\mathrm{Ev}_{1,w}) \cap \mathrm{Ker}(\mathrm{Ev}_{2,w}) = \emptyset$ *iff* $c_1 + c_2 > 1$.
(b) If $\mathrm{Ev}_1 < \mathrm{Ev}_2$ *then* $\mathrm{Ker}(\mathrm{Ev}_{1,w})^1\subset \mathrm{Ker}(\mathrm{Ev}_{2,w})$.

PROOF: (a) easily implies from (6), (7) and (8) and the above definition.
(b) immediately from the definition of $<$. □

This lemma clarifies why we have put $c > 0.5$ in (10). If the kernels of neighboring extensions were overlapping then the intuitive meaning of them as interpretation of the meaning of evaluating linguistic expressions would become dubious.

We will now select several adverbs, namely

> *extremely (Ex), significantly (Si), very (Ve), more or less (ML), roughly (Ro), quite roughly (QR), very roughly (VR)*

as basic linguistic hedges (in the brackets are shorts used below).

Furthermore, we will choose numbers $a_0, b_0, c_0 \in [0,1]$ and assign the abstract hedge $\nu_{a_0,b_0,c_0} \in \mathbf{Hf}^{quad}$ to the "empty hedge". Then we choose three abstract hedges $\nu_{Ex}, \nu_{Si}, \nu_{Ve} \in \mathbf{Hf}^{quad}$, for which

$$\nu_{Ex} < \nu_{Si} < \nu_{Ve} < \nu_{a_0,b_0,c_0}$$

holds and four abstract hedges $\nu_{ML}, \nu_{Ro}, \nu_{QR}, \nu_{VR} \in \mathbf{Hf}^{quad}$, for which

$$\nu_{a_0,b_0,c_0} < \nu_{ML} < \nu_{Ro} < \nu_{QR} < \nu_{VR}.$$

Finally we assign these hedges to the above selected words and understand the former as the *meaning* of the latter. This procedure enables to construct the meaning of each evaluating expression, which is the intension and if given a possible world, also their extension. Let us remark that in fact, we should introduce also a special grammar which would prevent, e.g. combination of narrowing linguistic hedge with medium (e.g. "very medium" has no sense). We will not care for this problem in this paper.

Note that we can also define other kinds of modifiers not belonging to the above group, for example "rather". This should be assigned the abstract hedge $\nu_{a,b,c}$ with $a > a_0$, $b \leq b_0$ and $c < c_0$. Thus, the above theory encompasses a large class of linguistic hedges.

The experimentally found values of the parameters a, b, c of the discussed linguistic hedges are given on Table 1.

Linguistic hedge	a	b	c
Extremely	0.5	0.75	0.95
Significantly	0.47	0.6	0.8
Very	0.35	0.58	0.83
empty	0.27	0.5	0.8
Rather	0.4	0.5	0.8
More or less	0.23	0.45	0.76
Roughly	0.2	0.4	0.7
Quite roughly	0.15	0.32	0.65
Very roughly	0.09	0.2	0.6

Table 1. Experimentally found values of parameters a, b, c of some linguistic hedges.

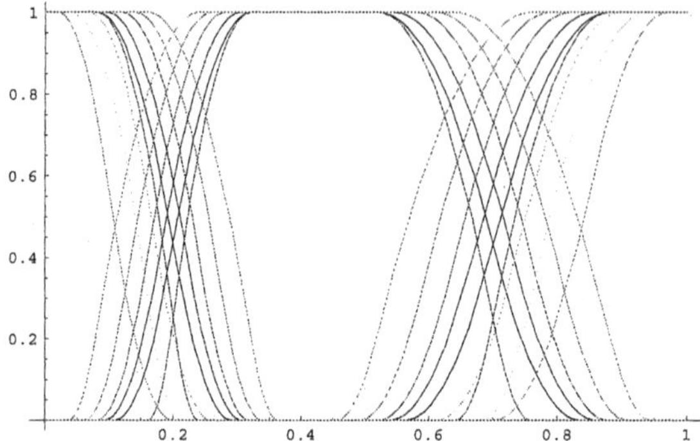

Fig. 2. Membership functions of extensions of selected evaluating expressions from Table 1.

Shapes of the membership functions of extensions of the experimentally set pure evaluating expressions are depicted on Figure 2

Let us also remark that linear abstract hedges could be used when a significant simplification (for the linguistic purposes non well suitable) is needed. However, from the linguistic point of view, they seem to be oversimplification of the problem. These hedges might be reasonable for the fuzzy numbers, which form a quite specific class of the linguistic expressions. Note that $\nu_{0,1}^{lin}$ gives extension equal to the original linear function (6), (7), (8) and (9).

4 Fuzzy Relation Equations

4.1 General solution

In this section, we will focus on fuzzy relation equations. Recall that this problem has been initiated by E. Sanchez ([14]) and studied further by many researchers (cf. Di Nola et. al. [3], S. Gottwald [5], I. Perfilieva and A. Tonis [11]). The initial situation is the following.

We are given fuzzy data, which is a finite set of tuples

$$\langle A_{11}, \ldots, A_{1n}, B_1 \rangle$$
$$\ldots\ldots\ldots\ldots\ldots\ldots\ldots \tag{13}$$
$$\langle A_{m1}, \ldots, A_{mn}, B_m \rangle \tag{14}$$

where $A_{j,i} \underset{\sim}{\subset} U_i$, $B_j \underset{\sim}{\subset} V$, $j = 1, \ldots, m$, $i = 1, \ldots, n$ are fuzzy sets. We will often suppose that $U_i, V \subset \mathbb{R}$ are a compact sets.

The imprecise information leading to fuzzy data is quite often given using linguistic expressions. Thus, the first information we are given takes the form

$$\begin{aligned}&\langle \mathcal{A}_{11}, \dots, \mathcal{A}_{1n}, \mathcal{B}_1\rangle \\ &\dots\dots\dots\dots\dots\dots \\ &\langle \mathcal{A}_{m1}, \dots, \mathcal{A}_{mn}, \mathcal{B}_m\rangle\end{aligned} \tag{15}$$

where $\mathcal{A}_{ji}, \mathcal{B}_j$ are evaluating linguistic predications. Then using the results of the previous section, (15) leads to the fuzzy data of the form

$$\begin{aligned}&\langle \mathrm{Ev}_{11,w_1}, \dots, \mathrm{Ev}_{1n,w_n}, \mathrm{Ev}_{1(n+1),w_{n+1}}\rangle \\ &\dots\dots\dots\dots\dots\dots\dots\dots\dots\dots \\ &\langle \mathrm{Ev}_{m1,w_1}, \dots, \mathrm{Ev}_{mn,w_n}, \mathrm{Ev}_{m(n+1),w_{n+1}}\rangle\end{aligned} \tag{16}$$

In (16), Ev_{ji,w_i}, $j = 1, \dots, m$, $i = 1, \dots, n+1$ are extensions of the linguistic predications from (15) in some possible worlds w_i, which take the role of the universes $U_1, \dots, U_n, V$ considered above (these are compact sets by Definition 3).

The problem is to find a fuzzy relation $R \underset{\sim}{\subset} U_1 \times \cdots \times U_n \times V$ such that

$$B_j = (A_{j1} \times \cdots \times A_{jn}) \circ R \tag{17}$$

holds for all $j = 1, \dots, m$ where the membership function of B_j in (17) is given by

$$B_j(v) = \bigvee_{\substack{u \in U_i \\ i=1,\dots,n}} ((A_{j1} \times \cdots \times A_{jn})(u_1, \dots, u_n) \otimes R(u, v)), \qquad v \in V \tag{18}$$

and $\otimes$ is a t-norm. If such relation exists, then we say that the system of fuzzy relation equations (17) is *solvable*. We will usually suppose that the composition in (18) is defined w.r.t. a continuous (left continuous) t-norm.

Let us denote

$$\hat{R} = \bigwedge_{j=1}^{m} ((A_{j1} \times \cdots \times A_{jn}) \ominus B_j). \tag{19}$$

Then the general solution is given in the following theorem.

Theorem 1. *Given the data (13). Then the system (17) is solvable iff*

$$B_k(v) = \bigvee_{\substack{u \in U_i \\ i=1,\dots,n}} ((A_{k1} \times \cdots \times A_{kn})(u_1, \dots, u_n) \otimes \hat{R}(u_1, \dots, u_n, v)), \qquad v \in V \tag{20}$$

holds for every $k = 1, \dots, m$. If (20) holds then $\hat{R}$ is the greatest solution of (17).

Note that it follows from this theorem that cartesian product $(A_{j1} \times \cdots \times A_{jn})$ in (19) and (20) can be defined using arbitrary t-norm, or even by a more general operation. In the sequel, we will for simplicity take $n = 1$.

Given a couple of fuzzy data $\langle A, B\rangle$, we will put

$$X(v) := \{u \in U \mid A(u) \geq B(v)\} \subseteq U, \qquad v \in V. \tag{21}$$

In case that $A := \mathrm{Ev}_{1,w}$, $B := \mathrm{Ev}_{2,w}$ for some possible world and intensions $\mathrm{Ev}_1, \mathrm{Ev}_2$, we will denote (21) by $X_{E_1E_2}$, or, in more details, $X_{SS}, X_{SM}, \ldots$ if $\mathrm{Ev}_1 := \mathrm{Sm}$ and $\mathrm{Ev}_2 := \mathrm{Sm}$ or $\mathrm{Ev}_2 := \mathrm{Me}$, respectively.

The following theorem is proved in [11].

Theorem 2 (Perfilieva and Tonis). *Let all the membership functions of the fuzzy data (13) be continuous. Then the system (17) is solvable iff for each i, $i = 1, \ldots, m$ and every $v \in V$ there exists $u \in X_i(v)$ such that $z(B_j(v)) < B_i(v)$ implies*

$$A_i(u) \rightarrow B_i(v) \leq A_j(u) \rightarrow B_j(v) \tag{22}$$

for each $j \neq i$ and $j = 1, \ldots, m$.

4.2 Pure evaluating expressions in fuzzy relations

In this subsection we will discuss the situation when the data are given by pure evaluating expressions.

Lemma 5. *Let us consider a couple of data*

$$\langle \mathrm{Ev}_{1,w}, \mathrm{Ev}_{2,w'}\rangle$$

for some possible worlds $w, w' \in W$ where $\mathrm{Ev}_1, \mathrm{Ev}_2$ are of the same type and $\mathrm{Ev}_1 < \mathrm{Ev}_2$. If $v \in \mathrm{Ker}(\mathrm{Ev}_{2,w'})$ then $X_{E_1E_2}(v) = \mathrm{Ker}(\mathrm{Ev}_{1,w})$.

PROOF: Immediately from the definition (21). □

Theorem 3. *Let the fuzzy data (13) contain couples*

$$\langle \mathrm{Ev}_{1,w}, \mathrm{Ev}_{2,w'}\rangle$$
$$\langle \mathrm{Ev}_{2,w}, \mathrm{Ev}_{3,w'}\rangle$$

where $\mathrm{Ev}_1, \mathrm{Ev}_2 \in \mathbf{Ev}^{pure}$, $\mathrm{Ev}_1 < \mathrm{Ev}_2$, are of the same type and $\mathrm{Ev}_3 \in \mathbf{Ev}$ is of the type different from the former. Then the system (17) is not solvable.

PROOF: We will use Theorem 2. We must show that there is $v \in w'$ such that for every $u \in X_{E_1E_2}(v)$, if $z(\mathrm{Ev}_{3,w'}(v)) < \mathrm{Ev}_{2,w'}(v)$ then

$$\mathrm{Ev}_{1,w}(u) \rightarrow \mathrm{Ev}_{2,w'}(v) > \mathrm{Ev}_{2,w}(u) \rightarrow \mathrm{Ev}_{3,w'}(v).$$

Let us take $v \in \mathrm{Ker}(\mathrm{Ev}_{2,w'}) - \mathrm{Ker}(\mathrm{Ev}_{1,w'})$. Then $z(\mathrm{Ev}_{3,w'}(v)) \le \mathrm{Ev}_{3,w'}(v) < \mathrm{Ev}_{2,w'}(v)$ because $\mathrm{Ker}(\mathrm{Ev}_{2,w'}) \cap \mathrm{Ker}(\mathrm{Ev}_{3,w'}) = \emptyset$. By Lemma 5, $X_{E_1E_2}(v) = \mathrm{Ker}(\mathrm{Ev}_{1,w})$.

Then for every $u \in X_{E_1E_2}(v)$, $\mathrm{Ev}_{1,w}(u) \to \mathrm{Ev}_{2,w'}(v) = 1$ and $\mathrm{Ev}_{2,w}(u) = 1$ because $\mathrm{Ker}(\mathrm{Ev}_{1,w}) = X_{E_1E_2}(v) \subset \mathrm{Ker}(\mathrm{Ev}_{2,w})$. However, $\mathrm{Ev}_{3,w'}(v) < 1$ because $\mathrm{Ker}(\mathrm{Ev}_{2,w'}) \cap \mathrm{Ker}(\mathrm{Ev}_{3,w'}) = \emptyset$. □

This theorem demonstrates that if we insist on the use of pure evaluating expressions then we are very limited provided we want the corresponding system of fuzzy relation equations to be solvable. In practice this means that for this case, pure linguistic expressions are inappropriate and thus, we should better take the fuzzy data to be fuzzy numbers.

Let us remark that this does not disqualify pure evaluating expressions from any use but only from the attempt to solve fuzzy relation equations with them. These expressions are, on the other hand, much more appropriate when deriving a conclusion using the logical deduction (cf., e.g., [1]).

4.3 Pure evaluating expressions in fuzzy IF-THEN rules

Let us remark that, in general, logical deduction leads to firing of one fuzzy IF-THEN rule. Let us analyze, what can be obtained by such a rule.

Recall that fuzzy IF-THEN rules are linguistically a special kind of conditional clause. Its semantics will be defined as follows.

Definition 6. Let the fuzzy IF-THEN rule $\mathcal{R}$ have the form (4). Then its meaning*) is

$$\mathrm{Int}\,\mathcal{R} := \mathrm{Ev}_1 \Rightarrow \mathrm{Ev}_2$$

where $\Rightarrow$ is a connective interpreted by some implication operation $\to$. Its extension is defined in a couple of possible worlds $w, w' \in W$ by

$$\mathrm{Ext}_{\langle w,w'\rangle}(\mathcal{R}) := \mathrm{Ev}_{1,w} \to \mathrm{Ev}_{2,w'} \tag{23}$$

where (23) is a fuzzy relation defined pointwisely.

Let us now define a defuzzification operation DEE for the evaluating linguistic expressions as follows: let $\mathrm{Ev} \in \mathbf{Ev}$ be an evaluating expression and $w \in W$ a possible world. Then the defuzzification operation DEF is

$$\mathrm{DEE}(\mathrm{Ev}_w) = \begin{cases} \mathrm{LOM}(\mathrm{Sm}_w), & \text{if } \mathrm{Ev} \in \mathbf{Sm}, \\ \mathrm{COG}(\mathrm{Ev}_w), & \text{if } \mathrm{Ev} \in \mathbf{Me} \text{ or } \mathrm{Ev} \in \mathrm{Fn}, \\ \mathrm{FOM}(\mathrm{Bi}_w), & \text{if } \mathrm{Ev} \in \mathbf{Bi} \end{cases} \tag{24}$$

where LOM is the *Least of Maxima*, FOM is the *First of Maxima* and COG is the *Center of Gravity* method.

---

*) In accordance with Remark 2, we semantically do not distinguish evaluating linguistic predications from the expressions inside them

Let $w = \langle v_L, v_S, v_R \rangle$ be a possible world. Then

$$\mathrm{LOM}(\mathrm{Sm}_{\nu,w}) = c v_L + (1-c) v_S, \tag{25}$$

$$\mathrm{FOM}(\mathrm{Bi}_{\nu,w}) = c v_R + (1-c) v_S \tag{26}$$

where $c \in (0.5, 1]$ is the parameter of the corresponding linguistic hedge ν.

Let $\mathcal{R}$ be a fuzzy IF-THEN rule (4) with the extension (23) in some possible worlds w, w'. Then it determines a function f_R by

$$f_R(u) = \mathrm{DEE}(\mathrm{Ev}_{1,w}(u) \to \mathrm{Ev}_{2,w'}). \tag{27}$$

To distinguish more subtly various kinds of the function f_R dependingly on the used evaluating expressions in the rule (4), we will write $f_{S_1 S_2}$ if it is obtained from the rule of the form

$$\mathcal{R} := \ \textsf{IF } X \textsf{ is } \mathit{Small} \textsf{ THEN } Y \textsf{ is } \mathit{Small},$$

$f_{S_1 B_2}$ if it is obtained from the rule $\mathcal{R} :=$ IF X is *Small* THEN Y is *Big*, etc.

A rather technical and routine proof gives us the following theorem.

Theorem 4. *Let $w, w' \in W$ be possible worlds and $\mathcal{R}$ a fuzzy IF-THEN rule (4). Then the function f_R from (27) is continuous. More specifically, f_{S_1,S_2} is nondecreasing, f_{B_1,B_2} is nondecreasing, f_{S_1,B_2} is nonincreasing, f_{B_1,S_2} is nonincreasing, f_{S_1,M_2} is nonincreasing until $v'_S \in w'$ and nondecreasing further and f_{B_1,M_2} is opposite.*

Let us define a *pseudoinverse* of $\nu_{a,b,c}$ in (10) by

$$\nu_{a,b,c}^{(-1)}(z) = \begin{cases} \nu^{-1}(z) & \text{if } z \in (0,1), \\ c & \text{if } z = 1, \\ a & \text{if } z = 0. \end{cases}$$

Then, for example, the function $f_{S_1 S_2}$ is given by the formula

$$f_{S_1 S_2}(u) = \nu_2^{(-1)}(\mathrm{Sm}_{1,w}(u)) v_L + (1 - \nu_2^{(-1)}(\mathrm{Sm}_{1,w}(u))) v_S$$

where $v_L, v_S \in w'$ and ν_2 is an abstract hedge inside Sm_2.

It follows from this theorem that by logical deduction we can obtain a piecewise continuous and monotonous function. An example of the behavior of the linguistic description (rule base) consisting of a set of monotonous rules such as

IF X is $ExSm$ THEN Y is $ExSm$
IF X is $SiSm$ THEN Y is $SiSm$
IF X is $VeSm$ THEN Y is $VeSm$
IF X is Sm THEN Y is Sm
IF X is $RoSm$ THEN Y is $RoSm$
etc.

is depicted on Figure 3

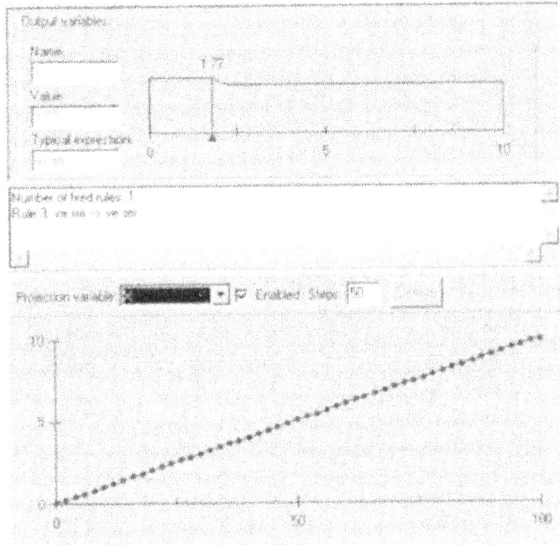

Fig. 3. Example of the construction of linguistic expressions so that the monotonous linguistic description leads to almost linear course.

4.4 Fuzzy numbers in fuzzy relations

As discussed in the previous section, pure evaluating expressions lead to mostly unsolvable fuzzy relation equations. Therefore, we we will confine only to fuzzy numbers in this subsection.

By a simple computation we get the following lemma.

Lemma 6. *Let the data (13) contain only fuzzy numbers, i.e. they are of the form*

$$\langle \mathrm{Fn}_{\nu_1,x_0,w}, \mathrm{Fn}_{\nu_2,y_0,w'} \rangle \tag{28}$$

then the set (21) is

$$X_1(v) = [x_0 - h_{L,1}(1-\nu_1^{(-1)}(\mathrm{Fn}_{\nu_2,y_0,w'})(v)), x_0 + h_{R,1}(1-\nu_1^{(-1)}(\mathrm{Fn}_{\nu_2,y_0,w'})(v))].$$

Moreover, if $\mathrm{Fn}_{\nu_2,y_0,w'}(v) \geq \mathrm{Fn}_{\nu_2,y_0,w'}(v')$ *then* $X_1(v) \subseteq X_2(v')$.

(Recall that the values $x_0 \in w$ and $y_0 \in w'$ are kernels of the fuzzy numbers $\mathrm{Fn}_{\nu_1,x_0,w}$ and $\mathrm{Fn}_{\nu_2,y_0,w'}$, respectively).

Theorem 5. *Let the fuzzy data (13) consist of fuzzy numbers of the form (28) and the corresponding system of fuzzy relations be solvable. Let*

$$\langle A := \mathrm{Fn}_{\nu_1,x_0,w}, B := \mathrm{Fn}_{\nu_2,y_0,w'} \rangle$$

be a new data. If $A_j(x_0) = 0$ for all $j = 1, \ldots, m$ then the new system is solvable.

PROOF: By Theorem 2: Let $v = y_0$. Then $B(y_0) = 1$ and thus, $X(y_0) = \{u \in w \mid A(u) = 1\}$, i.e. $x_0 \in X(y_0)$. Let $z(B_j(y_0)) < 1 = B(y_0)$. Then there must exist $u \in X(y_0)$ such that

$$A(u) \rightarrow 1 \leq A_j(u) \rightarrow B_j(y_0). \tag{29}$$

By the assumption, $A_j(x_0) = 0$ and thus (29) is fulfilled. Since by Lemma 6, $X(v)$ forms a nested system, $x_0 \in X(v)$ for every $v \in w'$ and thus (29) is always fulfilled. □

To finish this discussion, we will present the following theorem, which first occurred in [7] and holds generally in BL-logic.

Theorem 6. *Let the fuzzy data (13) consist of fuzzy numbers of the form (28). The solution of the corresponding system of fuzzy relations is the fuzzy relation*

$$R(u, v) = \bigvee_{j=1}^{m} (\mathrm{Fn}_{\nu_j, x_{0j}, w}(u) \otimes \mathrm{Fn}_{\nu_j, y_{0j}, w'}(v)), \qquad u \in w, v \in w', \tag{30}$$

iff

$$\bigvee_{u \in w} (\mathrm{Fn}_{\nu_i, x_{0i}, w}(u) \otimes \mathrm{Fn}_{\nu_j, x_{0j}, w}(u)) \leq \bigwedge_{v \in w'} (\mathrm{Fn}_{\nu_j, y_{0i}, w}(v) \leftrightarrow \mathrm{Fn}_{\nu_j, y_{0j}, w'}(v))$$

holds for every $i, j = 1, \ldots, m$.

The intersection of fuzzy sets in the following corollary is taken w.r.t. the t-norm $\otimes$.

Corollary 1. *(a) If $\mathrm{Fn}_{\nu_i, x_{0i}, w} \cap \mathrm{Fn}_{\nu_j, x_{0j}, w} = \emptyset$ then the fuzzy relation R in (30) is a solution.*

(b) If $\mathrm{Ker}(\mathrm{Fn}_{\nu_i, y_{0i}, w'}) \cap \mathrm{Ker}(\mathrm{Fn}_{\nu_j, y_{0j}, w'}) = \emptyset$ and the fuzzy relation R in (30) is a solution then $\mathrm{Fn}_{\nu_i, x_{0i}, w} \cap \mathrm{Fn}_{\nu_j, x_{0j}, w} = \emptyset$.

This corollary demonstrates that if the data consist of fuzzy numbers then the system of fuzzy relation equations can be solved by a fuzzy relation in the form of Mamdani-Assilian formula (30) provided that the data can be seen as a fuzzy function — a set of couples of fuzzy sets such that the closer are first components, the more closer are the second ones. Note that this follows also from Theorem 2.

5 Conclusion

In this paper we have developed a systematic theory of the evaluating linguistic expressions and their semantics. In the second part, we have focused on fuzzy relation equations derived from the fuzzy data given by these expressions. We have demonstrated that pure expressions (i.e. the expressions containing the adjectives "small, medium, big") mostly cannot lead to solvable fuzzy relation equations. Thus, we should use fuzzy numbers. If the structure of the data corresponds to a function then the solution can even be in the form of the widely used Mamdani-Assilian formula.

References

1. Bělohlávek, R. and V. Novák: *Learning Rule Base of the Linguistic Expert Systems.* Soft Computing **7**(2002), No. 2, 79–88.
2. Carnap, R.: **Meaning and Necessity: a Study in Semantics and Modal Logic**, University of Chicago Press, Chicago 1947.
3. Di Nola, A., Sessa, S., Pedrycz, W. and E. Sanchez: **Fuzzy Relation Equations and Their Applications to Knowledge Engineering**. Kluwer, Dordrecht 1989.
4. Gallin, D.: Intensional and Higher-Order Modal Logic (With Applications to Montague Semantics). North-Holland, Amsterdam 1975
5. Gottwald, S. **Fuzzy Sets and Fuzzy Logic**. Vieweg, Wiesbaden 1993.
6. Hájek, P.: **Metamathematics of fuzzy logic**. Kluwer, Dordrecht 1998.
7. Klawonn, F.: Fuzzy Points, Fuzzy Relations and Fuzzy Functions. In: [8], 431–453.
8. Novák, V. and I. Perfilieva (eds.): **Discovering World With Fuzzy Logic**. Springer-Verlag, Heidelberg 2000, 555 pp. (Studies in Fuzziness and Soft Computing, Vol. 57)
9. Novák, V., Perfilieva I. and J. Močkoř: **Mathematical Principles of Fuzzy Logic**. Kluwer, Boston 1999.
10. Novák, V.: *Antonyms and Linguistic Quantifiers in Fuzzy Logic.* Fuzzy Sets and Systems **124**(2001), 335-351.
11. Perfilieva, I. and A. Tonis: *Compatibility of systems of fuzzy relation equations.* Int. J. General Systems **29**(2000), 511–528.
12. Perfilieva, I.: *Normal Forms for Fuzzy Logic Functions and Their Approximation Ability.* Fuzzy Sets and Systems 124(2001), 371–384.
13. Perfilieva, I.: *Neural Nets and Normal Forms from Fuzzy Logic Point of View.* Neural Network World **6**(2001), 627–638.
14. Sanchez, E.: Resolution of composite fuzzy relation equations. *Information and Control*, **30**(1976), 38–48.
15. Vopěnka, P.: **Mathematics In the Alternative Set Theory**. Teubner, Leipzig 1979.
16. Vopěnka, P.: **Meditations on principles of science**, Práh, Prague 2001 (in Czech).
17. Zadeh, L.A. . *The concept of a linguistic variable and its application to approximate reasoning I, II, III*, Inf. Sci., **8**(1975), 199–257, 301–357; **9**(1975), 43–80.

A Normative View on Possibility Distributions

Christer Carlsson[1], Robert Fullér[1,2], and Péter Majlender[3,1,2]

[1] Institute for Advanced Management Systems Research, Åbo Akademi University, Lemminkäinengatan 14C, FIN-20520 Åbo, Finland, christer.carlsson@abo.fi,
[2] Department of Operations Research, Eötvös Loránd University, Pázmány Péter sétány 1C, P. O. Box 120, H-1518 Budapest, Hungary, robert.fuller@abo.fi,
[3] Turku Centre for Computer Science, Åbo Akademi University, Lemminkäinengatan 14B, FIN-20520 Åbo, Finland, peter.majlender@abo.fi

Abstract

In this paper we will consider fuzzy numbers from a normative point of view and will illustrate the concepts of possibilistic mean value, covariance, variance and correlation by several examples. We will show that zero correlation does not always imply non-interactivity. We will also discuss the limitations of direct definitions of joint possibility distributions (for example, when one simple aggregates the membership values of two fuzzy numbers by a triangular norm).

1 Probability and Possibility

The concept of independence (dependence) has been studied in depth in possibility theory, for good surveys see, e.g. (Campos and Huete 1999). Dubois and Prade (1987) defined an interval-valued expectation of fuzzy numbers, viewing them as consonant random sets. They also showed that this expectation remains additive in the sense of addition of fuzzy numbers. Carlsson and Fullér (2001) introduced the possibilistic mean value, variance and covariance of fuzzy numbers. Fullér and Majlender (2003) introduced the notations of crisp weighted possibilistic mean value, variance and covariance of fuzzy numbers, that are consistent with the extension principle. Fullér and Majlender (2003a) introduced a measure of interactivity between marginal distributions of a joint possibility distribution C as the expected value of their interactivity relation on C. Carlsson et al (2003) showed a possibilistic analog of the probabilistic Cauchy-Schwarz inequality.

In this paper we will illustrate some important feautures of possibilistic mean value, covariance, variance and correlation by several examples.

In probability theory, the dependency between two random variables can be characterized through their joint probability density function. Namely, if X and Y are two random variables with probability density functions $f_X(x)$ and $f_Y(y)$, respectively, then the density function, $f_{X,Y}(x,y)$, of their joint random variable (X,Y), should satisfy the following properties

$$\int_{\mathbb{R}} f_{X,Y}(x,t)dt = f_X(x), \quad \int_{\mathbb{R}} f_{X,Y}(t,y)dt = f_Y(y), \tag{1}$$

for all $x, y \in \mathbb{R}$. Furthermore, $f_X(x)$ and $f_Y(y)$ are called the the marginal probability density functions of random variable (X,Y). The covariance between X and Y is defined as $\text{Cov}(X,Y) = E(XY) - E(X)E(Y)$, where E is the expected value operator, and if X and Y are independent then $\text{Cov}(X,Y) = 0$. For any random variables X and Y and real numbers λ and μ the following relationship holds

$$\text{Var}(\lambda X + \mu Y) = \lambda^2\text{Var}(X) + \mu^2\text{Var}(Y) + 2\lambda\mu\text{Cov}(X,Y),$$

where $\text{Var}(X)$ denotes the variance of X. The correlation coefficient between X and Y is defined by

$$\rho(X,Y) = \frac{\text{Cov}(X,Y)}{\sqrt{\text{Var}(X)\text{Var}(Y)}},$$

and it is clear that $-1 \leq \rho(X,Y) \leq 1$.

A fuzzy set A in $\mathbb{R}$ is said to be a fuzzy number if it is normal, fuzzy convex and has an upper semi-continuous membership function of bounded support. The family of all fuzzy numbers will be denoted by $\mathcal{F}$. A γ-level set of a fuzzy set A in $\mathbb{R}^m$ is defined by $[A]^\gamma = \{x \in \mathbb{R}^m : A(x) \geq \gamma\}$ if $\gamma > 0$ and $[A]^\gamma = \text{cl}\{x \in \mathbb{R}^m : A(x) > \gamma\}$ (the closure of the support of A) if $\gamma = 0$. If $A \in \mathcal{F}$ is a fuzzy number then $[A]^\gamma$ is a convex and compact subset of $\mathbb{R}$ for all $\gamma \in [0, 1]$. Fuzzy numbers can be considered as possibility distributions (Zadeh 1978).

A fuzzy set B in $\mathbb{R}^m$ is said to be a joint possibility distribution of fuzzy numbers $A_i \in \mathcal{F}, i = 1, \ldots, m$, if it satisfies the relationship

$$\max_{x_j \in \mathbb{R},\ j \neq i} B(x_1, \ldots, x_m) = A_i(x_i),\ \forall x_i \in \mathbb{R}, i = 1, \ldots, m.$$

Furthermore, A_i is called the i-th marginal possibility distribution of B, and the projection of B on the i-th axis is A_i for $i = 1, \ldots, m$. If $A_i \in \mathcal{F}$, $i = 1, \ldots, m$, and B is their joint possibility distribution then the relationships $B(x_1, \ldots, x_m) \leq \min\{A_1(x_1), \ldots, A_m(x_m)\}$ and $[B]^\gamma \subseteq [A_1]^\gamma \times \cdots \times [A_m]^\gamma$ hold for all $x_1, \ldots, x_m \in \mathbb{R}$ and $\gamma \in [0, 1]$.

Fuzzy numbers $A_i \in \mathcal{F}, i = 1, \ldots, m$, are said to be non-interactive if their joint possibility distribution, B, is given by

$$B(x_1, \ldots, x_m) = \min\{A_1(x_1), \ldots, A_m(x_m)\},$$

or, equivalently, $[B]^\gamma = [A_1]^\gamma \times \cdots \times [A_m]^\gamma$, for all $x_1, \ldots, x_m \in \mathbb{R}$ and $\gamma \in [0, 1]$.

Note 1. If $A, B \in \mathcal{F}$ are non-interactive then their joint membership function is defined by $A \times B$. It is clear that in this case any change in the membership function of A does not effect the second marginal possibility distribution and vice versa. On the orther hand, if A and B are said to be interactive then they can not take their values independently of each other (Dubois, Prade 1988).

Note 2. Marginal probability distributions are determined from the joint one by the principle of 'falling integrals' and marginal possibility distributions are determined from the joint possibility distribution by the principle of 'falling shadows'.

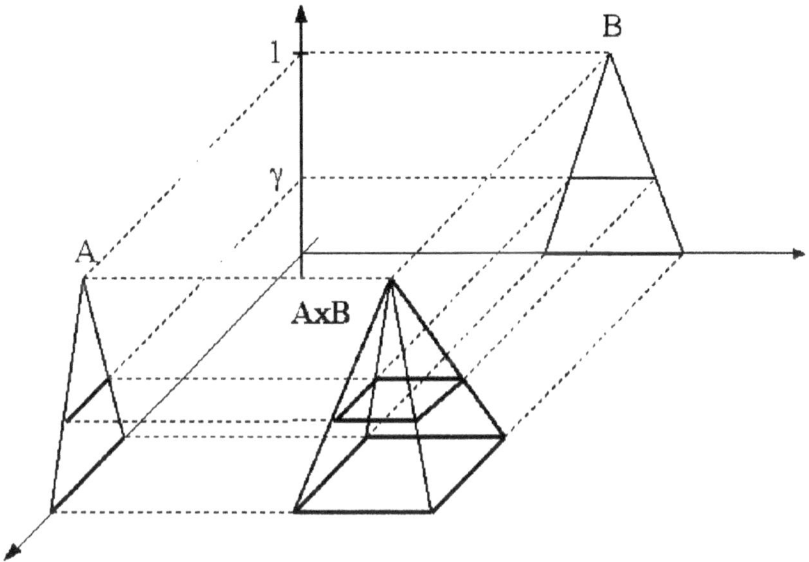

Fig. 1. Non-interactive possibility distributions.

Let $A \in \mathcal{F}$ be fuzzy number with $[A]^\gamma = [a_1(\gamma), a_2(\gamma)]$, $\gamma \in [0, 1]$. A function $f\colon [0, 1] \to \mathbb{R}$ is said to be a weighting function if f is non-negative, monotone increasing and satisfies the following normalization condition

$$\int_0^1 f(\gamma) d\gamma = 1. \tag{2}$$

Different weighting functions can give different (case-dependent) importances to γ-levels sets of fuzzy numbers.

2 Weighted Possibilistic Mean Value and Variance

The notations of weighted lower possibilistic and upper possibilistic mean values, crisp possibilistic mean value and variance of fuzzy numbers were introduced in (Fullér, Majlender 2003). Namely, they introduced the f-weighted possibilistic mean value of fuzzy number A as

$$E_f(A) = \int_0^1 \frac{a_1(\gamma) + a_2(\gamma)}{2} f(\gamma) d\gamma. \tag{3}$$

We note that if $f(\gamma) = 2\gamma$, $\gamma \in [0, 1]$ then

$$E_f(A) = \int_0^1 \frac{a_1(\gamma) + a_2(\gamma)}{2} 2\gamma d\gamma = \int_0^1 [a_1(\gamma) + a_2(\gamma)] \gamma d\gamma.$$

That is the f-weighted possibilistic mean value defined by (3) can be considered as a generalization of possibilistic mean value introduced in (Carlsson, Fullér 2001). From the definition of a weighting function it can be seen that $f(\gamma)$ might be zero for certain (unimportant) γ-level sets of A. So by introducing different weighting functions we can give different (case-dependent) importances to γ-levels sets of fuzzy numbers.

Example 1 *Let $A = (a, b, \alpha, \beta)$ be a fuzzy number of trapezoidal form with peak $[a, b]$, left-width $\alpha > 0$ and right-width $\beta > 0$, and let $f(\gamma) = (n + 1)\gamma^n$, $n \geq 0$. The γ-level of A is computed by $[A]^\gamma = [a - (1 - \gamma)\alpha, b + (1 - \gamma)\beta]$, $\forall \gamma \in [0, 1]$. Then the weighted possibilistic mean values of A are computed by*

$$E_f(A) = \frac{a + b}{2} + \frac{\beta - \alpha}{2(n + 2)}.$$

So,

$$\lim_{n \to \infty} E_f(A) = \lim_{n \to \infty} \left(\frac{a + b}{2} + \frac{\beta - \alpha}{2(n + 2)} \right) = \frac{a + b}{2}.$$

Let $A, B \in \mathcal{F}$ and let f be a weighting function. In (Fullér, Majlender 2003) the f-weighted possibilistic variance of A was introduced as

$$\mathrm{Var}_f(A) = \int_0^1 \left(\frac{a_2(\gamma) - a_1(\gamma)}{2} \right)^2 f(\gamma) d\gamma, \tag{4}$$

and the f-weighted covariance of A and B was defined by

$$\mathrm{Cov}_f(A, B) = \int_0^1 \frac{a_2(\gamma) - a_1(\gamma)}{2} \cdot \frac{b_2(\gamma) - b_1(\gamma)}{2} f(\gamma) d\gamma. \tag{5}$$

It should be noted that if $f(\gamma) = 2\gamma$, $\gamma \in [0, 1]$ then

$$\begin{aligned}\mathrm{Var}_f(A) &= \int_0^1 \left(\frac{a_2(\gamma) - a_1(\gamma)}{2}\right)^2 2\gamma d\gamma \\ &= \frac{1}{2}\int_0^1 [a_2(\gamma) - a_1(\gamma)]^2 \gamma d\gamma = \mathrm{Var}(A),\end{aligned}$$

and

$$\begin{aligned}\mathrm{Cov}_f(A, B) &= \int_0^1 \frac{a_2(\gamma) - a_1(\gamma)}{2} \cdot \frac{b_2(\gamma) - b_1(\gamma)}{2} 2\gamma d\gamma \\ &= \frac{1}{2}\int_0^1 (a_2(\gamma) - a_1(\gamma)) \cdot (b_2(\gamma) - b_1(\gamma)) \gamma d\gamma \\ &= \mathrm{Cov}(A, B).\end{aligned}$$

Where $\mathrm{Var}(A)$ and $\mathrm{Cov}(A, B)$ denote the possibilistic variance and covariance introduced by (Carlsson, Fullér 2001). That is the f-weighted possibilistic variance and covariance defined by (4) and (5) can be considered as a generalization of the concepts introduced by (Carlsson, Fullér 2001).

It can easily be verified that the weighted covariance is a symmetrical bilinear operator.

Example 2 *Let $A = (a, b, \alpha, \beta)$ be a trapezoidal fuzzy number and let $f(\gamma) = (n+1)\gamma^n$ be a weighting function. Then,*

$$\begin{aligned}\mathrm{Var}_f(A) &= (n+1)\int_0^1 \left[\frac{a_2(\gamma) - a_1(\gamma)}{2}\right]^2 \gamma^n d\gamma \\ &= \left[\frac{b-a}{2} + \frac{\alpha + \beta}{2(n+2)}\right]^2 + \frac{(n+1)(\alpha+\beta)^2}{4(n+2)^2(n+3)}.\end{aligned}$$

Example 3 *Let* $A = (a, b, \alpha, \beta)$ *and* $B = (a', b', \alpha', \beta')$ *be fuzzy numbers of trapezoidal form. Let* $f(\gamma) = (n+1)\gamma^n$, $n \geq 0$, *be a weighting function, then the power-weighted covariance of* A *and* B *is computed by*

$$\mathrm{Cov}_f(A, B) = \left[\frac{b-a}{2} + \frac{\alpha+\beta}{2(n+2)}\right]\left[\frac{b'-a'}{2} + \frac{\alpha'+\beta'}{2(n+2)}\right] + \frac{(n+1)(\alpha+\beta)(\alpha'+\beta')}{4(n+2)^2(n+3)}.$$

So,

$$\lim_{n\to\infty} \mathrm{Cov}_f(A, B) = \frac{b-a}{2} \cdot \frac{b'-a'}{2}.$$

If $a = b$ *and* $a' = b'$, *i.e. we have two triangular fuzzy numbers, then their covariance becomes*

$$\mathrm{Cov}_f(A, B) = \frac{(\alpha+\beta)(\alpha'+\beta')}{2(n+2)(n+3)}.$$

The following theorem (Fullér, Majlender 2003) shows that the variance of linear combinations of fuzzy numbers can easily be computed (in a similar manner as in probability theory).

Theorem 1 *Let* f *be a weighting function, let* A *and* B *be fuzzy numbers and let* λ *and* μ *be real numbers. Then the following properties hold,*

$$\mathrm{Cov}_f(\lambda A + \mu B, C) = |\lambda|\mathrm{Cov}_f(A, C) + |\mu|\mathrm{Cov}_f(B, C),$$

and

$$\mathrm{Var}_f(\lambda A + \mu B) = \lambda^2 \mathrm{Var}_f(A) + \mu^2 \mathrm{Var}_f(B) + 2|\lambda||\mu|\mathrm{Cov}_f(A, B),$$

where the operations additions and multiplication by scalar of fuzzy numbers are defined by the sup-min extension principle (Zadeh 1965).

3 Expected Value of Functions on Fuzzy Sets

The main drawback of definition (5) is that $\mathrm{Cov}_f(A, B)$ is always non-negative for any pair of fuzzy numbers, A and B. So even though definition (5) is consistent with the extension principle it may not always be consistent with the principle of 'falling shadows'. In this section we will use the concept of average values of well-chosen real-valued function on γ-level sets of a joint possibility distribution to measure covariance between γ-level sets of its marginal distributions.

Let B be a joint possibility distribution in $\mathbb{R}^n$, let $\gamma \in [0,1]$ and let $g\colon \mathbb{R}^n \to \mathbb{R}$ be a function. It is well-known from analysis that the average value of function g on $[B]^\gamma$ can be computed by

$$\begin{aligned} \mathcal{C}_{[B]^\gamma}(g) &= \frac{1}{\displaystyle\int_{[B]^\gamma} dx} \int_{[B]^\gamma} g(x)dx \\ &= \frac{1}{\displaystyle\int_{[B]^\gamma} dx_1 \dots dx_n} \int_{[B]^\gamma} g(x_1, \dots, x_n) dx_1 \dots dx_n. \end{aligned}$$

Following (Fullér, Majlender 2003a) will call C as the central value operator.

If $g\colon \mathbb{R} \to \mathbb{R}$ is a single-variable function and $A \in \mathcal{F}$ is a fuzzy number then the average value of function g on $[A]^\gamma$ is defined by

$$\mathcal{C}_{[A]^\gamma}(g) = \frac{1}{\displaystyle\int_{[A]^\gamma} dx} \int_{[A]^\gamma} g(x)dx.$$

Especially, if $g(x) = x$, for all $x \in \mathbb{R}$ is the identity function and $A \in \mathcal{F}$ is a fuzzy number with $[A]^\gamma = [a_1(\gamma), a_2(\gamma)]$, then the average value of the identity function on $[A]^\gamma$ is computed by

$$\mathcal{C}_{[A]^\gamma}(\mathrm{id}) = \frac{1}{\displaystyle\int_{[A]^\gamma} dx} \int_{[A]^\gamma} x dx = \frac{a_1(\gamma) + a_2(\gamma)}{2}.$$

Because $C_{[A]^\gamma}(\mathrm{id})$ is nothing else, but the center of $[A]^\gamma$ we will use the shorter notation $C([A]^\gamma)$ for $C_{[A]^\gamma}(\mathrm{id})$.

It is clear that $C_{[B]^\gamma}$ is linear for any arbitrarily fixed joint possibility distribution B and for any $\gamma \in [0,1]$. Let us denote the projection functions on $\mathbb{R}^2$ by π_x and π_y, that is, $\pi_x(u,v) = u$ and $\pi_y(u,v) = v$ for $u, v \in \mathbb{R}$. The following two theorems (Fullér, Majlender 2003a) show two important properties of central value operator.

Theorem 2 *If $A, B \in \mathcal{F}$ are non-interactive and $g = \pi_x + \pi_y$ is the addition operator on $\mathbb{R}^2$ then*

$$C_{[A\times B]^\gamma}(\pi_x + \pi_y) = C_{[A]^\gamma}(\mathrm{id}) + C_{[B]^\gamma}(\mathrm{id}) = C([A]^\gamma) + C([B]^\gamma),$$

for all $\gamma \in [0,1]$.

Theorem 3 *If $A, B \in \mathcal{F}$ are non-interactive and $p = \pi_x\pi_y$ is the multiplication operator on $\mathbb{R}^2$ then*

$$C_{[A\times B]^\gamma}(\pi_x\pi_y) = C_{[A]^\gamma}(\mathrm{id}) \cdot C_{[B]^\gamma}(\mathrm{id}) = C([A]^\gamma) \cdot C([B]^\gamma),$$

for all $\gamma \in [0,1]$.

The following defintion (Fullér, Majlender 2003a) is crucial for the theory of possibilistic dependencies.

Definition 1 *Let C be the joint possibility distribution with marginal possibility distributions $A, B \in \mathcal{F}$, and let $\gamma \in [0,1]$. The measure of interactivity between the γ-level sets of A and B is defined by*

$$\mathcal{R}_{[C]^\gamma}(\pi_x, \pi_y) = C_{[C]^\gamma}\big((\pi_x - C_{[C]^\gamma}(\pi_x))(\pi_y - C_{[C]^\gamma}(\pi_y))\big).$$

Using the definition of central value we have

$$\begin{aligned}\mathcal{R}_{[C]^\gamma}(\pi_x, \pi_y) &= \frac{1}{\displaystyle\int_{[C]^\gamma} dxdy} \int_{[C]^\gamma} (x - C_{[C]^\gamma}(\pi_x))(y - C_{[C]^\gamma}(\pi_y))dxdy \\ &= C_{[C]^\gamma}(\pi_x\pi_y) - C_{[C]^\gamma}(\pi_x) \cdot C_{[C]^\gamma}(\pi_y),\end{aligned}$$

for all $\gamma \in [0,1]$.

Note 3. The interactivity relation computes the average value of the interactivity function

$$g(x,y) = (x - C_{[C]^\gamma}(\pi_x))(y - C_{[C]^\gamma}(\pi_y)),$$

on $[C]^\gamma$.

We can also use the principle of central values to introduce the notation of expected value of functions on fuzzy sets. Let $g\colon \mathbb{R} \to \mathbb{R}$ be a function and $A \in \mathcal{F}$. Let us consider again the average value of function g on $[A]^\gamma$

$$C_{[A]^\gamma}(g) = \frac{1}{\displaystyle\int_{[A]^\gamma} dx} \int_{[A]^\gamma} g(x)dx.$$

In (Fullér, Majlender 2003a) the expected value of function g on A with respect to a weighting function f was defined by

$$E_f(g;A) = \int_0^1 C_{[A]^\gamma}(g)f(\gamma)d\gamma = \int_0^1 \frac{1}{\displaystyle\int_{[A]^\gamma} dx} \int_{[A]^\gamma} g(x)dx f(\gamma)d\gamma.$$

Especially, if $g(x) = x$, for all $x \in \mathbb{R}$ is the identity function then we get

$$E_f(\text{id};A) = E_f(A) = \int_0^1 \frac{a_1(\gamma)+a_2(\gamma)}{2} f(\gamma)d\gamma,$$

which is thc$f$-weighted possibilistic mean value of A introduced in (Fullér, Majlender 2003).

Note 4. The expected value of a function on a fuzzy number A is nothing else but the expected value of its average values on all gamma level sets of A.

The variance of A was defined in (Fullér, Majlender 2003a) as the expected value of function $g(x) = (x - C([A]^\gamma))^2$ on A. That is,

$$\text{Var}_f(A) = E_f(g;A) = \int_0^1 \frac{(a_2(\gamma)-a_1(\gamma))^2}{12} f(\gamma)d\gamma.$$

The covariance between marginal possibility distributions was defined in (Fullér, Majlender 2003a) as the expected value of their interactivity function on their joint possibility distribution. Namely, if C is a joint possibility distribution in $\mathbb{R}^2$ and $A, B \in \mathcal{F}$ denote its marginal possibility distributions then the covariance of A and B with respect to a weighting function f (and with repect to their joint possibility distributioin C) was defined by

$$\mathrm{Cov}_f(A, B) = \int_0^1 \mathcal{R}_{[C]^\gamma}(\pi_x, \pi_y) f(\gamma) d\gamma$$
$$= \int_0^1 \left[\mathcal{C}_{[C]^\gamma}(\pi_x \pi_y) - \mathcal{C}_{[C]^\gamma}(\pi_x) \cdot \mathcal{C}_{[C]^\gamma}(\pi_y)\right] f(\gamma) d\gamma. \quad (6)$$

The next theorem (Fullér, Majlender 2003a) states the bilinearity of the interactivity relation operator.

Theorem 4 *Let C be a joint possibility distribution in $\mathbb{R}^2$, and let $\lambda, \mu \in \mathbb{R}$. Then*

$$\mathcal{R}_{[C]^\gamma}(\lambda\pi_x + \mu\pi_y, \lambda\pi_x + \mu\pi_y) =$$
$$\lambda^2 \mathcal{R}_{[C]^\gamma}(\pi_x, \pi_x) + \mu^2 \mathcal{R}_{[C]^\gamma}(\pi_y, \pi_y) + 2\lambda\mu \mathcal{R}_{[C]^\gamma}(\pi_x, \pi_y).$$

The following theorem (Carlsson et al 2003) shows a very important property of the correlation operator.

Theorem 5 *Let $A, B \in \mathcal{F}$ be fuzzy numbers with joint possibility distribution C. Then, their correlation coefficient defined by*

$$\rho_f(A, B) = \frac{\mathrm{Cov}_f(A, B)}{\sqrt{\mathrm{Var}_f(A)\mathrm{Var}_f(B)}},$$

satisfies the relationship

$$-1 \leq \rho_f(A, B) \leq 1$$

for any weighting function f.

4 Illustrations of Possibilistic Correlation

Let us consider several interesting cases. In (Carlsson et al 2003) we proved that if A and B are non-interactive, that is, their joint possibility distribution is $A \times B$ then $\rho_f(A, B) = 0$.

Example 1 *Consider now the case depicted in Fig. 2. It can be shown (Carlsson et al 2003) that in this case* $\rho_f(A, B) = 1$.

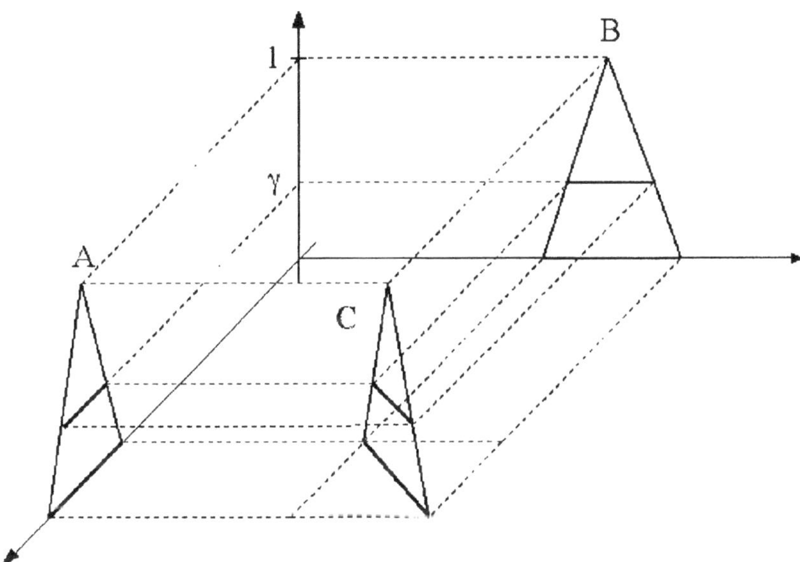

Fig. 2. The case of $\rho_f(A, B) = 1$.

Note 5. In this case $\rho_f(A, B) = 1$ for any weighting function f; and if $A(u) = \gamma$ for some $u \in \mathbb{R}$ then there exists a unique $v \in \mathbb{R}$ that B can take (see Fig. 2), furthermore, if u is moved to the left (right) then the corresponding value (that B can take) will also move to the left (right). Loosely speaking, in this case *the shadows move in tandem* that is, if one moves A to the left (right) then B will also move to the left (right).

Example 2 *Consider now the case depicted in Fig. 3. It can be shown (Carlsson et al 2003) that in this case* $\rho_f(A, B) = -1$.

Note 6. In this case $\rho_f(A, B) = -1$ *for any weighting function* f*; and if* $A(u) = \gamma$ *for some* $u \in \mathbb{R}$ *then there exists a unique* $v \in \mathbb{R}$ *that* B *can take (see Fig. 3), furthermore, if* u *is moved to the left (right) then the corresponding value (that* B *can take) will move to the right (left). Loosely speaking, in this case* the shadows move oppositively *that is, if one moves* A *to the left (right) then* B *will move to the right (left).*

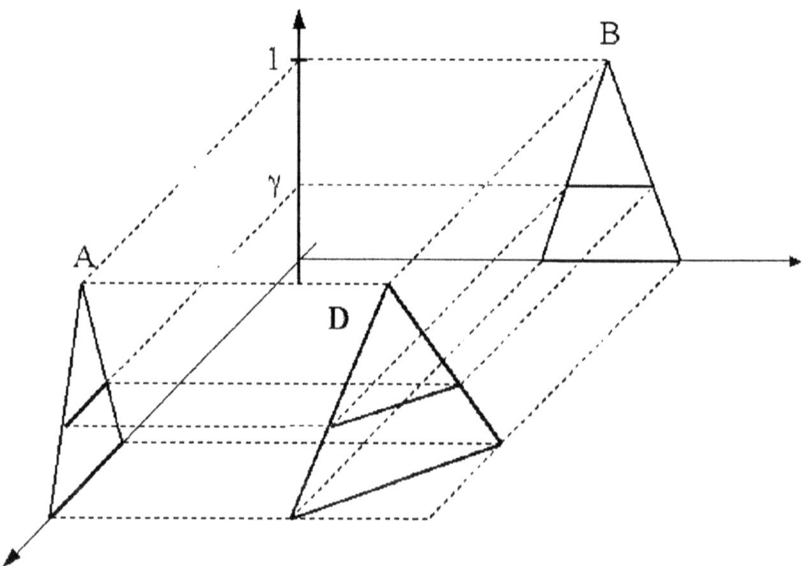

Fig. 3. The case of $\rho_f(A, B) = -1$.

Example 3 *Let* $A(x) = B(x) = x \cdot \chi_{[0,1]}(x)$ *for* $x \in \mathbb{R}$*. Thus, the* γ*-level sets of* A *and* B *are* $[A]^\gamma = [B]^\gamma = [\gamma, 1]$ *for* $\gamma \in [0, 1]$*. Let their joint possibility distribution* I *be defined by*

$$I(x, y) = (x + y - 1) \cdot \chi_T(x, y),$$

where

$$T = \{(x, y) \in \mathbb{R}^2 | x \le 1, y \le 1, x + y \ge 1\}.$$

It is easy to see that,

$$\begin{aligned}[I]^\gamma &= \mathrm{cl}\{(x, y) \in \mathbb{R}^2 | I(x, y) > \gamma\} \\ &= \{(x, y) \in \mathbb{R}^2 | x \le 1, y \le 1, x + y \ge 1 + \gamma\}.\end{aligned}$$

This situation is depicted on Fig. 4, where we have shifted the fuzzy sets

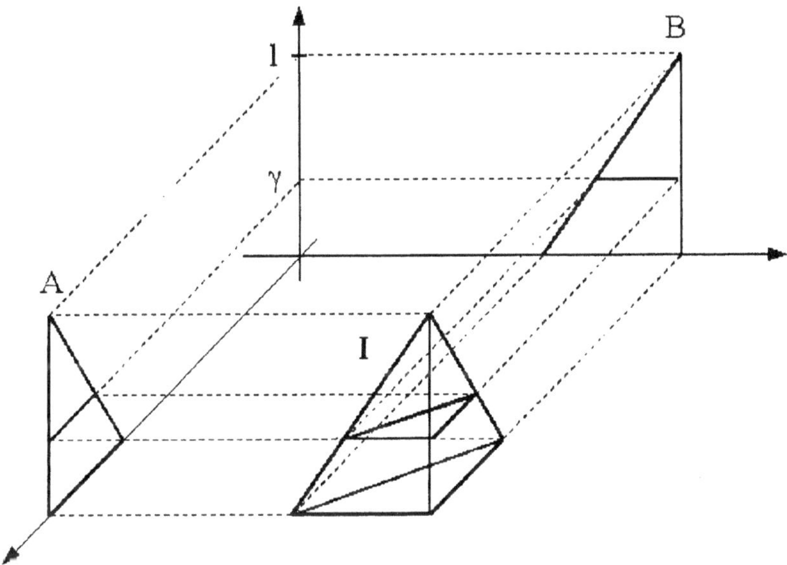

Fig. 4. The case of $\rho_f(A, B) = -1/3$.

to get a better view of the situation.

We shall compute the correlation coefficient between A and B. From the equations,

$$\int_{[I]^\gamma} dxdy = \frac{(1-\gamma)^2}{2}, \int_{[I]^\gamma} xydxdy = \frac{(1-\gamma)^2(1+\gamma)(5+\gamma)}{24},$$

$$\int_{[I]^\gamma} xdxdy = \int_{[I]^\gamma} ydxdy = \frac{1}{2}\int_\gamma^1 (1-x^2)dx = \frac{(1-\gamma)^2(2+\gamma)}{6},$$

and

$$\int_{[A]^\gamma} dx = 1-\gamma, \int_{[A]^\gamma} x^2 dx = \frac{(1-\gamma)(1+\gamma+\gamma^2)}{3},$$
$$\int_{[A]^\gamma} x dx = \frac{(1-\gamma)(1+\gamma)}{2}$$

we get

$$\mathrm{Cov}_f(A,B) = -\frac{1}{36}\int_0^1 (1-\gamma)^2 f(\gamma) d\gamma,$$

and

$$\mathrm{Var}_f(A) = \mathrm{Var}_f(B) = \frac{1}{12}\int_0^1 (1-\gamma)^2 f(\gamma) d\gamma,$$

Therefore,

$$\rho_f(A,B) = -1/3,$$

for any weighting function f.

Example 4 *Let $A(x) = B(1-x) = x \cdot \chi_{[0,1]}(x)$ for $x \in \mathbb{R}$. So, the γ-level set of A and B are computed by $[A]^\gamma = [\gamma, 1]$ and $[B]^\gamma = [0, 1-\gamma]$ for $\gamma \in [0,1]$. Let*

$$H(x,y) = (x-y) \cdot \chi_V(x,y),$$

where

$$V = \{(x,y) \in \mathbb{R}^2 | x \leq 1, y \geq 0, x-y \geq 0\}.$$

This situation is depicted on Fig. 5, where we have shifted the fuzzy sets to get a better view of the situation. After some calculations we get

$$\rho_f(A,B) = 1/3,$$

for any weighting function f.

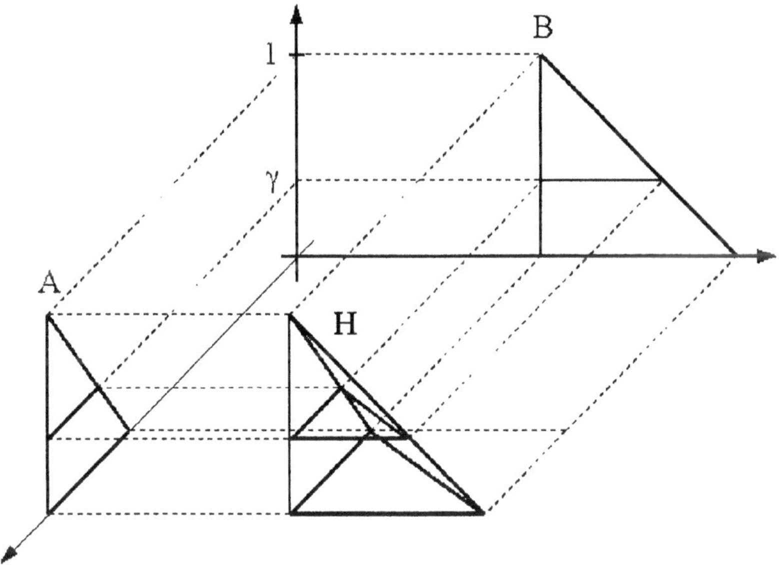

Fig. 5. The case of $\rho_f(A, B) = 1/3$.

5 Interactive Fuzzy Numbers can be Uncorrelated

Zero correlation does not always imply non-interactivity. Let $A, B \in \mathcal{F}$ be fuzzy numbers, let C be their joint possibility distribution, and let $\gamma \in [0, 1]$. Suppose that $[C]^\gamma$ is symmetrical, i.e. there exists $a \in \mathbb{R}$ such that

$$C(x, y) = C(2a - x, y),$$

for all $x, y \in [C]^\gamma$ (hence, line defined by $\{(a, t)|t \in \mathbb{R}\}$ is the axis of symmetry of $[C]^\gamma$). We shall show that in this case the interactivity relation of $[A]^\gamma$ and $[B]^\gamma$ vanishes, i.e. $\mathcal{R}_{[C]^\gamma}(\pi_x, \pi_y) = 0$. Indeed, let

$$H = \{(x, y) \in [C]^\gamma | x \leq a\},$$

then

$$\int_{[C]^\gamma} xydxdy = \int_H (xy + (2a - x)y)dxdy = 2a \int_H ydxdy,$$

$$\int_{[C]^\gamma} x dx dy = \int_H (x + (2a - x)) dx dy = 2a \int_H dx dy,$$

$$\int_{[C]^\gamma} y dx dy = 2 \int_H y dx dy, \quad \int_{[C]^\gamma} dx dy = 2 \int_H dx dy,$$

therefore, we obtain

$$\mathcal{R}_{[C]^\gamma}(\pi_x, \pi_y) = \frac{1}{\displaystyle\int_{[C]^\gamma} dx dy} \int_{[C]^\gamma} xy dx dy$$

$$- \frac{1}{\displaystyle\int_{[C]^\gamma} dx dy} \int_{[C]^\gamma} x dx dy \frac{1}{\displaystyle\int_{[C]^\gamma} dx dy} \int_{[C]^\gamma} y dx dy = 0.$$

So, if all γ-level sets of joint possibility distribution C are symmetrical

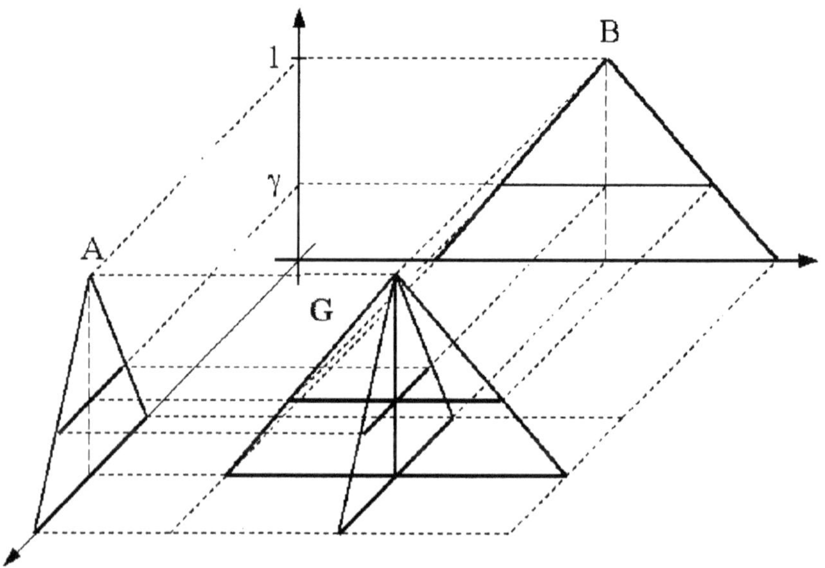

Fig. 6. The case of $\rho_f(A, B) = 0$ for interactive fuzzy numbers.

then

$$\mathrm{Cov}_f(A, B) = 0,$$

for any weighting function f.

6 Explicitly Given Joint Possibility Distributions

In many papers authors consider joint possibility distributions that are derived from given marginal distributions by aggregating their membership values. Namely, let $A, B \in \mathcal{F}$. We will say that their joint possibility distribution C is *directly defined* from its marginal distributions if

$$C(x, y) = T(A(x), B(y)),\ x, y \in \mathbb{R},$$

where $T\colon [0,1] \times [0,1] \to [0,1]$ is a function satisfying the properties

$$\max_y T(A(x), B(y)) = A(x), \forall x \in \mathbb{R}, \tag{7}$$

and

$$\max_x T(A(x), B(y)) = B(y), \forall y \in \mathbb{R}, \tag{8}$$

for example a triangular norm.

Note 7. In this case the joint distribution depends barely on the membership values of its marginal distributions.

We will show that in this case the covariance (and, consequently, the correlation) between its marginal distributions will be zero whenever at least one of its marginal distributions is symmetrical.

Theorem 6 *Let $A, B \in \mathcal{F}$ and let their joint possibility distribution C be defined by*

$$C(x, y) = T(A(x), B(y)),$$

for $x, y \in \mathbb{R}$, where T is a function satisfying conditions (7) and (8). If A is a symmetrical fuzzy number then

$$\mathrm{Cov}_f(A, B) = 0,$$

for any fuzzy number B, aggregator T, and weighting function f.

Proof. If A is a symmetrical fuzzy number with center a such that $A(x) = A(2a - x)$ for all $x \in \mathbb{R}$ then,

$$C(x, y) = T(A(x), B(y)) = T(A(2a - x), B(y)) = C(2a - x, y),$$

that is, C is symmetrical. Hence, considering the results obtained above we have

$$\mathcal{R}_{[C]^\gamma}(\pi_x, \pi_y) = 0,$$

and, therefore,

$$\mathrm{Cov}_f(A, B) = 0,$$

for any weighting function f. Which ends the proof.

Summary

We have illustrated some important feautures of possibilistic mean value, covariance, variance and correlation by several examples. We have shown that zero correlation does not always imply non-interactivity. We have also shown the limitations of direct definitions of joint possibility distributions from individual fuzzy numbers, for example, when one simply aggregates the membership values of two fuzzy numbers by a triangular norm

References

Carlsson C, Fullér R (2001) On possibilistic mean value and variance of fuzzy numbers, *Fuzzy Sets and Systems*, 122(2001) 315-326.

Carlsson C, Fullér R, Majlender P (2003) On possibilistic Cauchy-Schwarz inequality *Fuzzy Sets and Systems*, (submitted).

Dubois D, Prade H (1987) The mean value of a fuzzy number, *Fuzzy Sets and Systems*, 24(1987) 279-300.

Dubois D, Prade H (1988) *Possibility Theory: An Approach to Computerized Processing of Uncertainty*, Plenum Press, New York, 1988.

Fullér R, Majlender P (2003) On weighted possibilistic mean and variance of fuzzy numbers, *Fuzzy Sets and Systems*, (to appear).

Fullér R, Majlender P (2003a) On interactive fuzzy numbers, *Fuzzy Sets and Systems*, (submitted).
Campos LM, Huete JF (1999) Independence concepts in possibility theory: Part I and II, *Fuzzy Sets and Systems*, 103(1999) 127-152, 487-505.
Zadeh LA (1965) Fuzzy Sets, *Information and Control*, 8 (1965) 338-353.
Zadeh LA (1978) Fuzzy sets as a basis for a theory of possibility, *Fuzzy Sets and Systems*, 1(1978) 3-28.

FREs: the ODEs and PDEs of the Fuzzy Modelling Paradigm

Bernard De Baets

Department of Applied Mathematics, Biometrics and Process Control
Ghent University, Coupure links 653, B-9000 Gent, Belgium
E-mail: Bernard.DeBaets@rug.ac.be

Abstract

Fuzzy relational equations are without doubt the most important inverse problems arising from fuzzy set theory, and in particular from fuzzy relational calculus. Indeed, the calculus of fuzzy relations is a powerful one, with applications in fuzzy control and fuzzy systems modelling in general, approximate reasoning, relational databases, clustering, etc. In this paper, fuzzy relational equations are approached from an order-theoretical point of view. It is shown how all inverse problems can be reduced to systems of polynomial lattice equations. The exposition is limited to the description of exact solutions of systems of sup-$\mathcal{T}$ equations, and analytical ways are presented for obtaining the complete solution set when working in a broad and interesting class of distributive lattices.

1 Introduction

Mathematical relations, just as human relations, are invaluable to any scientific researcher. As Goguen (1967) writes, "Science is, in a sense, the discovery of relations between observables. Zadeh has shown the study of relations to be equivalent to the general study of systems (a system is a relation between an "input" and an "output" space)." However, the typical behaviour of a mathematician to describe relationships in a black-or-white manner is not suitable for every-day problems. Fuzzy relations, however, allow for a description of gradual relationships. Depending on the context, fuzzy relations can be interpreted in two ways: a disjunctive way, or a conjunctive way, see e.g. De Baets (1996) and Dubois and Prade (1992).

In the *disjunctive interpretation*, a fuzzy relation is considered as an elastic restriction on the more or less possible values of a couple of variables, each of which takes exactly one value in reality. In this interpretation, fuzzy relations are extensively used for modelling systems of which the input-output behaviour is known only through linguistic descriptions. The basic operations on fuzzy relations are the *direct image* and the *composition*. The generalized modus ponens and the generalized modus tollens, the key operations behind forward and backward chaining inference mechanisms in fuzzy rule-based systems, perfectly fit within this disjunctive interpretation and are immediate applications of the basic operations. The design and deployment of such systems require the solution of inverse problems connected with the basic operations: from known input-output behaviour a system model is built through the solution of fuzzy relational equations (FREs).

The notion of *mutually exclusive values* is no longer present in the conjunctive interpretation. In a diagnostic problem, for instance, a single cause may lead to multiple defects. In a decision problem, alternatives are evaluated on a number of criteria, of which, of course, more than one may be satisfied. Although most attention has been directed towards the conjunctive interpretation, this disjunctive interpretation also has an incredible potential. The basic operations from the conjunctive interpretation are still applicable in the disjunctive interpretation. However, Bandler and Kohout (1980a, 1980b) realized that in the conjunctive interpretation additional operations play an important role. The most important such operations are their *triangular compositions.* The contributions of Bandler and Kohout are, without doubt, among the most fundamental in the study of fuzzy relations. However, the solution of inverse problems related to these 'new' operations has received only little attention.

In this paper, we present a uniform framework for solving the inverse problems arising in both interpretations. Remarkably, particular solutions in the disjunctive interpretation, for instance, can be described by means of operations from the conjunctive one. This demonstrates that it is really opportune to treat these problems in one common framework. However, due to space limitations, only the inverse problems arising in the disjunctive interpretation will be discussed. This paper is organized as follows. In Section 2, we briefly introduce the reader to the basic operations of fuzzy relational calculus, followed by an identification of the corresponding inverse problems in Section 3. The most important fuzzy relational equations are related to systems of sup-$\mathcal{T}$ equations. Their solution procedures are explained in Sections 4 and 5.

2 Images and Compositions

2.1 Relational Calculus and Boolean Equations

In essence, there exist two basic types of operations in relational calculus: *images* and *compositions.* Consider a relation R from X to Y. The *afterset* xR of $x \in X$ is the subset of Y defined as $xR = \{y \in Y \mid (x, y) \in R\}$. The *foreset* Ry of $y \in Y$ is the subset of X defined as $Ry = \{x \in X \mid (x, y) \in R\}$. Given a subset A of X, the *direct image* $R(A)$, *subdirect image* $R_{\triangleleft}(A)$ and *superdirect image* $R_{\triangleright}(A)$ of A under R are the subsets of Y defined as:

$$\begin{aligned} R(A) &= \{y \in Y \mid (\exists x \in X)(x \in A \wedge (x, y) \in R)\} \\ &= \{y \in Y \mid A \cap Ry \neq \emptyset\} \end{aligned}$$

$$R_{\triangleleft}(A) = \{y \in Y \mid (\forall x \in X)(x \in A \Rightarrow (x,y) \in R)\}$$
$$= \{y \in Y \mid A \subseteq Ry\}$$

$$R_{\triangleright}(A) = \{y \in Y \mid (\forall x \in X)((x,y) \in R \Rightarrow x \in A)\}$$
$$= \{y \in Y \mid Ry \subseteq A\}.$$

Similarly, given a second relation S from Y to Z, the *(round) composition* $R \circ S$, *subcomposition* $R \triangleleft S$ and *supercomposition* $R \triangleright S$ of R and S are the relations from X to Z defined as:

$$R \circ S = \{(x,z) \in X \times Z \mid (\exists y \in Y)((x,y) \in R \wedge (y,z) \in S)\}$$
$$= \{(x,z) \in X \times Z \mid xR \cap Sz \neq \emptyset\}$$

$$R \triangleleft S = \{(x,z) \in X \times Z \mid (\forall y \in Y)((x,y) \in R \Rightarrow (y,z) \in S)\}$$
$$= \{(x,z) \in X \times Z \mid xR \subseteq Sz\}$$

$$R \triangleright S = \{(x,z) \in X \times Z \mid (\forall y \in Y)((y,z) \in S \Rightarrow (x,y) \in R)\}$$
$$= \{(x,z) \in X \times Z \mid Sz \subseteq xR\}.$$

Strictly speaking, for the sub- and superdirect images, and the sub- and supercompositions, one should also take into account some non-emptiness conditions, see De Baets and Kerre (1993). In the context of relational equations, however, the above formulations are more suitable. Sub- and supercompositions of relations have been studied extensively by Bandler and Kohout (1980a, 1980b) under the name *triangle relational products.*

It is now not difficult to imagine what types of inverse problems one can consider:

(i) *image equations*, i.e. equations of the type $R(A) = B$, $R_{\triangleleft}(A) = B$ or $R_{\triangleright}(A) = B$ in the unknown set A or in the unknown relation R;
(ii) *composition equations*, i.e. equations of the type $R \circ S = T$, $R \triangleleft S = T$ or $R \triangleright S = T$ in the unknown relation R or in the unknown relation S.

Identifying sets and relations with their characteristic mapping, these inverse problems, called relational equations, can be translated into Boolean equations (see e.g. Rudeanu (1974)). Note that the characteristic mapping χ_A of a crisp subset A of X is defined by

$$\chi_A(x) = \begin{cases} 1\ , & \text{if } x \in A \\ 0\ , & \text{elsewhere} \end{cases} .$$

2.2 Fuzzy Relational Calculus

The above images and compositions are extended to fuzzy relations in a by now traditional manner: supremum and infimum are playing the role of the existential and universal quantifiers and a conjunctor and implicator are modelling pointwise intersection and inclusion. The most natural mathematical framework for dealing with these operations is without doubt the lattice-theoretic one (see e.g. Birkhoff (1967), Davey and Priestley (1990)). Hence, we will work with so-called

L-fuzzy relations, not to artificially increase the level of mathematical abstractness, but because lattice theory provides the proper framework for formulating and solving fuzzy relational equations.

Consider a complete lattice $(L, \leq)$ with smallest element 0 and greatest element 1. Denote the meet and join operation by $\frown$ and $\smile$. An L-fuzzy set A in a universe X is simply an $X \to L$ mapping; an L-fuzzy relation R from X to Y is an $X \times Y \to L$ mapping, see Goguen (1967). From here on, we will omit the prefix L if no confusion can occur. The binary Boolean logical operations $\wedge$ and $\to$, for instance, can be extended in the following obvious way:

(i) a *conjunctor* $\mathcal{C}$ is a binary operation on L with order-preserving partial mappings such that $\mathcal{C}(0,1) = \mathcal{C}(1,0) = 0$ and $\mathcal{C}(1,1) = 1$;
(ii) an *implicator* $\mathcal{I}$ is a binary operation on L with order-reversing first partial mappings and order-preserving second partial mappings such that $\mathcal{I}(1,0) = 0$ and $\mathcal{I}(0,0) = \mathcal{I}(1,1) = 1$.

Obviously, additional properties can be imposed on these operators. For instance, if a conjunctor is also required to be commutative, associative and to have 1 as neutral element, then we are in the setting of t-norms, see De Cooman and Kerre (1994) and Schweizer and Sklar (1983). For the sake of simplicity and in view of their familiarity to the reader, we will mainly restrict to t-norms from here on. Most results, however, can be stated much more generally. The three most important continuous t-norms on the real unit interval are the minimum operator M, the (algebraic) product P and the Łukasiewicz t-norm W defined by $W(x,y) = \max(x+y-1,0)$.

We are now ready to introduce the basic operations of fuzzy relational calculus. Consider a fuzzy relation R from X to Y. The *afterset* xR of $x \in X$ is the fuzzy set in Y defined by $xR(y) = R(x,y)$. The *foreset* Ry of $y \in Y$ is the fuzzy set in X defined by $Ry(x) = R(x,y)$. Let $\mathcal{C}$ be a conjunctor and $\mathcal{I}$ be an implicator. Given a fuzzy set A in X, the *direct image* $R(A)$, *subdirect image* $R_{\lhd}(A)$ and *superdirect image* $R_{\rhd}(A)$ of A under R are the fuzzy sets in Y defined by:

$$
\begin{aligned}
R(A)(y) &= \sup_{x \in X} \mathcal{C}(A(x), R(x,y)) \\
R_{\lhd}(A)(y) &= \inf_{x \in X} \mathcal{I}(A(x), R(x,y)) \\
R_{\rhd}(A)(y) &= \inf_{x \in X} \mathcal{I}(R(x,y), A(x)).
\end{aligned}
$$

Given a second fuzzy relation S from Y to Z, the *(round) composition* $R \circ S$, *subcomposition* $R \lhd S$ and *supercomposition* $R \rhd S$ of R and S are the fuzzy relations from X to Z defined by:

$$
\begin{aligned}
R \circ S(x,z) &= \sup_{y \in Y} \mathcal{C}(R(x,y), S(y,z)) \\
R \lhd S(x,z) &= \inf_{y \in Y} \mathcal{I}(R(x,y), S(y,z)) \\
R \rhd S(x,z) &= \inf_{y \in Y} \mathcal{I}(S(y,z), R(x,y)).
\end{aligned}
$$

For an in-depth study of these operations and their applications, we refer to De Baets and Kerre (1995).

3 Types of Inverse Problems

As in the Boolean case, we can consider the following types of inverse problems:

(i) *image equations*, i.e. equations of the type $R(A) = B$, $R_{\triangleleft}(A) = B$ or $R_{\triangleright}(A) = B$ in the unknown fuzzy set A or in the unknown fuzzy relation R;
(ii) *composition equations*, i.e. equations of the type $R \circ S = T$, $R \triangleleft S = T$ or $R \triangleright S = T$ in the unknown fuzzy relation R or in the unknown fuzzy relation S.

These inverse problems are what one usually calls *fuzzy relation(al) equations*. They can all be formulated as systems of particular lattice equations. Consider a binary operator $\mathcal{O}$ on L with monotone partial mappings, then we can distinguish the following four basic types of equations, given a family $(a_i)_{i\in I}$ in L and b in L:

- left sup-$\mathcal{O}$ equation: $\sup\limits_{i\in I} \mathcal{O}(x_i, a_i) = b$;
- right sup-$\mathcal{O}$ equation: $\sup\limits_{i\in I} \mathcal{O}(a_i, x_i) = b$;
- left inf-$\mathcal{O}$ equation: $\inf\limits_{i\in I} \mathcal{O}(x_i, a_i) = b$;
- right inf-$\mathcal{O}$ equation: $\inf\limits_{i\in I} \mathcal{O}(a_i, x_i) = b$.

These equations can be considered as *polynomial lattice equations*. Finite index sets will be denoted I_n, containing the first n integers. A first observation is that, due to the monotonicity of the operator $\mathcal{O}$, the solution set of any of the above equations is *order convex*, i.e. if $(x_i)_{i\in I}$ and $(z_i)_{i\in I}$ are two solutions of one of the above equations, then so is $(y_i)_{i\in I}$, whenever $x_i \leq y_i \leq z_i$, for any $i \in I$. This allows us to focus on extremal solutions, i.e. maximal (greatest) and minimal (smallest) solutions. In case of a commutative operator $\mathcal{O}$, there is, of course, no need to talk about left or right equations.

Let us consider the image equations $R(A) = B$ and composition equations $R \circ S = T$:

- the equation $R(A) = B$ in the unknown fuzzy relation R, called image equation of *type 1*, is equivalent to a family of independent sup-$\mathcal{C}$ equations in the foresets Ry, $y \in Y$:

$$\sup_{x\in X} \mathcal{C}(A(x), Ry(x)) = B(y);$$

- the equation $R(A) = B$ in the unknown fuzzy set A, called image equation of *type 2*, is equivalent to a system of sup-$\mathcal{C}$ equations, $y \in Y$:

$$\sup_{x\in X} \mathcal{C}(A(x), R(x, y)) = B(y);$$

- the composition equation $R \circ S = T$ in the unknown fuzzy relation R or S can be reformulated as a system of image equations of type 1 or as a family of independent image equations of type 2.

Of course, the same observations hold for the other images and compositions.

In this paper we will only deal with sup-$\mathcal{T}$ equations, with $\mathcal{T}$ a t-norm, and the above corresponding fuzzy relational equations. Note that the associativity of the t-norm $\mathcal{T}$ plays no role, while the commutativity is only imposed for reducing the number of residual operators associated with it (see Section 4). It would therefore be sufficient to consider a t-seminorm, see De Cooman and Kerre (1994). Our treatment of fuzzy relational equations is built up in the following way: we will gradually impose conditions on the t-norm involved and restrict the class of lattices considered, in order to go from a description of the greatest solution, and necessary and sufficient solvability conditions, to a full description of the solution set. Conditions imposed on the t-norm are typically continuity conditions, or in the lattice-theoretic framework, *morphism* conditions. Recall that an $L \to L$ mapping is called a *sup-morphism (inf-morphism)* if for any non-empty subset A of L it holds that $f(\sup A) = \sup f(A)$ $(f(\inf A) = \inf f(A))$. A mapping is called a *homomorphism* if it is both a sup-morphism and an inf-morphism. For instance, in the more familiar case of the unit interval, sup-morphims, resp. inf-morphisms, are nothing else but increasing left-continuous, resp. right-continuous, mappings.

4 Sup-$\mathcal{T}$ Equations

4.1 The Equation $\mathcal{T}(a, x) = b$

The most basic equation we consider is the equation $\mathcal{T}(a, x) = b$ in the unknown x, with $\mathcal{T}$ a t-norm on a complete lattice $(L, \leq)$. In order to describe the solution set of this equation, we associate to a given t-norm $\mathcal{T}$ two binary operators $\mathcal{I}_{\mathcal{T}}$ and $\mathcal{L}_{\mathcal{T}}$ on L, called *residual operators*, defined by

$$\mathcal{I}_{\mathcal{T}}(x, y) = \sup\{z \in L \mid \mathcal{T}(x, z) \leq y\}$$

$$\mathcal{L}_{\mathcal{T}}(x, y) = \inf\{z \in L \mid \mathcal{T}(x, z) \geq y\}.$$

The first operator is called the *residual implicator* of $\mathcal{T}$ and satisfies $(\forall(x, y) \in L^2)$ $(x \leq y \Rightarrow \mathcal{I}_{\mathcal{T}}(x, y) = 1)$, the second one has no particular interpretation.

Since the solution set of the equation $\mathcal{T}(a, x) = b$ can be seen as the intersection of the solution sets of the inequalities $\mathcal{T}(a, x) \leq b$ and $\mathcal{T}(a, x) \geq b$, we start by investigating the latter.

On a complete lattice. The following observations hold true:

- If the partial mapping $\mathcal{T}(a, \cdot)$ is a sup-morphism, then the solution set of the inequality $\mathcal{T}(a, x) \leq b$ is given by $[0, \mathcal{I}_{\mathcal{T}}(a, b)]$.

- If the partial mapping $\mathcal{T}(a,\cdot)$ is an inf-morphism, then the solution set of the inequality $\mathcal{T}(a,x) \geq b$ is given by

$$\begin{cases} [\mathcal{L}_\mathcal{T}(a,b),1] & \text{, if } b \leq a \\ \emptyset & \text{, elsewhere} \end{cases}.$$

- If the partial mapping $\mathcal{T}(a,\cdot)$ is a homomorphism, then the solution set of the equation $\mathcal{T}(a,x) = b$ is given by

$$\begin{cases} [\mathcal{L}_\mathcal{T}(a,b),\mathcal{I}_\mathcal{T}(a,b)] & \text{, if } b \in \operatorname{rng}(\mathcal{T}(a,\cdot)) \\ \emptyset & \text{, elsewhere} \end{cases};$$

  if the partial mapping $\mathcal{T}(a,\cdot)$ is a maximally surjective homomorphism (i.e. if $\operatorname{rng}(\mathcal{T}(a,\cdot)) = [0,a]$), then the solution set of the equation $\mathcal{T}(a,x) = b$ is given by

$$\begin{cases} [\mathcal{L}_\mathcal{T}(a,b),\mathcal{I}_\mathcal{T}(a,b)] & \text{, if } b \leq a \\ \emptyset & \text{, elsewhere} \end{cases}.$$

On the real unit interval. A t-norm $\mathcal{T}$ on $[0,1]$ is continuous if and only if all of its partial mappings are homomorphisms. Moreover, the partial mappings of a continuous t-norm $\mathcal{T}$ on $[0,1]$ are maximally surjective. Hence, if $\mathcal{T}(a,\cdot)$ is continuous, then the solution set of the equation $\mathcal{T}(a,x) = b$ is given by

$$\begin{cases} [\mathcal{L}_\mathcal{T}(a,b),\mathcal{I}_\mathcal{T}(a,b)] & \text{, if } b \leq a \\ \emptyset & \text{, elsewhere} \end{cases}.$$

Therefore, a necessary and sufficient solvability condition is given by $b \leq a$.

4.2 Greatest Solution – Solvability Conditions

Let $(a_i)_{i\in I}$ be an arbitrary family in L and $b \in L$, then we want to determine the solution set of the equation

$$\sup_{i\in I} \mathcal{T}(a_i, x_i) = b$$

in the family of unknowns $(x_i)_{i\in I}$ in L. The families $(a_i)_{i\in I}$ and $(x_i)_{i\in I}$ can be seen as L-fuzzy sets in the index set I. The problem can therefore be reformulated as follows. Given $A \in \mathcal{F}_L(I)$ (with $\mathcal{F}_L(I)$ the set of L-fuzzy sets in I) and $b \in L$, determine the solution set of the equation

$$\sup_{i\in I} \mathcal{T}(A(i), X(i)) = b \tag{1}$$

in the unknown L-fuzzy set X in I.

Recall that $(\mathcal{F}_L(I), \subseteq)$ is a complete lattice, with order relation $\subseteq$ (inclusion) defined by

$$A \subseteq B \Leftrightarrow (\forall i \in I)(A(i) \leq B(i)),$$

and as infimum and supremum the intersection and union defined by

$$\left(\bigcap_{j\in J} A_j\right)(i) = \inf_{j\in J} A_j(i) \quad \text{and} \quad \left(\bigcup_{j\in J} A_j\right)(i) = \sup_{j\in J} A_j(i).$$

The solution set of equation (1) then is a subset of the complete lattice $(\mathcal{F}_L(I), \subseteq)$ and can be seen as the intersection of the solution sets of the inequalities

$$\sup_{i\in I} \mathcal{T}(A(i), X(i)) \leq b \tag{2}$$

and

$$\sup_{i\in I} \mathcal{T}(A(i), X(i)) \geq b. \tag{3}$$

Obviously, due to monotonicity considerations, inequality (2) always has a solution, namely $\chi_\emptyset$, while inequality (3) has a solution if and only if χ_I is a solution, i.e. if and only if

$$b \leq \sup_{i\in I} A(i). \tag{4}$$

In general, inequality (3) is a lot harder to solve than inequality (2). For the moment, we are only concerned with the greatest solution of equation (1).

On a complete lattice. The following statements can be verified:

- If the partial mappings of $\mathcal{T}$ are sup-morphisms, then the solution set of inequality (2) is given by $[\chi_\emptyset, G]$ with G defined by

$$G(i) = \mathcal{I}_\mathcal{T}(A(i), b). \tag{5}$$

  In particular, this result shows us the solution set for equation (1) in case $b = 0$.
- If the partial mappings of $\mathcal{T}$ are sup-morphisms, then the solution set of equation (1) is not empty if and only if the fuzzy set G defined by (5) is a solution. Moreover, if G is a solution, then it is the greatest solution.

Necessary and sufficient solvability conditions. Notice that, in general, no simple necessary and sufficient solvability condition for equation (1) can be given: one has to construct the potential greatest solution and verify that it is indeed a solution. In some cases, however, such a simpler condition can be formulated:

- In any case, condition (4) is a necessary condition for the existence of a solution to equation (1). If L is a complete chain and the partial mappings of $\mathcal{T}$ are maximally surjective homomorphisms, then condition (4) becomes also a sufficient condition. Moreover, in case $\mathcal{T}$ is the meet operation, then L need not be totally ordered (in this case, we are working in a complete Brouwerian lattice), see Zhao (1987).

- If the partial mappings of $\mathcal{T}$ are maximally surjective homomorphisms, then the condition
$$(\exists k \in I)(b \leq A(k)), \tag{6}$$
a stronger version of condition (4), is a sufficient condition for solvability.

On the real unit interval. For a left-continuous t-norm $\mathcal{T}$ on $[0,1]$, the solution set of equation (1) is not empty if and only if the fuzzy set G defined by (5) is the greatest solution. Moreover, if $\mathcal{T}$ is continuous, then condition (4) is a necessary and sufficient solvability condition.

4.3 Complete Solution Set

For most practical considerations, it is sufficient to know the greatest solution of equation (1). From a mathematical point of view, one is, of course, highly interested in the complete solution set. Before we can discuss this solution set, we need to develop an appropriate representation for it. Obviously, it would be desirable to find a representation that is closed under (arbitrary) intersections, since that would learn us how to solve systems of equations at the same time. Fortunately, such a representation exists.

Root systems. A subset R of an ordered set $(P, \leq)$ is called a *root system* if there exists an element σ in P and an antichain O in $\downarrow \sigma = \{x \in P \mid x \leq \sigma\}$ such that
$$R = \bigcup_{\omega \in O} [\omega, \sigma].$$
For a root system R, the corresponding element σ and antichain O are unique. The element σ is called the *stem* and the elements of the antichain O are called the *offshoots* of the root system. A root system is called *finitely generated* if the set of offshoots is finite. The stem is the greatest element and the offshoots are the minimal elements of the root system. Hence, a root system clearly is an order convex subset that is completely determined by its greatest element and its minimal elements.

The following results are of extreme importance in the present discussion, see De Baets (1995b):

- Let $(R_i)_{i \in I}$ be a family of finitely generated root systems of a complete lattice $(L, \leq)$ with stem σ_i and set of offshoots O_i. If the intersection $\bigcap_{i \in I} R_i$ is not empty, then it is a root system with stem $\sigma = \inf_{i \in I} \sigma_i$ and as offshoots the minimal elements of the set
$$\{\sup_{i \in I} \omega_i \mid (\forall i \in I)(\omega_i \in O_i \wedge \omega_i \leq \sigma)\}. \tag{7}$$
- If the intersection of a finite family of finitely generated root systems of a complete lattice is not empty, then it is a finitely generated root system. These offshoots can be determined in an exhaustive way, or following an iterative algorithm, see De Baets (1995b). We will call this method the pen-and-paper method.

If the root systems are not finitely generated, then, in general, nothing can be said about their intersection.

On a complete lattice. So far, we know the greatest solution of equation (1), if there exists a solution at all. Under very general conditions, we can always find a root system of $(\mathcal{F}_L(I), \subseteq)$ that is contained in the solution set, see De Baets (1995b). Indeed, if the partial mappings of $\mathcal{T}$ are maximally surjective homomorphisms and condition (6) holds, then the solution set of equation (1) contains the root system with stem G defined by (5) and as set of offshoots the set $\{M_k \mid b \leq A(k)\}$, where M_k is defined by

$$M_k(i) = \begin{cases} \mathcal{L}_\mathcal{T}(A(k), b) & \text{, if } i = k \\ 0 & \text{, elsewhere} \end{cases} . \tag{8}$$

Moreover, these offshoots are minimal solutions. If the index set I is finite, then this root system is, of course, finitely generated. This result shows how particular minimal solutions, that could be considered as Dirac-solutions or point solutions, can be constructed. In general, it is not known how other minimal solutions can be determined. As pointed out further on, in some cases, the above root system coincides with the solution set. For instance, under the same conditions on $\mathcal{T}$ and condition (6), but for a complete chain L and a finite index set I, the solution set coincides with the above root system, which is then finitely generated.

On a distributive, complete lattice with join-irreducible and join-decomposable elements. So far, the above discussion reveals only the complete solution set in case of a complete chain and under some additional conditions. However, there is an interesting class of lattices, in particular from a practical point of view, in which the complete solution set can be determined. Let us first recall some order-theoretic notions, see Birkhoff (1967) and Davey and Priestley (1990):

(i) A lattice $(L, \leq)$ is called *distributive* if the following property holds:

$$(\forall (a, b, c) \in L^3)(a \frown (b \smile c) = (a \frown b) \smile (a \frown c)).$$

For instance, any chain is distributive and also the Cartesian product of distributive lattices is distributive.

(ii) An element a of a lattice $(L, \leq)$ is called *join-irreducible* if

$$(\forall (b, c) \in L^2)(b \smile c = a \Rightarrow (b = a \vee c = a)),$$

and *join-decomposable* if there exists a set A of join-irreducible elements of L, with $\#A \geq 2$, such that $a = \sup A$.

For instance, all elements of a chain are join-irreducible, and all elements of a finite distributive lattice or a Cartesian product of a finite number of chains are join-irreducible or join-decomposable.

In order to describe the complete solution set of equation (1), we now direct our attention towards inequality (3), inequality (2) being solved already on a complete lattice. The following results hold, for a distributive, complete lattice $(L, \leq)$ and a finite index set I, see De Baets (1995a):

- **The case of a join-irreducible right-hand side.** If b is join-irreducible, the partial mappings of $\mathcal{T}$ are inf-morphisms and the solution set of inequality (3) is not empty (see condition (4)), then it is a finitely generated root system with stem χ_I and as set of offshoots the set $\{M_k \mid b \leq A(k)\}$, where M_k is defined by (8).
- **The case of a join-decomposable right-hand side.** If b has a join-decomposition $b = \sup_{j \in J} b_j$, the partial mappings of $\mathcal{T}$ are inf-morphisms and the solution set of inequality (3) is not empty (see condition (4)), then it is a root system with stem χ_I and as offshoots the *minimal* elements of the set

$$\{\bigcup_{j \in J} M_k^j \mid b_j \leq A(k)\},$$

where M_k^j is defined by

$$M_k^j(i) = \begin{cases} \mathcal{L}_\mathcal{T}(A(k), b_j) & , \text{if } i = k \\ 0 & , \text{elsewhere} \end{cases} . \tag{9}$$

If the join-decomposition is finite, then this root system is also finitely generated.

Combining this information with our knowledge of the solution set of inequality (2) leads to a full description of the solution set of equation (1), again for a distributive, complete lattice and a finite index set I, see De Baets (1995a):

- **The case of a join-irreducible right-hand side.** If b is join-irreducible, the partial mappings of $\mathcal{T}$ are homomorphisms and the solution set of equation (1) is not empty, then it is a finitely generated root system with stem G defined by (5) and as set of offshoots the set $\{M_k \mid b \leq A(k) \wedge M_k \subseteq G\}$, where M_k is defined by (8); moreover, if the partial mappings of $\mathcal{T}$ are maximally surjective, then the set of offshoots can be written as $\{M_k \mid b \leq A(k)\}$. In the case of a complete chain, the latter result was already mentioned above.
- **The case of a join-decomposable right-hand side.** If b has a join-decomposition $b = \sup_{j \in J} b_j$, the partial mappings of $\mathcal{T}$ are homomorphisms and the solution set of equation (1) is not empty, then it is a root system with stem G defined by (5) and as offshoots the *minimal* elements of the set

$$\{\bigcup_{j \in J} M_k^j \mid b_j \leq A(k) \wedge M_k^j \subseteq G\},$$

where M_k^j is defined by (9). If the join-decomposition is finite, then this root system is also finitely generated, and if $\mathcal{L}_\mathcal{T} \leq \mathcal{I}_\mathcal{T}$, then the offshoots are the *minimal* elements of the set

$$\{\bigcup_{j \in J} M_k^j \mid b_j \leq A(k)\}.$$

Note that if L is a complete chain and the partial mappings of $\mathcal{T}$ are maximally surjective, then the inequality $\mathcal{L}_\mathcal{T} \leq \mathcal{I}_\mathcal{T}$ always holds.

These results have been obtained by Zhao (1987) in the case of a complete Brouwerian lattice, i.e. when choosing as t-norm the meet operation, of which all elements have a finite irredundant decomposition in join-irreducible elements. They have been generalized and rephrased in the terminology of root systems by De Baets (1995a).

On the real unit interval. The real unit interval is a distributive, complete lattice of which all elements are join-irreducible. Hence, the above results can be applied. For a continuous t-norm $\mathcal{T}$ and a finite index set I, the solution set of equation (1) is not empty if and only if condition (6) holds. This also follows from the previous subsection, since for finite I conditions (4) and (6) are equivalent. If the solution set is not empty, then it is the finitely generated root system with stem G defined by (5) and as set of offshoots the set $\{M_k \mid b \leq A(k)\}$, where M_k is defined by (8).

On the real unit hypercube. A very attractive lattice satisfying the above requirements is the unit hypercube $([0,1]^m, \leq)$. From an implementational point of view, the real unit hypercube is the ultimate ordinal and numerical working environment; it allows for incomparability and offers at the same time the possibility to fall back on the underlying real unit interval. Moreover, it is a distributive, complete lattice and all elements of it are either join-irreducible or have a finite join-decomposition. Consider $a \in [0,1]^m$, $a \neq (0,\ldots,0) = 0_m$, then a can be decomposed as

$$a = \sup_{a(i) \neq 0} \underline{a}_i$$

with $\underline{a}_i = (0,\ldots,a(i),\ldots,0)$.

From the above discussion, it is clear that we should consider a t-norm $\mathcal{T}$ on $[0,1]^m$ with partial mappings that are homomorphisms. However, this implies that the t-norm $\mathcal{T}$ is the direct product of m t-norms $\mathcal{T}_i$, $i = 1,\ldots,m$, on $[0,1]$, i.e. for any x and y in $[0,1]^m$ it holds that

$$\mathcal{T}(x,y) = (\mathcal{T}_1(x(1),y(1)),\ldots,\mathcal{T}_m(x(m),y(m))).$$

The latter follows even under weaker conditions, see De Baets and Mesiar (1999). For the sake of simplicity, let us consider a t-norm $\mathcal{T}$ on $[0,1]$ and the t-norm $\mathcal{T}^{(m)}$ on $[0,1]^m$ defined by:

$$\mathcal{T}^{(m)}(x,y) = (\mathcal{T}(x(1),y(1)),\ldots,\mathcal{T}(x(m),y(m))).$$

The corresponding operators $\mathcal{I}_{\mathcal{T}^{(m)}}$ and $\mathcal{L}_{\mathcal{T}^{(m)}}$ are then given by, with $1_m = (1,\ldots,1)$, see De Baets (1995a):

$$\mathcal{I}_{\mathcal{T}^{(m)}}(x,y) = (\mathcal{I}_{\mathcal{T}}(x(1),y(1)),\ldots,\mathcal{I}_{\mathcal{T}}(x(m),y(m)))$$

and

$$\mathcal{L}_{\mathcal{T}^{(m)}}(x,y) = (\mathcal{L}_{\mathcal{T}}(x(1),y(1)),\ldots,\mathcal{L}_{\mathcal{T}}(x(m),y(m)))$$

if $(\forall i \in I_m)(y(i) \leq x(i))$, and $\mathcal{L}_{\mathcal{T}^{(m)}}(x,y) = 1_m$ elsewhere.

Combining the above results on join-decomposable and join-irreducible right-hand sides, we can state the next result. The most important difference is that now the offshoots can be written down *immediately*, and do not have to be determined as minimal elements of an auxiliary set. If $\mathcal{T}$ is continuous, I is finite and the solution set of the equation

$$\sup_{i\in I} \mathcal{T}^{(m)}(A(i), X(i)) = b \tag{10}$$

is not empty, then it is a finitely generated root system with stem G defined by

$$G(i) = \mathcal{I}_{\mathcal{T}^{(m)}}(A(i), b)$$

and as set of offshoots the set

$$\{ \bigcup_{j\in J \wedge b(j)\neq 0} M_k^j \mid b(j) \leq A(k)(j)\}$$

with M_k^j defined by

$$M_k^j(i) = \begin{cases} \mathcal{L}_{\mathcal{T}^{(m)}}(A(k), \underline{b}_j) & \text{, if } i = k \\ 0_m & \text{, elsewhere} \end{cases}.$$

Note that $\mathcal{L}_{\mathcal{T}^{(m)}}(A(k), \underline{b}_j) = (0, \ldots, \mathcal{L}_{\mathcal{T}}(A(k)(j), b(j)), \ldots, 0)$.

4.4 Systems of Sup-$\mathcal{T}$ Equations

It should be clear from the above that in many important cases the solution set of a sup-$\mathcal{T}$ equation, if not empty, is a finitely generated root system. Recalling that the intersection of a family of finitely generated root systems, if not empty, is again a root system, leads immediately to the description of solution procedures for systems of sup-$\mathcal{T}$ equations.

The problem we are concerned with is the following: given a complete lattice $(L, \leq)$ and a family $(A_s)_{s\in S}$ in $\mathcal{F}_L(I)$ and a family $(b_s)_{s\in S}$ in L, determine the solution set of the system of equations $(E_s)_{s\in S}$

$$\sup_{i\in I} \mathcal{T}(A_s(i), X(i)) = b_s \tag{11}$$

in the unknown L-fuzzy set X in I.

Greatest solution. If the partial mappings of $\mathcal{T}$ are sup-morphisms, then the solution set of system (11) is not empty if and only if the fuzzy set $G = \bigcap_{s\in S} G_s$, with G_s the potential greatest solution of equation E_s, i.e.

$$G(i) = \inf_{s\in S} \mathcal{I}_{\mathcal{T}}(A_s(i), b_s), \tag{12}$$

is a solution. Moreover, if G is a solution, then it is the greatest solution.

In general, no simpler necessary and sufficient solvability conditions are known. Let us therefore turn our attention to the description of the complete solution set.

On a distributive, complete lattice with join-irreducible or finitely join-decomposable elements. Consider a finite index set I and assume that the right-hand sides of system (11) are join-irreducible or finitely join-decomposable. If the partial mappings of $\mathcal{T}$ are homomorphisms and the solution set of system (11) is not empty, then it is a root system; if the system is finite, then this root system is finitely generated. Suppose that the solution is not empty, and that the solution sets of the equations E_s are the root systems with stem G_s and set of offshoots O_s, then the solution set of the system $(E_s)_{s \in S}$ is the root system with stem $G = \bigcap_{s \in S} G_s$ and as set of offshoots the set

$$\{\bigcup_{s \in s} N_s \mid N_s \in O_s \wedge N_s \subseteq G\}.$$

A necessary and sufficient solvability condition therefore is that G is a solution. The case of a complete Brouwerian lattice was again discussed by Zhao (1987), the above more general case by De Baets (1998).

5 Fuzzy Relational Equations

We are now in a position to tackle what are called fuzzy relational equations. For the sake of brevity, we will restrict ourselves to the real unit interval; the reader can easily generalize the results to the more general setting of distributive, complete lattices with join-irreducible or finitely join-decomposable elements.

Image equations of type 1. Consider a fuzzy set A in X and a fuzzy set B in Y. As mentioned before, the image equation $R(A) = B$ in the unknown fuzzy relation R is equivalent to a family of independent sup-$\mathcal{T}$ equations in the foresets Ry, $y \in Y$. This equation is an important identification problem in fuzzy rule modelling: determine a relational correspondence given the input A and the output B. Clearly, the solution set of a family of independent equations is isomorphic to the Cartesian product of the solution sets of the individual equations. Since the Cartesian product of root systems obviously also is a root system, our knowledge of sup-$\mathcal{T}$ equations is sufficient for solving this type of image equation. Note that the set of offshoots of a Cartesian product of root systems is nothing else but the Cartesian product of the corresponding sets of offshoots. For instance, the solution set is not empty if and only if the solution sets of all individual equations are not empty.

We then have that for a left-continuous t-norm $\mathcal{T}$, the solution set of the above image equation is not empty if and only if the fuzzy relation G defined by

$$G(x, y) = \mathcal{I}_{\mathcal{T}}(A(x), B(y)) \tag{13}$$

is the greatest solution. In case of a complete Brouwerian lattice and the meet operation, this result is known as one of the very first results on fuzzy relational equations, see Sanchez (1977). The case of the real unit interval was discussed by Di Nola, Sessa, Pedrycz and Sanchez (1989). For a continuous t-norm $\mathcal{T}$, a necessary and sufficient solvability condition is given by

$$\sup_{y \in Y} B(y) \leq \sup_{x \in X} A(x), \tag{14}$$

or, in words, the height of A is not smaller than the height of B. In the case of the real unit interval, this condition can also be found in the work of Gottwald and Pedrycz (1988) and Gottwald (1994).

Using the above framework of sup-$\mathcal{T}$ equations and root systems, the complete solution set can be written down immediately. If the universe X is finite, $\mathcal{T}$ is continuous and the solution set of the equation $R(A) = B$ is not empty, then it is a root system with stem G defined by (13) and as set of offshoots the set

$$O = \{M \mid (\forall y \in Y)(My \in O_y)\},$$

where

$$O_y = \{M_u^y \mid B(y) \leq A(u)\}$$

with M_u^y defined by

$$M_u^y(x) = \begin{cases} \mathcal{L}_{\mathcal{T}}(A(u), B(y)) & \text{, if } x = u \\ 0 & \text{, elsewhere} \end{cases}.$$

If the universe Y is also finite, then this root system is finitely generated.

The complete solution set was described first by Sanchez (1977) for the case of the real unit interval, finite underlying universes and the minimum operator as t-norm. The generalization to continuous t-norms has been done by Di Nola, Sessa, Pedrycz and Sanchez (1989). In both sources, explicit proofs are given.

Image equations of type 2. Consider a fuzzy relation R from X to Y and a fuzzy set B in Y. The image equation $R(A) = B$ in the unknown fuzzy set A is equivalent to a system of sup-$\mathcal{T}$ equations. This inverse problem plays a crucial role in backward chaining in fuzzy rule-based systems. Here, we can immediately apply the results from the previous section.

For a left-continuous t-norm $\mathcal{T}$, the solution set of this equation is not empty if and only if the fuzzy set G defined by

$$G(x) = \inf_{y \in Y} \mathcal{I}_{\mathcal{T}}(R(x, y), B(y)) \tag{15}$$

is the greatest solution. Denoting the converse (or transpose) of a fuzzy relation R by R^t, i.e. $R^t(y, x) = R(x, y)$, we can rewrite G as follows:

$$G = (R^t)_{\triangleright}(B),$$

using as implicator the residual implicator $\mathcal{I}_{\mathcal{T}}$. In view of the importance of this equation, we also mention the complete solution set. If the universe X is finite, $\mathcal{T}$ is continuous and the solution set of the equation $R(A) = B$ is not empty, then it is a root system with stem G defined by (15) and as offshoots the minimal elements of the set

$$\{\bigcup_{y\in Y} M_u^y \mid B(y) \leq R(u,y) \wedge M_u^y \subseteq G\}$$

where M_u^y is defined by

$$M_u^y(x) = \begin{cases} \mathcal{L}_{\mathcal{T}}(R(u,y), B(y)) & \text{, if } x = u \\ 0 & \text{, elsewhere} \end{cases}. \tag{16}$$

This type of image equation has received a lot of attention in the literature. Most attempts concern finite universes, the real unit interval (or a complete chain) and mostly the minimum operator. The case of a (left-)continuous t-norm is also discussed in Pedrycz (1985), Di Nola, Pedrycz and Sessa (1987) and Di Nola, Sessa, Pedrycz and Sanchez (1989). However, using our knowledge about the intersection of finitely generated root systems, the above results can be been derived easily and more generally.

Composition equations. Consider a fuzzy relation S from Y to Z and a fuzzy relation T from X to Z. The composition equation $R \circ S = T$ in the unknown fuzzy relation R is equivalent to a family of independent image equations $S(xR) = xT$ in the unknown aftersets xR, for all $x \in X$. The solution set therefore is isomorphic to the Cartesian product of the solution sets of these image equations, each of which is nothing else but a system of sup-$\mathcal{T}$ equations. Of course, the equation $R \circ S = T$ in the unknown fuzzy relation S is solved at the same time, since it can be reformulated as follows: $S^t \circ R^t = T^t$.

Therefore, for a left-continuous t-norm $\mathcal{T}$, the solution set of this equation is not empty if and only if the fuzzy relation G defined by

$$G(x,y) = \inf_{z\in Z} \mathcal{I}_{\mathcal{T}}(S(y,z), T(x,z)) \tag{17}$$

is the greatest solution. We can rewrite G as follows:

$$G = T \triangleright S^t,$$

using as implicator the residual implicator $\mathcal{I}_{\mathcal{T}}$. In the case of a complete Brouwerian lattice and the meet operation as t-norm, this is again one the first results of Sanchez (1977); in this case, there also exists an alternative necessary and sufficient solvability condition by Di Nola (1990). In the case of finite universes, a completely distributive complete lattice $(L, \leq)$ and the meet operation as t-norm, the solution set is also a root system. Indeed, Di Nola (1987, 1990) has shown that for any solution at least one underlying minimal solution can be constructed; however, not all minimal solutions can be constructed explicitly. This result complements the results obtained in our general framework. It is clear that for specific operators and in specific situations, additional results can be obtained. Of course, in case of join-irreducibility and join-decomposability, we are again in our framework.

However, there is another way of dealing with the equation $R \circ S = T$. Indeed, it can be seen as a system of image equations $S(xR) = xT$ in the unknown fuzzy relation S, for all $x \in X$. At first sight, the latter approach seems to be the most practical one, in particular in view of determining the minimal solutions. Indeed, since each of the image equations $S(xR) = xT$ in the unknown fuzzy relation S stands for a set of independent sup-$\mathcal{T}$ equations, their minimal solutions can be written down immediately and the pen-and-paper method has to be applied only once to find the minimal solutions of the composition equation $R \circ S = T$. In the first approach, the minimal solutions of each of the independent image equations $S(xR) = xT$ in the unknown aftersets xR have to be found using the pen-and-paper method. The minimal solutions of the composition equation $R \circ S = T$ are then found immediately. However, in practice it is difficult to say which method is the most efficient one. In the second approach, the pen-and-paper method has to be applied only once, but then possibly on a huge number of *fuzzy relations*, while for each of the independent image equations the minimal solutions have to be found among a smaller number of *fuzzy sets*. The second approach is the one usually followed in the literature, see Di Nola, Sessa, Pedrycz and Sanchez (1989), while we favour the first one, in particular from an implementational point of view.

6 Conclusion

When the input-output behaviour of a system is (only) described linguistically by an expert, the fuzzy systems modelling approach can be called on. A system model is then built by solving fuzzy relational equations. In this paper, we have explained how these inverse problems can be solved in the framework of polynomial lattice equations and have presented a uniform framework in which various types of image and composition equations can be dealt with. Future work will consist of providing a similar framework for approximate solution methods in case of voidness of the solution set.

References

Bandler W. and Kohout L. (1980a). Fuzzy relational products as a tool for analysis and synthesis of the behaviour of complex natural and artificial systems, *Fuzzy Sets: Theory and Application to Policy Analysis and Information Systems* (Wang P. and Chang S., eds.), Plenum Press, 341–367.

Bandler W. and Kohout L. (1980b). Semantics of implication operators and fuzzy relational products, *Int. J. Man-Machine Studies*, **12**, 89–116.

Birkhoff G. (1967). *Lattice Theory*, AMS Colloquium Publications, Volume XXV, American Mathematical Society.

Davey B. and Priestley H. (1990). *Introduction to Lattices and Order*, Cambridge University Press.

De Baets B. (1995a). An order-theoretic approach to solving sup-$\mathcal{T}$ equations, *Fuzzy Set Theory and Advanced Mathematical Applications* (Ruan D., ed.), Kluwer Academic Publishers, 67–87.

De Baets B. (1995b). *Oplossen van vaagrelationele vergelijkingen: een ordetheoretische benadering*, Ph.D. dissertation, University of Gent, 389 p.

De Baets B. (1996). Disjunctive and conjunctive fuzzy modelling, *Proc Symposium on Qualitative System Modelling, Qualitative Fault Diagnosis and Fuzzy Logic and Control* (Budapest and Balatonfüred, Hungary), 63–70.

De Baets B. (1998). Sup-$\mathcal{T}$ equations: state of the art, *Computational Intelligence: Soft Computing and Fuzzy-Neural Integration with Applications* (Kaynak O., Zadeh L., Turksen B. and Rudas I., eds.), NATO ASI Series F: Computer and Systems Sciences, Vol. 162, Springer Verlag, 80–93.

De Baets B. and Kerre E. (1993). A revision of Bandler–Kohout compositions of relations, *Math. Pannon.*, **4**, 59–78.

De Baets B. and Kerre E. (1995). Fuzzy relations and applications, *Advances in Electronics and Electron Physics* (Hawkes P., ed.), **89**, Academic Press, 255–324.

De Baets B. and Mesiar R. (1999). Triangular norms on product lattices, *Fuzzy Sets and Systems*, **104**, 61-75.

De Cooman G. and Kerre E. (1994). Order norms on bounded partially ordered sets, *J. Fuzzy Math.*, **2**, 281–310.

Di Nola A. (1987). Fuzzy equations in infinitely distributive lattices, *Proc Second IFSA World Congress* (Tokyo, Japan), 533–534.

Di Nola A. (1990). On solving relational equations in Brouwerian lattices, *Fuzzy Sets and Systems*, **34**, 365–376.

Di Nola A., Pedrycz W. and Sessa S. (1987). Fuzzy relation equations under lsc and usc t-norms and their Boolean solutions, *Stochastica*, **11**, 151–183.

Di Nola A., Sessa S., Pedrycz W. and Sanchez E. (1989). *Fuzzy Relation Equations and their Applications to Knowledge Engineering,* Theory and Decision Library. Series D. System Theory, Knowledge Engineering and Problem Solving, Kluwer Academic Publishers.

Dubois D. and Prade H. (1992). Upper and lower images of a fuzzy set induced by a fuzzy relation: applications to fuzzy inference and diagnosis, *Inform. Sc.*, **64**, 203–232.

Goguen J. (1967). *L*-fuzzy sets, *J. Math. Anal. Appl.*, **18**, 145–174.

Gottwald S. (1994). Approximately solving fuzzy relational equations: some mathematical results and some heuristic proposals, *Fuzzy Sets and Systems*, **66**, 175–193.

Pedrycz W. (1985). Applications of fuzzy relational equations for methods of reasoning in presence of fuzzy data, *Fuzzy Sets and Systems*, **16**, 163–175.

Pedrycz W. (1988). Approximate solutions of fuzzy relational equations, *Fuzzy Sets and Systems*, **28**, 183–202.

Rudeanu S. (1974). *Boolean Functions and Equations*, North-Holland.

Sanchez E. (1977). Solutions in composite fuzzy relation equations: application to medical diagnosis in Brouwerian logic, *Fuzzy Automata and Decision Processes* (Gupta M., Saridis G. and Gaines B., eds.), North-Holland, New York, 221–234.

Schweizer B. and Sklar A. (1983). *Probabilistic Metric Spaces*, Elsevier.

Zhao C. (1987). On matrix equations in a class of complete and completely distributive lattices, *Fuzzy Sets and Systems*, **22**, 303–320.

Equations and Inequalities With BK-Products of Relations

L.J. Kohout

Dept. of Computer Science, Florida State University,
Tallahassee, Florida 32306-4530, USA
E-mail: kohout@cs.fsu.edu URL: http://cs.fsu.edu/~ kohout

Abstract. The paper presents an algebraic setting for dealing with fuzzy relational inequalities and equations. The algebraic apparatus that provides constructive means for formulating the relational algorithns is based on pseudoassociative compositions of fuzzy relations called BK-products in the literature. BK-products allow us to develop the theory of generalized morphisms which is an extension and substantial generalization of crisp and fuzzy relational homomorphisms. This extension yields morphisms of two kinds: (1) Generalized Morphisms (or forward compatibility) and (2) Generalized Proteromorphisms (backward compatibility). When both properties (1) and (2) hold, we obtain Generalized Ampimorphisms (both-ways compatibility). Stadard crisp or fuzzy relational homomorphisms are special cases of Generalized Amphimorphisms.

This theory of BK-product based generalized morphisms yields algebraic conditions for solvability of fuzzy relational inequalities and equations. The essential mechanism for formulating constructive formulas for solution is the relatinship between the standard associative relational product $\circ$ and the pseudoassociative BK-subproduct $\triangleleft$ and the BK-superproduct $\triangleright$.

For computationa reasons distinguish four different ways of defining relational products: (i) by quantifiers, (ii) by foreset-, afterset notation, (iii) by matrix products representation, and (iv) by tensor representation.

1 Introduction: Relations in Distributed Intelligent Systems

Relations play very important unifying role in specification, design and utilization of Intelligent Systems, and indeed in computer science in general. Fuzzy relations are important for effective representation of knowledge and user friendly computing, essential in scientific applications, medicine, engineering and commerce. Fuzzy relations are indispensable in *unified* represwentation and manipulation of knowledge in soft computing [47], computing with words [41] and computational theory of perceptions [46].

In all these applications of relations, relational equations and inequalities form the essential backbone of mathematical theory and computational algorithms. We show in this chapter, how BK-products of fuzzy relations can significantly advance both theory and practical applications of fuzzy relations. As crisp relations are a special case of fuzzy relations, the chapter presents result that are also highly relevant to this special case.

1.1 An Overview of Relational Methodology and Goals

Development of generalized morphisms and relational inequalities is part of a broader program the aim of which is to advance the theory and practice of fuzzy relational computing. Relational computing methods and structures make it possible to perform knowledge representation as well as to express requirements for computations in distributed information/intelligent systems in a unified way. In order to apply the insights obtained by means of relational methods in practice, the mathematics and logic of fuzzy systems has to be aided by an adequate methodology that provides foundations for analysis and synthesis of intelligent systems. For this purpose we have been developing a methodology of *Activity Structures* [29],[38] – the conceptual and methodological tools that can capture the diverse features of adaptive activities in natural an artificial systems, and of interaction of these systems with their environment. Activity Structures have three facets:

- conceptual,
- methodological,
- formal and algorithmic.

Relational inequalities and fuzzy generalized morphisms provide essential algebraic tools for imposing elastic constraints as well as for approximate matching of Activity Structures.

Our goal is to advance the theory as well as practical applications of generalized morphisms and relational inequalities based on BK-products of fuzzy relations. We work simultaneously on

- mathematical development of relational theories and algorithms,
- computational, knowledge engineering and scientific applications exploring BK-Products of relations.

1. Mathematical Development consists of:
   (a) Extending the mathematical theory of Generalized Morphisms and of associated relational inequalities from BL systems to general monoidal logic based systems.
   (b) Development of conceptual algebras of semiotic descriptors based on (1a). Such algebras could serve as a foundation of fuzzy AI ontologies that would be compatible with Zadeh's computational theory of perception and computing with words methodologies.
   (c) Examining extendability of Tarski's algebras of crisp relations to t-norm based algebras axiomatized in Hájek's BL predicate calculus.
2. Computational, Knowledge Engineering & Scientific Applications Exploring BK-Products:

(a) Development of Knowledge Elicitation methods compatible with the framework of Zadeh's *Computational Theory of Perceptions* and Bandler-Kohout-Mancini's *Train-of-Thought* cognitive paradigm [15], [26]. Using fuzzy generalized morphisms for comparison and evaluation of elicited cognitive structures.
(b) Development of distributed agents based knowledge-based systems exploring Fuzzy Generalized Morphisms and accompanying relational inequalities.
(c) Development of BK-product based relational architectures for computing with words.
(d) Reengineering extant BK-product based computational tools into an open-ended distributed architecture that could coperatively compute over the Internet using the OpenMath protocols [1],[53].
(e) Developing the Internet OpenMath compatible relational architectures for *Computing with Words.*
(f) Development of high level relational representation of scientific knowledge for Partial Differential Equations Problem Solving Environments exploring fuzzy relational knowledge representaton schemes.

1.2 Distinguishing Features of Relational Calculus Enriched with BK-products

There are six distinguishing features of the BK-product systems of relations that facilitate the unification of different many-valued systems of fuzzy relations and enhance their practical applicability [31]:

1. Non-associative BK-products are introduced and used both in definitions of relational properties and in computations. These products are defined not only for homogeneous but also for heterogeneous relations.
2. Homomorphisms between relations are extended from mappings used in the literature to general relations. This yields *generalized morphisms* important for practical solving of relational inequalities and equations.
3. Relational properties are not only global but also local [12] (important for applications).
4. The unified treatment of computational algorithms by means of matrix notation is used which is equally applicable to both crisp and fuzzy relations.
5. The theory unifying crisp and fuzzy relations in some distinguished logics makes it possible to represent a whole *finite nested family* of crisp relations with special properties as a *single* cutworthy fuzzy relation for the purpose of computation. After completing the computations, the resulting fuzzy relation is again converted by α-cuts to a nested family of crisp relations, thus increasing the computing performance considerably [13].
6. Relations in their predicate forms are distinguished from their satisfaction sets; foresets and aftersets of relations are used in addition to relational

predicates. This makes it possible to introduce interpretable linguistic labels (semiotic descriptors) that have a clearly defined meaning within the domains of their applications. Then one can develop an algebra of meaning defined by equations and ineqalities that defines a computational basis for forming of ontologies in knowledge engineering applications as well as in *computing with words*.

2 BK-Products of Relations

2.1 Enhancing Expressive Power of Calculus of Relations

Relational representation of knowledge makes it possible to perform all the computation and decision making by means of *special relational compositions* called triangle and square products [39],[31],[46],[9]. These were first introduced by Bandler and Kohout in 1977 [2],[11],[5] and are referred to as the BK-products in the literature [21],[16],[19],[18],[20],[36].

The representational and computational power of BK-products $\triangleright$ and $\triangleleft$ resides in their algebraic properties. The following mixed pseudo-associativities hold for $\triangleleft$ and $\triangleright$:

$$Q \triangleleft (R \triangleright S) = (Q \triangleleft R) \triangleright S,$$

$$Q \triangleleft (R \triangleleft S) = (Q \circ R) \triangleleft S$$

$$Q \triangleright (R \triangleright S) = Q \triangleright (R \circ S).$$

The interplay of $\circ, \triangleleft, \triangleright$ that is afforded by relaxing the property of full associativity is essential for enriching the expressive power of the calculus of relations. The mutual interaction of these three relational compositions plays a crucial role in defining the key inequalities of relational calculus.

One such set of inequalities called *Residuation bootstrap of BK-products* that plays a crucial role in the development of fuzzy relational calculi [33] will be proved and used extensively in this chapter. It consists of the following relational inequalities that hold for arbitrary $V \in \mathcal{B}(\mathcal{A} \rightarrow \mathcal{C})$:

$$R \circ S \sqsubseteq V \text{ iff } R \sqsubseteq V \triangleright S^T \text{ iff } S \sqsubseteq R^T \triangleleft V$$

These inequalities were discovered by Bandler and Kohout in 1977 and play an important role in solving relational inequalities over generalized morphisms as will be shown in the sequel. Using these inequalities significantly simplifies relational inference.

2.2 A Brief Overview of BK-Products

Mathematical definitions. Where R is a relation from X to Y, and S a relation from Y to Z, a *product relation* $R@S$ is a relation from X to Z, determined by R and S. There are several types of product used to produce product-relations [11], [39]. Each product type performs a **different logical action** on the intermediate sets, as *each logical type* of the product enforces a *distinct specific meaning* on the resulting product-relation $R@S$. In the following definitions of the products, R_{ij}, S_{jk} represent the fuzzy degrees to which the respective statements $x_i R y_j$, $y_j S z_k$ are true.

Table 1. Set-based definitions of fuzzy BK-products

PRODUCT TYPE	SET-BASED DEFINITION		
Circle product:	$x(R \circ S)z$	$\Leftrightarrow$	xR intersects Sz
Triangle Subproduct	$x(R \triangleleft S)z$	$\Leftrightarrow$	$xR \tilde{\subseteq} Sz$
Triangle Superprod.	$x(R \triangleright S)z$	$\Leftrightarrow$	$xR \tilde{\supseteq} Sz$
Square product:	$x(R \square S)z$	$\Leftrightarrow$	$xR \cong Sz$

Table 2. Other logically equivalent notations for BK-products of relations

PRODUCT TYPE	MANY-VALUED LOGIC BASED DEFINITION	TENSOR NOTATION
Circle product:	$(R \circ S)_{ik} = \bigvee_j (R_{ij} \& S_{jk})$	$(R \circ S)_{ik} = R_{ij} \circ S^{jk}$
Triangle Subproduct	$(R \triangleleft S)_{ik} = \bigwedge_j (R_{ij} \rightarrow S_{jk})$	$(R \triangleleft S)_{ik} = R_{ij} \triangleleft S^{jk}$
Triangle Superprod.	$(R \triangleright S)_{ik} = \bigwedge_j (R_{ij} \leftarrow S_{jk})$	$(R \triangleright S)_{ik} = R_{ij} \triangleright S^{jk}$
Square product:	$(R \square S)_{ik} = \bigwedge_j (R_{ij} \equiv S_{jk})$	$(R \square S)_{ik} = R_{ij} \square S^{jk}$

There, however, exist several other different notational forms in which BK-products can be expressed:

1. the notation shown in Table 1 using the concept of fuzzy set inclusion and equality [4],[5].
2. many-valued logic(MVL) based notation, which uses the logic connectives $\bigwedge$, &, $\rightarrow$ or $\equiv$which is displayed in Table 2.
3. The tensor notation also shown in Table 2.
4. The fuzzy predicate calculus form, discussed in Sec. 4 below.

These four different forms of relational compositions are logically equivalent under some reasonable logic assumptions, producing the same mathematical results. Distinguishing these forms is, however, important when constructing fast and efficient computational algorithms [39].

The tensor notation in its presentation abstracts from the display of the type of MVL connectives shown by logic-based notation. It preserves, on the other hand, the information about the way the BK-products were composed from their componets. This is important when we want to keep track of the ways in which several distinct, but logically equivalent streems of relational computation were constructed.

The logical symbols for the logic connectives AND &, both *implications* and the *equivalence* in the formulas shown in Table 2 represent connectives of some many-valued logic, **chosen** according to the logic properties of the products required. An important special case is when the AND connective & is represented semantically by a t-norm *. If the logics are residuated, then the implications are residua of the t-norm, and the equivalence is a biresiduum of the t-norm.

The generic formula

$$(R@S)_{ik} := \oplus_j(R_{ij}\#S_{jk}),$$

yields two types of fuzzy relational products. We can replace the outer connective $\bigoplus$ with $\bigwedge$(defined above) or with $\frac{1}{|J|}\sum$;

$$(R@S)_{ik} := \bigwedge_j (R_{ij}\#S_{jk}): \textit{Harsh} \text{ product},$$

$$(R@S)_{ik} := \frac{1}{|J|}\sum_j (R_{ij}\#S_{jk}): \textit{Mean} \text{ product}.$$

By choosing appropriate many-valued logic operations for the logic connectives, the crisp case extends to a wide variety of many-valued logic based (fuzzy) relational systems [39], [6],[9],[10],[43],[39]. While we often used in our applications the classical *min* and *max* for *t-norm* and *t-conorm*, respectively, we applied various MVL implication operators for the computation of BK-products. The details of choice of the appropriate many-valued connectives are discussed in [6],[9],[10],[43],[39].

2.3 Foresets and Aftersets of Crisp and Fuzzy Relations

The **afterset** of $x \in X$ is the *subset of* Y consisting of the elements $y \in Y$ to which x is related by R (where $\mu_A x = \delta\{xRy\}$, the degree to which x and y are R-related):

$$xR = \{y/\delta\{xRy\} \mid y \in Y \text{ and } \delta\{xRy\} > 0\}.$$

The **foreset** of $y \in Y$ is the *subset of* X consisting of all the elements $x \in X$ which are related by R to y (where $\mu_A y = \delta\{xRy\}$, the degree to which x

and y are R-related):

$$Ry = \{x/\delta\{xRy\} \mid x \in X \text{ and } \delta\{xRy\} > 0\}.$$

When we take the matrix representation of a relation R, aftersets are given by the rows, while the foresets are given by the columns of the matrix R_M. (Here, of course, we assume that the satisfaction set R_S corresponds under the axiom of extensionality to the relation given by its predicate form R_P).

The notions of afterset and foreset of an element can be extended to afterset and foreset of a set in (at least) two distinct but equally important ways: an *inclusive* or *exclusive* afterset / foreset (see [8]).

The **inclusive** after- and foresets are given by

$$A'R = A'_R \circ R, \quad RB' = R \circ B'_C$$

The **exclusive** after- and foresets are given by

$$A'R = A'_R \triangleleft R, \quad RB' = R \triangleright B'_C$$

To understand how this composition can be computed, one has to look at its component-wise definition that involves indexed elements. A_i is the membership (characteristic function) giving the degree to which the predicate $a_i \in A$ is TRUE; and R_{ij} is the degree to which the predicate $R_{ij} \in R$ is TRUE, where R_{ij} is an element of R.

$$C_k = (A \circ R) = \bigvee_i (A_i \wedge R_{ik})$$
$$C_k = (A \triangleleft R) = \bigwedge_i (A_i \to R_{ik})$$
$$C_k = (A \triangleright R) = \bigwedge_i (A_i \leftarrow R_{ik})$$

3 BK-Products in Predicate Calculus Form

After discussing the importance of computationally friendly notation for relational products, we briefly summarize some notions of Basic Logic (BL) predicate calculus of Hájek [22] which axiomatizes all logics in the infinite family of fuzzy logics based on continuous t-norms. That will provide us with the necessary technical tools for proving the basic BK-product inequalities that we call *The residuation Bootstrap of BK-products.* These inequalities are essential for the development of fuzzy relational calculus based on logics using residuated t-norms. Then in Section 4.2 we present a rigorous proof of these inequalities using BL predicate calculus.

3.1 The Usefulness of Different Notations for Computing with Relational Products

The triangle and square BK-products significantly extend the expressive products of relational calculi. The **foreset-afterset notation** and the many-valued logic based **MVL-notation** are advantageous when relations are manipulated by a computer, applying fuzzy relational computations to large sets of empirical data. They are semantic in their nature and computations with these are done numerically.

For abstract proofs, on the other hand, it is often advantageous to use the fourth notational form of the BK-products, namely its **predicate form** expressed symbolically within an appropriate system of formal logic using the universal quantifier. Manipulation of relational formulas expressed this way is syntactic/symbolic, using appropriate (syntactic) inference rules. These two different approaches, the semantic (computing numerically) and the syntactic/symbolic (computing with syntactic forms - manipulating strings symbolically) mutually complement each other.

The next definition expresses the relational products in a first order predicate logic form.

Zadeh's definition of fuzzy version of standard relationa composition (circle product) in predicate forms is

$$R \circ S = (\forall x)(\forall z)(\exists y)(xRy \ \& \ ySz).$$

BK-products can be expressed in the following way:

Definition 1 BK-PRODUCTS OF RELATIONS $\lhd, \rhd, \square$. *For arbitrary fuzzy relations in [0, 1], R from the set X to Y, S from Y to Z define:*

1. *The subproduct:* $R \lhd S = (\forall x)(\forall z)(\forall y)(xRy \rightarrow ySz)$
2. *The superproduct:* $R \rhd S = (\forall x)(\forall z)(\forall y)(xRy \leftarrow ySz)$;
3. *The square product:* $R \square S = (\forall x)(\forall z)(\forall y)(xRy \equiv ySz)$;

For crisp relations the logic connectives are Boolean, and $\forall, \exists$ are quantifiers of the standard (2-valued) predicate calculus. For fuzzy logics the above formulas cover the infinite spectrum of many different fuzzy relational calculi. Indeed, taking the *supremum* to interpret $\exists$ and the *infimum* to interpret $\forall$, any continuous t-norm to interpret & and the residuum of & to interpret $\rightarrow$ yields such a relational calculus. The fuzzy predicate calculus of this form is called *Basic Logic* (BL). It has been axiomatized by Hájek [22]. BL predicate calculus has been used by Kohout to axiomatize the Basic Algebra of BK-products of relations in t-norm fuzzy logics [33],[34]. BL predicate calculus subsumes classical predicate calculus as a special case. It might be advantageous to use BL logic instead of classical logic in courses of logic for computer science. The students will have obtained the background not only for classical computer science and software engineering, but also for approximate reasoning so important in soft computing and computational intelligence.

3.2 Basic Logic BL of Hájek

The unification and extension of the technical apparatus of relations from crisp to fuzzy, including a substantial generalization of BK-products, is supported by Hájek's BL-predicate calculus $PC(*)$ [22]. This axiomatization puts on a firm logical footing the whole family of the t-norm based fuzzy logics. In the sequel $PC(*)$ will denote a predicate calculus where * is an arbitrary t-norm of this family which represents the `AND` connective, from which the implication operator is produced by residuation.

$$x * z \leq y \quad \text{if and only if} \quad z \leq (x \rightarrow y)$$

Some formulas that are 1-tautologies of every $PC(*)$ for any continuous t-norm $(*)$ are chosen as the axioms of Basic Logic (BL). BL forms a common base of all the logics $PC(*)$.

Definition 2: The following formulas are axioms of the basic logic BL:

(A1) $(\varphi \rightarrow \psi) \rightarrow ((\psi \rightarrow \chi) \rightarrow (\varphi \rightarrow \chi))$
(A2) $(\varphi \& \psi) \rightarrow \varphi$
(A3) $(\varphi \& \psi) \rightarrow (\psi \& \varphi)$
(A4) $(\varphi \& (\varphi \rightarrow \psi)) \rightarrow (\psi \& (\psi \rightarrow \varphi))$
(A5a) $(\varphi \rightarrow (\psi \rightarrow \chi)) \rightarrow ((\varphi \& \psi) \rightarrow \chi)$
(A5b) $((\varphi \& \psi) \rightarrow \chi) \rightarrow (\varphi \rightarrow (\psi \rightarrow \chi))$
(A6) $((\varphi \rightarrow \psi) \rightarrow \chi) \rightarrow (((\psi \rightarrow \varphi) \rightarrow \chi) \rightarrow \chi)$
(A7) $\bar{0} \rightarrow \varphi$

The *deduction rule* of BL is modus ponens. Given this, the notions of a *proof* and of a *provable formula* in BL are defined in the obvious way (cf. [22]).

The following are *logical axioms on quantifiers:*

($\forall$1) $(\forall x)\varphi(x) \rightarrow \varphi(t)$ (t substitutable for x in $\varphi(x)$)
($\exists$1) $\varphi(t) \rightarrow (\exists x)\varphi(x)$ (t substitutable for x in $\varphi(x)$)
($\forall$2) $(\forall x)(\nu \rightarrow \varphi) \rightarrow (\nu \rightarrow (\forall x)\varphi)$ (x not free in ν)
($\exists$2) $(\forall x)(\varphi \rightarrow \nu) \rightarrow ((\exists x)\varphi \rightarrow \nu)$ (x not free in ν)
($\forall$3) $(\forall x)(\varphi \vee \nu) \rightarrow ((\forall x)\varphi \vee \nu)$ (x not free in ν)

Let $\mathcal{C}$ be a schematic extension of the basic propositional logic BL. We associate with $\mathcal{C}$ the corresponding predicate calculus $\mathcal{C}\forall$ (over a given predicate language $\mathcal{J}$) by taking as logical axioms

- all formulas resulting from the axioms of $\mathcal{C}$ by substituting arbitrary formulas of $\mathcal{J}$ for propositional variables, and
- the axioms ($\forall$1), ($\forall$2), ($\forall$3), ($\exists$1), ($\exists$2) for quantifiers
- taking as deduction rules modus ponens (from φ, $\varphi \rightarrow \psi$ inhfer ψ) and generalization (from φ infer $(\forall x)\varphi$).

Given this, the notions of proof, provability, theory, proof/provability in a theory over $\mathcal{C}\forall$ are obvious.

4 Fundamental Theorems of Enriched Calculus of Fuzzy Relations

4.1 Pseudo-associativity of BK-Triangle Products

The representational and computational power of BK-products, the triangle subproduct and the triangle superproduct resides in their algebraic properties. Although the triangle products are not associative, the following mixed pseudo-associative equation holds for and :

$$Q \lhd (R \rhd S) = (Q \lhd R) \rhd S, \quad \text{hence written} \quad Q \lhd R \rhd S.$$

This equation depends on very general properties of the implication operators, hence do not depend on residuation.

There is also a link between $\circ, \lhd, \rhd$ given by the following relational equations:

1. $Q \lhd (R \lhd S) = (Q \circ R) \lhd S$;
2. $Q \rhd (R \rhd S) = Q \rhd (R \circ S)$.

These equations also cover a broad range of fuzzy relational systems. They hold not only in all relational systems based on t-norm residuated fuzzy logics, but extend to the realm of more general relational relational systems [31] based on monoidal logics over residuated lattices [25].

4.2 Relational Inequalities with Fuzzy BK-Products

The interplay of relational compositions generates powerful inequalities that enrich the computational power of both crisp and fuzzy relational calculi. Important formulas of the BK-enriched relational calculus are given by the next theorem. This theorem which describes the interrelationship of $\circ, \lhd, \rhd$ plays a substantial role in further development of the theory of crisp and fuzzy relations. Because these formulas depend on residuation, they carry over into relational theories based on t-norms and corresponding residuated implication operators.

Theorem 2 *Residuation bootstrap of BK-products [33] in t-norm fuzzy logics.*
For arbitrary $V \in \mathcal{B}(\mathcal{A} \to \mathcal{C})$,

$$R \circ S \sqsubseteq V \textit{ iff } R \sqsubseteq V \rhd S^T \textit{ iff } S \sqsubseteq R^T \lhd V$$

where relations R^T *and* S^T *are transpose relations. Because of residuation, these relations are also inverses of* R *and* S*, sometime denoted by* R^{-1} *and* S^{-1}
Proof:
$BL \vdash (\forall x)(\varphi \to \nu) \equiv (\exists x)\varphi \to \nu$ *(by [22], T 5.1.4(2))*

The substitutions $\varphi := xRy\&ySz, \nu := xVz$ *yield*
$\vdash (\forall y)((xRy\&ySz) \to xVz) \equiv ((\exists y)(xRy\&ySz) \to xVz)$
$\vdash (\forall y)((xRy\&ySz) \to xVz) \to ((\exists y)(xRy\&ySz) \to xVz)$ *By [22], L 2.2.16(25)*
$\vdash (\forall x)(\forall z)(\forall y)((xRy\&ySz) \to xVz) \to (\forall x)(\forall z)((\exists y)(xRy\&ySz) \to xVz))$
T 5.1.16(5) and MP
$\vdash (\forall x)(\forall y)(xRy \to (\forall z)(xVz \leftarrow zS^Ty)) \to (\forall x)(\forall z)((\exists y)(xRy\&ySz) \to xVz))$ *By T 15.1.14(1)*
$(A) \vdash (\forall x)(\forall y)(xRy \to x(V \rhd S^T)y) \to (\forall x)(\forall z)(x(R \circ S)z \to xVz)$
Similarly, we derive (B)
$(B) \vdash (\forall x)(\forall y)(xRy \to x(V \rhd S^Ty) \leftarrow (\forall x)(\forall z)(x(R \circ S)z \to xVz)$
Substituting $\varphi := (A), \psi := (B)$ *into* $\vdash \varphi \to (\psi \to \varphi\&\psi)$,
applying MP twice and using the definition of $\equiv$ *yields*
$(B) \vdash (\forall x)(\forall y)(xRy \to x(V \rhd S^T)y) \equiv (\forall x)(\forall z)(x(R \circ S)z \to xVz)$
which is $\vdash (R \sqsubseteq V \rhd S^T) \equiv (R \circ S \sqsubseteq V)$

Similar proofs can be given for other equalities of Th. 1.

For further details and discussion of the importance of this theorem as the core statement in axiomatization of relations see [33],[34]. The theorem also plays a major role in constructive axiomatization of generalized morphisms [32]. Basic inequalities characterising generalized morphisms are discussed below, in section 5. Because this characterisation is constructive, not just existentional, it also provides direct computational methods that have been employed in design of distributed agent-based fuzzy intelligent systems [51] exploring the framework of computational theory of agent perception [30],[35].

4.3 Universal Representation in BL, of Some Relational Properties by Inequalities Using BK-Superproduct

BK-products have great representational power. For example, they can provide universal representation of relations with special properties. In this section we look at representation of preordered relations. Here we assume that $\circ, \lhd, \rhd$ are defined in BL.

The following inequalities and equation in BL provide universal representation of all reflexive, transitive and preordered relations respectively:

Theorem 3 *[14]*

1. *R is transitive if and only if* $R \sqsubseteq R \rhd R^{-1}$.
2. *R is reflexive if and only if* $R \rhd R^{-1} \sqsubseteq R$
3. *Hence, R is a preorder if and only if* $R = R \rhd R^{-1}$.
4. *Every preorder can be expressed this way.*

It has been shown by Kohout [31] that this can be generalized outside the realm of t-norm based fuzzy logics that have the truth functions in the interval

[0, 1] to monoidal logics with truth function in a more general lattice. This generalization obviously subsumes both BL systems (residuated systems with continuous t norms) as well as the residuated systems with left-continuous t-norms.

4.4 Solving Relational Equations: Some Well-Known Results Revisited within BL

Theorems concerning solutions of relational equations can be easily derived from the BK-residuation bootstrap inequalities as shown in this section. In this section we assume that $\circ, \lhd, \rhd$ are defined in BL. That is the & connective defining $\circ$ is a continuous t-norm, and the implications defining $\lhd, \rhd$ are residuated with the t-norm. Then the inequalities expressing *the Residuation Bootstrap of BK-products* hold as given in Section 4.2 above. $\mathcal{R}_F(X \rightsquigarrow Y)$ denotes the lattice of all fuzzy relations from X to Y.

Solving $R \circ S = V$ Let X, Y, Z be fuzzy sets and R, S, V relations such that $R \in \mathcal{R}_F(X \rightsquigarrow Y)$, $S \in \mathcal{R}_F(Y \rightsquigarrow Z)$, $V \in \mathcal{R}_F(X \rightsquigarrow Z)$. Further, let the relation equation $R \circ S = V$ be given. Then when solving this equation we have two cases:

1. In the equation $R \circ S = V$, the relations R, V are given, S is unknown. The task is to solve the equation in S.
2. In the equation $R \circ S = V$, the relations S, V are given, R is unknown. The task is to solve the equation in R.

Theorem 4 *Let $\mathcal{S}$ be a family of relations such that*

$$\mathcal{S}_{sol} = \{S \mid R \circ S = V\}$$

Then $\mathcal{S}_{sol}$ is non-empty iff $(R^T \lhd V) \in \mathcal{S}_{sol}$. Then $(R^T \lhd V)$ is the greatest element in $\mathcal{S}_{sol}$.

Proof:
This directly follows from $R \circ S \sqsubseteq V$ iff $R \sqsubseteq V \rhd S^T$ iff $S \sqsubseteq R^T \lhd V$. Indeed, if $R^T \lhd V \in \mathcal{S}_{sol}$ then $\mathcal{S}_{sol}$ is non-empty. If, on the other hand $\mathcal{S}_{sol}$ is nonempty, then for all S such that $S \in \mathcal{S}_{sol}$, it must hold that $S \sqsubseteq R^T \lhd V$. If, furthermore $S = R^T \lhd V$, then $R^T \lhd V$ must be the greatest solution because it is impossible for any $S \in \mathcal{S}_{sol}$ be greater than $R^T \lhd V$.

Theorem 5 *Let $\mathcal{S}$ be a family of relations such that*

$$\mathcal{R}_{sol} = \{R \mid R \circ S = V\}$$

Then $\mathcal{R}_{sol}$ is non-empty iff $(V \rhd S^T) \in \mathcal{R}_{sol}$. Then $(V \lhd S^T)$ is the greatest element in $\mathcal{R}_{sol}$.

Proof:
This directly follows from $R \circ S \sqsubseteq V$ iff $R \sqsubseteq V \rhd S^T$ iff $S \sqsubseteq R^T \lhd V$ in a similar fashion as the proof of Th.4.

Solving $R \lhd S = V$ Let X, Y, Z be fuzzy sets and R, S, V relations such that $R \in \mathcal{R}_F(X \rightsquigarrow Y)$, $S \in \mathcal{R}_F(Y \rightsquigarrow Z)$, $V \in \mathcal{R}_F(X \rightsquigarrow Z)$. Further, let the relation equation $R \lhd S = V$ be given. Then when solving this equation we have two cases:

1. In the equation $R \lhd S = V$, the relations R, V are given, S is unknown. The task is to solve the equation in S.
2. In the equation $R \lhd S = V$, the relations S, V are given, R is unknown. The task is to solve the equation in R.

Theorem 6 *Let $\mathcal{S}$ be a family of relations such that*

$$\mathcal{S}_{sol} = \{S \mid R \lhd S = V\}$$

Then $\mathcal{S}_{sol}$ is non-empty iff $(R^T \circ V) \in \mathcal{S}_{sol}$. Then $(R^T \circ V)$ is the smallest element in $\mathcal{S}_{sol}$.

Proof:
This directly follows from the inequalities $R^T \circ V \sqsubseteq S$ iff $R^T \sqsubseteq S \rhd V^T$ iff $V \sqsubseteq R \lhd S$ (*the Residuation Bootstrap of BK-products*). Indeed, if $R^T \lhd V \in \mathcal{S}_{sol}$ then $\mathcal{S}_{sol}$ is non-empty. If, on the other hand $\mathcal{S}_{sol}$ is nonempty, then for all S such that $S \in \mathcal{S}_{sol}$, it must hold that $S \sqsupseteq R^T \lhd V$. If, furthermore $S = R^T \lhd V$, then $R^T \lhd V$ must be the smallest solution because it is impossible for any $S \in \mathcal{S}_{sol}$ be smaler than $R^T \lhd V$.

Theorem 7 *Let $\mathcal{S}$ be a family of relations such that*

$$\mathcal{R}_{sol} = \{R \mid R \lhd S = V\}$$

Then $\mathcal{R}_{sol}$ is non-empty iff $(V \rhd S^T) \in \mathcal{R}_{sol}$. Then $(V \rhd S^T)$ is the greatest element in $\mathcal{R}_{sol}$.

Proof:
This directly follows from the inequalities $R^T \circ V \sqsubseteq S$ iff $R^T \sqsubseteq S \rhd V^T$ iff $V \sqsubseteq R \lhd S$ (*the Residuation Bootstrap of BK-products*) in a similar fashion as in the previous theorems.

Solving $R \rhd S = V$ Let X, Y, Z be fuzzy sets and R, S, V relations such that $R \in \mathcal{R}_F(X \rightsquigarrow Y)$, $S \in \mathcal{R}_F(Y \rightsquigarrow Z)$, $V \in \mathcal{R}_F(X \rightsquigarrow Z)$. Further, let the relation equation $R \rhd S = V$ be given. Then when solving this equation we have two cases:

1. In the equation $R \rhd S = V$, the relations R, V are given, S is unknown. The task is to solve the equation in S.
2. In the equation $R \rhd S = V$, the relations S, V are given, R is unknown. The task is to solve the equation in R.

Theorem 8 *Let $\mathcal{S}$ be a family of relations such that*

$$\mathcal{S}_{sol} = \{S \mid R \rhd S = V\}$$

Then $\mathcal{S}_{sol}$ is non-empty iff $(V^T \lhd R)^T \in \mathcal{S}_{sol}$. Then $(V^T \lhd R)^T$ is the greatest element in $\mathcal{S}_{sol}$.

Proof:
This directly follows from the inequalities $V \circ S^T \sqsubseteq R$ iff $V \sqsubseteq R \rhd S$ iff $S \sqsubseteq (V^T \lhd R)^T$ (*the Residuation Bootstrap of BK-products*). Indeed, if $(V^T \lhd R)^T \in \mathcal{S}_{sol}$ then $\mathcal{S}_{sol}$ is non-empty. If, on the other hand $\mathcal{S}_{sol}$ is nonempty, then for all S such that $S \in \mathcal{S}_{sol}$, it must hold that $S \sqsubseteq (V^T \lhd R)^T$. If, furthermore $S = (V^T \lhd R)^T$, then $(V^T \lhd R)^T$ must be the greatest solution because it is impossible for any $S \in \mathcal{S}_{sol}$ be greater than $(V^T \lhd R)^T$.

Theorem 9 *Let $\mathcal{S}$ be a family of relations such that*

$$\mathcal{R}_{sol} = \{R \mid R \lhd S = V\}$$

Then $\mathcal{R}_{sol}$ is non-empty iff $(V \circ S^T) \in \mathcal{R}_{sol}$. Then $V \circ S^T$ is the smallest element in $\mathcal{R}_{sol}$.

Proof:
This directly follows from the inequalities $V \circ S^T \sqsubseteq R$ iff $V \sqsubseteq R \rhd S$ iff $S \sqsubseteq (V^T \lhd R)^T$ (*the Residuation Bootstrap of BK-products*)in a similar fashion as in the previous theorems.

5 From Equations to Inequalities

We have shown in the previous section that the BK-relational products play a significant role in simplifying the methods of solution of relational equations. Relational equations are useful in providing equational definitions of other relational constructs, such as homomorphisms. The classical crisp theory of homomorphisms can be substantialy generalized by employing relational inequalities formed by means of crisp BK-products [2]. Similarly, fuzzy generalized morphisms can be substantially extended by means of fuzzy BK-products [32], as shown in this section.

5.1 Generalized Morphisms and Relational Inequalities

Homomorphisms play an important role in mathematics, general system theory and computing as well as in large number practical applications that require comparison of structures and their matching. Many diverse problems of compatibility of structures can be unified by generalizing the concept of a homomorphism. Homomorphisms have been successfully generalized and form one of the basic concepts of mathematics of fuzzy sets.

In 1977 Bandler and Kohout further extended and substantially generalized the notion of homomorphism. They introduced *generalized homomorphism, proteromorphism* and *amphimorphism*, forward and backward compatibility of relations, and non-associative and pseudoassociative products (compositions) of relations in crisp setting [2]. The proofs in the original papers of Bandler and Kohout were based on residuation without specific use of negation [8]. Hence they are generally valid in fuzzy relational calculi based on residuated fuzzy logics. Therefore the concepts of generalized morphisms, compatibility etc. can be rigorously extended to relations based on any system of fuzzy logic using continuous residuated t-norms. Rigorous proofs in the first order predicate calculus for BL family of fuzzy logics of Petr Hájek were given by Kohout [32]. Kohout [31] has also demonstrated that these relational calculi extend outside BL to the systems based on *left-continuous* t-norm family of fuzzy logics, and are even extendable to general lattice based monoidal logics pioneered by Höhle.

5.2 Motivation for Generalized Morphisms: Standard Homomorphisms

Let A, B, C, D be sets with relations R, S upon them – R from A to B and S from C to D, where each relation determines some structure. In addition, we have homomorphic mappings F and G. F is from A to C and G is from B to D. The usual way of visualizing this is by eans of arrow diagrams, as shown in Fig.1.

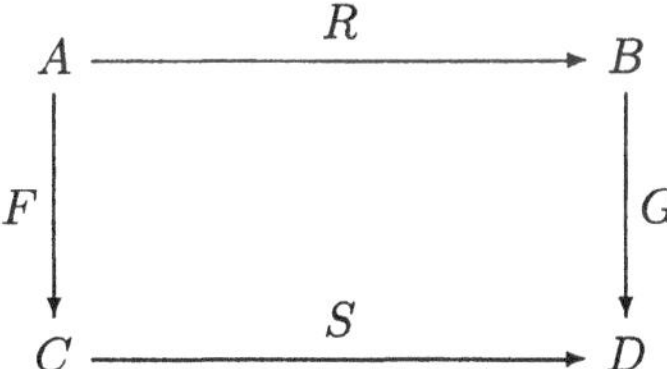

Fig. 1. A diagram for conventional homomorphisms. Here homomorphic maps F and G are functions, i.e. univalent and covering relations.

There are two points of departure that stem from this fundamental algebraic notion of *homomorphism*:

1. the design or checking mappings which will "preserve" or "respect" certain given relations, and on the other hand
2. the design or checking of relations which "absorb" or "validate" certain given mappings.

For example let $A = B$, $C = D$ and R, S be orders. Given A and C we wish to find one or all the mappings from A to C that preserve orders – this

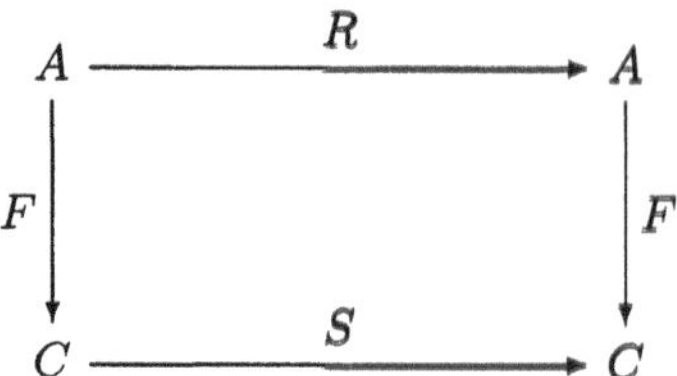

Fig. 2. Structure preserving mappings.

illustrates the case (1). This can be pictured by a commutative diagram - see Fig.2.

An example of (2) is, given a mapping from A to C, how to match the order on A given by R, with some other order on C, or vice versa – so that some given mapping will preserve or co-preserve them.

In this situation the conventional homomorphism yields a commuting diagram of arrows such that $R \circ G = F \circ S$, where of course, the morphisms F and G are the relations which are both covering and univalent (i.e. functional) as shown in Fig.1. To obtain the constructions that solve the problems (1) or (2) requires to solve the above **relational equation** with respect to one of the relations R, S, F or G.

5.3 Generalized Morphisms and Proteromorphisms

When the mappings (functional relations) F and G are *replaced by general relations*, the equation is no longer valid but has to be **replaced by two inequalities**. The notion of a homomorphism splits into two independent notions, generalized morphism and generalized proteromorphism.

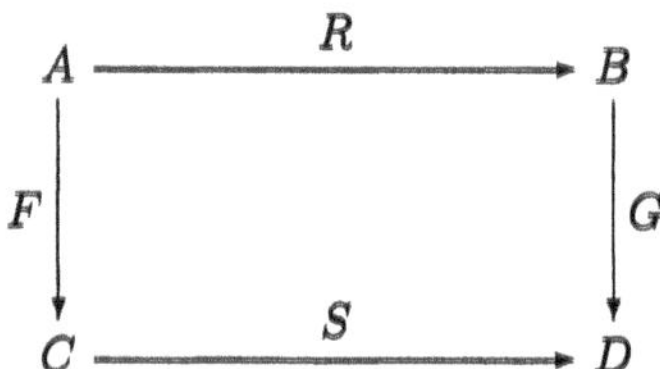

Fig. 3. A diagram for forward and backward compatibility. Here homomorphic maps F and G multivalued, i.e. are general relations, not just functions.

Definition 10 Generalised Morphisms.
Let F, R, G, S be heterogenous relations between the sets A, B, C, D such that $F \in \mathcal{R}(A \rightsquigarrow C)$, $R \in \mathcal{R}(A \rightsquigarrow B)$, $G \in \mathcal{R}(B \rightsquigarrow D)$, $S \in \mathcal{R}(C \rightsquigarrow D)$ (see Fig. 3).

The conditions that (for all $a \in A$, $b \in B$, $c \in C$, $d \in D$)

$$\left.\begin{array}{l} aFc \\ aRb \\ bGd \end{array}\right\} \Longrightarrow cSd,$$

will be expressed in any of the following ways:

1. *FRG: S are compatible forward or forward-compatible.*
2. *F, G respect R, S forwards*
3. *R, S absorbs F, G forwards*
4. *F, G are generalized homorphisms from R to S.*

Theorem 11 (Forward Compatibility)
FRG: S are forward-compatible iff
$F^{-1} \circ R \circ G \sqsubseteq S$

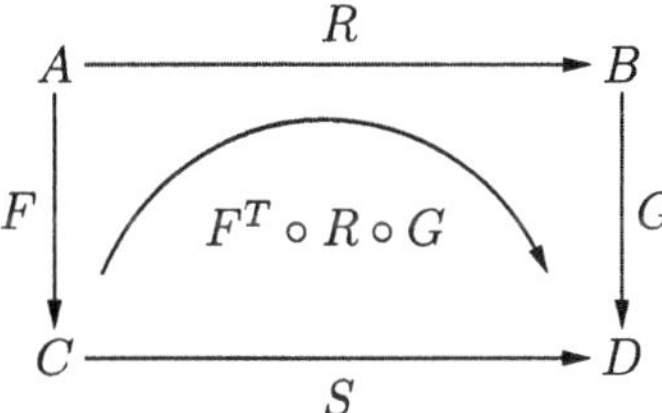

Fig. 4. Forward Compatibility

Definition 12 Generalized Proteromorphisms. *Let F, R, G, S be heterogenous relations between the sets A, B, C, D such that $F \in \mathcal{R}(A \leadsto C)$, $R \in \mathcal{R}(A \leadsto B)$, $G \in \mathcal{R}(B \leadsto D)$, $S \in \mathcal{R}(C \leadsto D)$ (see Fig. 4).*

The condition that (for all $a \in A$, $b \in B$, $c \in C$, $d \in D$)

$$\left.\begin{array}{l} aFc \\ cSd \\ bGd \end{array}\right\} \Longrightarrow aRb$$

will be expressed in any of the following ways:

1. *FRG: S are compatible backward or backward-compatible*
2. *F, G respect R, S backwards*
3. *R, S absorb F, G backwards*
4. *F, G are generalized proteromorphisms from R to S.*

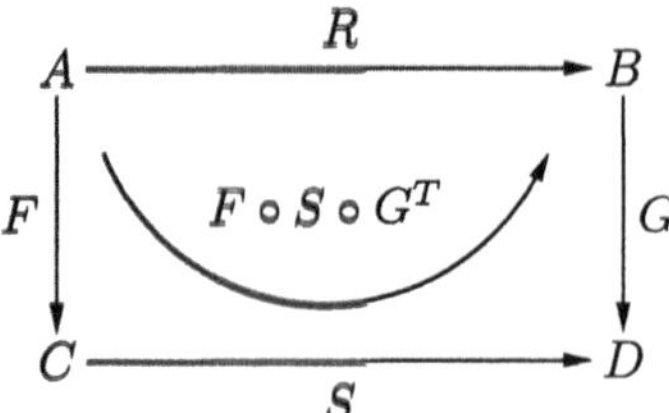

Fig. 5. Backward Compatibility

Theorem 13 (Backward Compatibility)

FRG:S are backward-compatible iff
$F \circ S \circ G^{-1} \sqsubseteq R$

The simultaneous fulfillment of the conditions in Definitions 10 and 12 will be expressed as compatibility, respect, absorption **both ways** and as **generalized amphimorphisms**

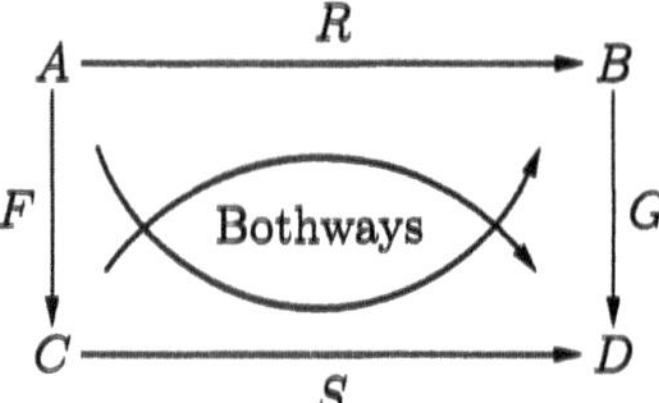

Fig. 6. Bothways Compatibility

6 Solution of Relational Inequalities Representing Generalized Morphisms

Inequalities defined by conventional circle products $\circ$ presented in the previous section are existentional definitions that would satisfy a mathematician interested in the existentional theories which are, however, not constructive. A computer scientist in addition to this general mahematical knowledge also wishes to know how to solve these inequalities comptationally. BK-products of relations provide rigorous mahematical specification of such solutions that are constructive. Hence, these allow us to write efficient computer algorithms for computing with generalized morphisms.

6.1 An Overview of Solutions in t-Norm Residuated Fuzzy Logics

Theorem 14 COMPATIBILITY SOLUTION.
1. $FRG : S$ *are forward compatible* $\iff F^T \circ R \circ G \sqsubseteq S \iff R \sqsubseteq F \lhd S \rhd G^T)$

2. $FRG : S$ *are backward compatible* $\iff F \circ S \circ G^T \sqsubseteq R \iff S \sqsubseteq F^T \lhd R \rhd G$

$FRG : S$ are both-ways compatible iff they are both forward and backward compatible. The R's of forward compatibility constitute a lower ideal, while those of backward compatibility constitute an upper ideal or filter; the bothways problem has a solution iff their intersection is non-empty. The conventional homomorphism is a special case of *Both-ways* compatibility. The approach encompasses a broader range of inequaltites than would superficially apear. Deriving (by proofs in BL) the residuated bootstrap inequality in BK-product first, one obtains algebraic inequalities that can be used for derivation of other relational equations and inequallities purely symbolicaly in algebra of BK-products, bypassing more cumbersome derivations in BL.

Theorem 15 (Criteria for F and G)

1. $FRG : S$ *are forward-compatible iff*
   (a) $F \sqsubseteq R \lhd (G \lhd S^{-1})$
   or equivalently
   (b) $G \sqsubseteq R^{-1} \lhd (F \lhd S)$
2. $FRG : S$ *are backward-compatible iff*
   (a) $F \sqsubseteq (R \rhd G) \rhd S^{-1}$
   or equivalently
   (b) $G \sqsubseteq (R^{-1} \rhd F) \rhd S$
3. $FRG : S$ *are accordingly bothways-compatible iff*
   (a) $F \sqsubseteq R \lhd (G \lhd S^{-1}) \sqcap (R \rhd G) \rhd S^{-1}$
   or equivalently
   (b) $G \sqsubseteq R^{=1} \lhd (F \lhd S) \sqcap (R^{-1} \rhd F) \rhd S$

4. *The relations which appear in these criteria have the following alternative forms:*

$$R \lhd (G \lhd S^{-1}) = (R \circ G) \lhd S^{-1}$$
$$R^{-1} \lhd (F \lhd S) = (R^{-1} \circ F) \lhd S$$
$$(R \rhd G) \rhd S^{-1} = R \rhd (G \circ S^{-1})$$
$$(R^{-1} \rhd F) \rhd S = R^{-1} \rhd (F \circ S).$$

6.2 The Range of Validity of Solutions

The results presented hold not only for t-norm based residuated relations but are broader. They extend also to relational systems based on left-continuous t-norm residuated fuzzy logics, and even to general theory of relations based on fuzzy monoidal logics. The proofs in BL predicate calculus are rigorous but tedious. The right strategy is to proof the basic inequalities such as

$$R \circ S \sqsubseteq V \text{ iff } R \sqsubseteq V \rhd S^T \text{ iff } S \sqsubseteq R^T \lhd V$$

(as we have done in section 4.2) and proceed with further derivations algebraically. We have shown how this works in section 4.4 above for relational equations. Similarly, all the inequaltites of previous sections can be obtained by double application [8] of residuation botstrap inequalities of Theorem 2 in section 4.2 above.

7 Applications

This section presents brief discusssion and references to papers on some selected applications. A more extensive surveys with more extensive list of references can be found in [35],[36],[41],[46],[40],[47].

7.1 BK-Products in Data Analysis and Knowledge Representation

$\lhd, \rhd, \square$ products add the expressive power to the mathematics of relations. Universal representation of preorders and similaritites is provided by residuated BK-products defined above. Testing relational properties, discovering local and global fuzzy preorders, similarities and equivalences and approximating these by fuzzy closures and interiors is the important technique for data analysis that we have developed [36],[35]. Another important relational property is *classivalence* (partial difunctionality). Elements *aRb, aRb', a'Rb, a'Rb'* of a relation are *classivalent* or *partial difunctional* if they satisfy the logic expression *aRb&a'Rb&a'Rb'* $\rightarrow$ *aRb'*. Classivalence is a generalization of equivalence and replaces fuzzy similarity when we have to match structures comming from two different contexts. Classivalence (like preorders and similarities) can also be characterised by fuzzy relational inequalities and equations containing BK-products [37].

The essential facet of the fuzzy approach is the fact that fuzzy relations can manipulate semantic information that is carried by linguistic labels. For this purpose we introduce semiotic descriptors that can be defined by relational equations. Their purpose is to capture the ontology and epistemology of the relational model in which the knowledge is represented, for example *"physics"* in mechanical engineering or geodynamics. The basic ontological categories of semiotic descriptors are: *objects, qualities and relationships.* Each category may contain various kinds of semiotic descriptors [46].

7.2 BK-products in Engineering, Medicine and Sciences

Triangle relational products together with fast fuzzy relational algorithms [7],[12] have been applied to various practical problems in a number of scientific fields: computer protection and AI [29], medicine, information retrieval, handwriting classification, [42], architecture and urban studies, investment and control fields [38]. A promising recent application is concerned with generating efficient search strategies for resolution-based theorem proving. See the survey in [39] with a list of 50 selected references on the theory and applications and [36],[35] for updates. The relational methods combining linguistic labels with BK-products give a natural conceptual framework for knowledge representation and inference from imprecise, incomplete, or not totally reliable information in a consistent manner. All these approaches can extended to the realm of interval computations [40],[27]. For example or knowledge-based medical system CLINAID [28] combines fuzzy relations, with methods of interval inference [40].

BK-triangle products are essential for the characterization of granularity, knowledge representation and relational computations in fuzzy systems for *Computing with Words* [41]. They are also used in relational interval computations in Intelligent Systems [49],[28]. Application of Generalized Morphisms to design and implementation of Communicating Agents in Distributed Intelligent Systems is particularly promising [30],[51]. Other applications include diagnostic data and patient management, [3], [6],[9], medical sign and symptom comparison [17]; information retrieval [45], handwriting classification, [44], natural language understanding [52], cognitive structure analysis, elicitation of cognitive maps of knowledge of human experts and extracting the consensus structures of a group of experts/engineers; also in generating efficient search strategies for resolution-based theorem proving [48], analysis of protection and security of computing systems [29]. Important application is relational value analysis applied to industrial engineering (see [47],[46] for references).

In all these applications the unifying thread runs through the Activity Structures methodology. The significance of relational equalities, inequalities and generalized morphisms expressed by means of fuzzy BK-products of relations lies in the fact that these provide mathematical and logic methods that unify the conceptual, methodological, and formal/algorithmic facets of Activity Structures methodology in a pragmatic workable framework (see chapter 5 of [38] and [29],[35]).

7.3 Current Development

Interesting current work in progress is in the area Geographic Information Systems, in analysis and contextual matching of remotely sensed data obtained from satalite images of the landscape and its transformation to symbolic relational representation. GUHA and relational analysis with BK-products, two mutually complementary methods are used in this work. Fuzzy relational computations with BK-products extract clusters and hierarchies latently contained in data. GUHA method for statistical exploratory analysis by logic method that has been under developed by Dr. Hájek and his team at the

Czech Academy of Sciences, Prague [24] systematically creates all hypotheses interesting from the point of view specified by logic means from the base of given data. GUHA produces poly-factorial hypotheses, not only relating one variable with another, but also expressing relationships among single variables, pairs, triples, quadruples and larger combination of variables. The generalized GUHA quantifiers as used by Hájek at al. are interestingly related to the checklist paradigm of Bandler and Kohout that provides semantics for fuzzy logic connectives used by BK-products [23].

An application of BK-products and relational equations is envisaged in Partial Differential Equations Problem Solving environments [50].

7.4 Relevance to Geosciences

Characteristic Analysis used in Geosciences (e.g. by Blobol, Chaves etc.) can be viewed as a special case of GUHA methods. BK-products and GUHA techniques used for analysis of remote sensed data is applicable to evaluating and classifying features of geological maps. Explorationists learn through personal experience and the collective experience of the industry. The rationale in applying geology to exploration is based on applicability and transfer of knowledge characterising the old oil fields to the new old fields that are similar geologically. If experience is quantified and formalized, then the experience gained in the old oil fields can help in assessing new region by application of relationships linking geology and oil occurrence. Here GUHA method may be of considerable interest. BK-product methods and Fast Fuzzy relational algorithms can also be used for eliciting and comparing subjective knowledge of experts [46],[47].

References

1. J. Abbott, A. van Leeuwen, and A. Strotmann. OpenMath: Communicating mathematical information between co-operating agents in a knowledge network. *Journal of Intelligent Systems* **8**, No. 3–4, 1998.
2. W. Bandler and L.J. Kohout. *Mathematical Relations, their Products and Generalized Morphisms.* Tech. report, Man-Machine Systems Laboratory, EES-MMS-REL 77-3, Dept. of Electrical Eng., University of Essex, Colchester, Essex, U.K., 1977. Reprinted as Ch. 2 in: Kohout, L.J. & Bandler, W., Survey of Fuzzy and Crisp Relations, Lect. Notes in Fuzzy Mathematics and Computer Sci., Creighton Univ. Omaha (to appear).
3. W. Bandler and L.J. Kohout. The use of new relational products in clinical modelling. In B.R. Gaines, editor, *General Systems Research: A Science, a Methodology, a Technology (Proc. 1979 North American Meeting of the Society for General Systems Research).*, pages 240–246, Louisville KY, January 1979. Society for General Systems Research.
4. W. Bandler and L.J. Kohout. Fuzzy power sets and fuzzy implication operators. *Fuzzy Sets and Systems*, **4**, 13–30, 1980. Reprinted in: *Readings in Fuzzy*

Sets for Intelligent Systems, D. Dubois, H. Prade and R. Yager (eds.), Morgan Kaufmann Publishers, San Mateo, Calif., 1993, pages 88-96.
5. W. Bandler and L.J. Kohout. Fuzzy relational products as a tool for analysis and synthesis of the behaviour of complex natural and artificial systems. In P.P. Wang and S.K. Chang, editors, *Fuzzy Sets: Theory and Applications to Policy Analysis and Information Systems*, pages 341–367. Plenum Press, New York and London, 1980.
6. W. Bandler and L.J. Kohout. Semantics of implication operators and fuzzy relational products. *Internat. Journal of Man-Machine Studies*, **12**, 89–116, 1980. Reprinted in Mamdani, E.H. and Gaines, B.R. eds. *Fuzzy Reasoning and its Applications*. Academic Press, London, 1981, pages 219-246.
7. W. Bandler and L.J. Kohout. Fast fuzzy relational algorithms. In A. Ballester, D. Cardús, and E. Trillas, editors, *Proc. of the Second Internat. Conference on Mathematics at the Service of Man*, pages 123–131, Las Palmas, 1982. (Las Palmas, Canary Islands, Spain, 28 June - 3 July), Universidad Politechnica de las Palmas.
8. W. Bandler and L.J. Kohout. On the general theory of relational morphisms. *International Journal of General Systems*, **13**, 47–66, 1986.
9. W. Bandler and L.J. Kohout. A survey of fuzzy relational products in their applicability to medicine and clinical psychology. In L.J. Kohout and W. Bandler, editors, *Knowledge Representation in Medicine and Clinical Behavioural Science*, pages 107–118. Gordon and Breach Publ., London and New York, 1986.
10. W. Bandler and L.J. Kohout. Fuzzy implication operators. In M.G. Singh, editor, *Systems and Control Encyclopedia*, pages 1806–1810. Pergamon Press, Oxford, 1987.
11. W. Bandler and L.J. Kohout. Relations, mathematical. In M.G. Singh, editor, *Systems and Control Encyclopedia*, pages 4000 – 4008. Pergamon Press, Oxford, 1987.
12. W. Bandler and L.J. Kohout. Special properties, closures and interiors of crisp and fuzzy relations. *Fuzzy Sets and Systems*, **26**, 317–332, June 1988.
13. W. Bandler and L.J. Kohout. Cuts commute with closures. In B. Lowen and M. Roubens, editors, *Fuzzy Logic: State of the Art*, pages 161–167. Kluwer Academic, Boston and Dordrecht, 1993.
14. W. Bandler and L.J. Kohout. On the universality of the triangle superproduct and the square product of relations. *Internat. Journal of General Systems*, **25**, 399–403, 1997.
15. W. Bandler, V. Mancini, and L.J. Kohout. Investigating the structures of knowledge and thought. In G.E. Lasker, editor, *Advances in Computer Science*. International Institute for Advanced Studies in Systems Research and Cybernetic, Ontario, Canada, 1989. ISBN 0-921856-03-1.
16. R. Bělohlávek. Similarity relations and BK-relational products. *Information Sciences*, **126**, 287–295, 2000.
17. B. Ben-Ahmeida, L.J. Kohout, and W. Bandler. The use of fuzzy relational products in comparison and verification of correctness of knowledge structures. In L.J. Kohout, J. Anderson, and W. Bandler, editors, *Knowledge-Based Systems for Multiple Environments*, chapter 16. Ashgate Publ. (Gower), Aldershot, U.K., 1992.
18. B. DeBaets and E. Kerre. Fuzzy relational compositions. *Fuzzy Sets and Systems*, **60**, 109–120, 1993.

19. B. DeBaets and E. Kerre. A revision of Bandler-Kohout composition of relations. *Mathematica Pannonica*, **4** ,59–78, 1993.
20. B. DeBaets and E. Kerre. The cutting of compositions. *Fuzzy Sets and Systems*, **62**, 295–310, 1994.
21. P. Hájek. A remark on Bandler-Kohout products of relations. *Int. J. of General Systems*, **25** ,165–166, 1996.
22. P. Hájek. *Metamathematics of Fuzzy Logics*. Kluwer, Dordrecht, 1998.
23. P. Hájek and L.J. Kohout. Fuzzy implications and generalized quantifiers. *Internat. Journal of Uncertainty, Fuzziness and Knowledge Based Systems*, **4** ,225–233, 1996.
24. Zvárová J. Hájek P., Sochorová A. Guha for personal computers. *Computational Statistics & Data Analysis*, **19**, 149–153, 1995.
25. U. Höhle. Commutative, residuated l-monoids. In U. Höhle and E.P. Klement, editors, *Epistemological aspects of many-valued logics and fuzzy structures*, chapter 4, pages 53–106. Kluwer, Dordrecht and Boston, 1995.
26. B. Juliano and W. Bandler. *Tracing Chains-of-Thought: Fuzzy Methods in Cognitive Diagnosis*. Physica Verlag. A Springer-Verlag Company, Heidelberg, 1996.
27. L. J. Kohout and W. Bandler. Interval-valued systems for approximate reasoning based on the checklist paradigm. In P. Wang, Paul, editor, *Advances in Fuzzy Theory and Technology, vol. 1*, pages 167–193. Bookwrights Press, Durham, N.C., 1993.
28. L. J. Kohout, I. Stabile, W. Bandler, and J. Anderson. CLINAID: Medical knowledge-based system based on fuzzy relational structures. In M. Cohen and D. Hudson, editors, *Comparative Approaches in Medical Reasoning*, pages 1–25. World Scientific, 1995.
29. L.J. Kohout. *A Perspective on Intelligent Systems: A Framework for Analysis and Design*. Chapman and Hall & Van Nostrand, London & New York, 1990. In 1991, received an international prize from The International Institute for Advanced Studies in Systems Research: *"The best book of the year in the area of AI Systems"*.
30. L.J. Kohout. Dynamic sharing of cognitive structures by means of generalized morphisms. In P.P. Wang, editor, *Proc. of the 2nd Joint Conf. on Information Sciences (Sept. 28 - Oct. 1, Wrightsville Beach, NC, USA)*, pages 597–600. Duke University, 1995.
31. L.J. Kohout. Theory and applications of non-associative products of relations. In T. Childers, editor, *The Logica Yearbook 1997*, chapter 16, pages 152–164. Filosofia (Publisher of the Institute of Philosophy, Czech Academy of Sciences), Prague, 1998. (ISBN 80-7007-109-5).
32. L.J. Kohout. Generalized morphisms in BL-logics. *The Bulletin of Symbolic Logic*, **5**, 116–117, 1999.
33. L.J. Kohout. Basic algebra of BK products of relations in t-norm fuzzy logics. In R. Langari and J. Yen, editors, *Proc. of FUZZ-IEEE2000*, pages 599–604, Piscataway, NJ, 2000. IEEE Neural Network Council, IEEE.
34. L.J. Kohout. Extension of Tarski's axioms of relations to t-norm fuzzy logics. In P.P. Wang, editor, *Proc. of 5th Joint Conference on Information Sciences*, pages 44–47, Durham, NC, February 27 — March 3, 2000. Association for Intelligent Machinery. vol. 1.

35. L.J. Kohout. Foundations of knowledge engineering for systems with distributed intelligence: A relational approach. In A. Kent and J.G. Williams, editors, *Encyclopedia of Microcomputers, volume 24*, pages 127–158. Marcel Dekker Inc., New York and Basel, 2000.
36. L.J. Kohout. Boolean and fuzzy relations. In P.M. Pardalos and C.A. Floudas, editors, *The Encyclopedia of Optimization*, pages 189–202. Kluwer, Boston, 2001. vol.I, A-D.
37. L.J. Kohout. Classivalent and difunctional relations in the interval calculus of fuzzy BK–products. In *Proc. of IFSA and NAFIPS conferences.*, pages 1659–1664. IFSA & NAFIPS, IEEE, July 2001. ISBN: 0-7803-7079-1, IEEE Catalog Number: 01TH8569C.
38. L.J. Kohout, J. Anderson, and W. Bandler. *Knowledge-Based Systems for Multiple Environments*. Ashgate Publ. (Gower), Aldershot, U.K., 1992. A Scientific Monograph, 382 pages. Written by the principal author in collaboration with others. Awarded *"Outstanding Scholarly Contribution Award"* by the Systems Research Foundation in 1993.
39. L.J. Kohout and W. Bandler. Fuzzy relational products in knowledge engineering. In V. Novák et al., editor, *Fuzzy Approach to Reasoning and Decision Making*, pages 51–66. Academia and Kluwer, Prague and Dordrecht, 1992.
40. L.J. Kohout and W. Bandler. Fuzzy interval inference utilizing the checklist paradigm and BK-relational products. In R.B. Kearfott and V. Kreinovich, editors, *Applications of Interval Computations*, chapter 12, pages 291–335. Kluwer, Boston, 1996.
41. L.J. Kohout, B. Granville, and E. Kim. Granular relational computing with semiotic descriptors using BK-products of fuzzy relations. In P. Wang, Paul, editor, *Computing with Words*, chapter 4, pages 89–146. John Wiley, New York, 2001.
42. L.J. Kohout and M. Kallala. Evaluator of neurological patients' dexterity based on relational fuzzy products. In *Proc. of Second Expert Systems International Conference (London, October 1986)*, pages 1–12. Learned Information Inc., New Jersey, USA and Oxford, UK, 1986.
43. L.J. Kohout and M. Kallala. Choice of fuzzy optimal logics for pattern classifiers by means of measure analysis. In *Proc. of 7th International Congress on Cybernetics and Systems*, Imperial College, London, September 1987.
44. L.J. Kohout and M. Kallala. The use of fuzzy information retrieval in knowledge-based management of patients' clinical profiles. In B. Bouchon and R.R. Yager, editors, *Uncertainty in Knowledge-Based Systems (Lecture Notes in Computer Science vol. 286)*, pages 275–282. Springer Verlag, Berlin, 1987.
45. L.J. Kohout, E. Keravnou, and W. Bandler. Automatic documentary information retrieval by means of fuzzy relational products. In B.R. Gaines, L.A. Zadeh, and H.-J. Zimmermann, editors, *Fuzzy Sets in Decision Analysis*, pages 383–404. North-Holland, Amsterdam, 1984.
46. L.J. Kohout and E. Kim. Reasoning with cognitive structures of agents I: :Acquisition of rules for computational theory of perceptions by fuzzy relational products. In Da Ruan and E. Kerre, editors, *Fuzzy IF-THEN Rules in Computational Intelligence*, chapter 8, pages 161–188. Kluwer, Boston, 2000.
47. L.J. Kohout and E. Kim. The role of BK-products of relations in soft computing. *Soft Computing*, **6**, 92–115, 2002.

48. L.J. Kohout and Yong-Gi Kim. Generating control strategies for resolution-based theorem provers by means of fuzzy relational products and relational closures. In B. Lowen and M. Roubens, editors, *Fuzzy Logic: State of the Art*, pages 181–192. Kluwer Academic, Boston and Dordrecht, 1993.
49. L.J. Kohout and I. Stabile. Relational computations in medical KBS CLINAID: the means for integrating interval, symbolic logic and neural network techniques. In T. et al. Furuhashi, editor, *Proc. of the 1st On-Line Workshop of Soft Computing*, pages 228–233, Nagoya, 464-01, Japan, August 1996. The Society of Fuzzy Theory and Systems (SOFT), Nagoya University. http://www.bioele.nuee.nagoya-u.ac.jp/wsc1.
50. L.J. Kohout, A. Strotmann, and R. van Engelen. Knowledge engineering methods for climate models. In *2001 IEEE Internat. Conference on Systems, Man & Cybernetics: SMC 2001 Conf. Proc.*, pages 2193–2198. Systems, Man and Cybernetics Society of the IEEE, IEEE, October 2001. CD-ROM Proceedings, IEEE catalog number 01CH37236C.
51. J. Muhammad and L.J. Kohout. The role of generalized morphisms in cooperative activities of agents. In P.P. Wang, editor, *Joint Conference on Information Sciences - JCIS'98 Proceedings*, pages 56–59, Research Triangle Park, N.C., October 1998. Association for Intelligent Machinery. In vol.1.
52. S. Nagarajan and L.J. Kohout. Fuzzy syntactic and semantic analyzer. In R. Lowen and M. Roubens, editors, *Proceedings of the 4th IFSA World Congress IFSA91 Brussels, Vol. Artificial Intelligence*, pages 149–152. International Fuzzy Systems Association, 1991.
53. A. Strotmann and L.J. Kohout. OpenMath: Compositionality achieved at last. *ACM SIGSAM Bulletin*, (December), 2000. Special issue on OpenMath, Mike Dewar, ed.

Decomposition of Fuzzy Relations and Functional Relations

Elie Sanchez

University of Aix-Marseille II
Service Universitaire de Biomathématiques, Faculté de Médecine, 27 Bd Jean-Moulin, 13385 Marseille Cedex5, France, elie.sanchez@medecine.univ-mrs.fr

Abstract: Many general problems on the resolution of composite fuzzy relation equations have been investigated for nearly thirty years. This paper now shows how to solve a class of SUP-min fuzzy relational equations, such as $T = R o Q$, where Q is a functional relation. Such equations are related to systems of truth-qualified propositions and the concept of the decomposition of a fuzzy relation by a fuzzy set is introduced. Finally, the problem of *'decomposability'* of a fuzzy relation by a given fuzzy set is solved by combining these methodologies.

Introduction

Resolution of fuzzy relational equations was initiated in the mid-seventies (Sanchez 1974, 1976). The class of usual Zadeh's fuzzy sets (Zadeh 1965) was extended to L-fuzzy sets, where L was a complete Brouwerian lattice. The aim was to propose the largest class of lattice-valued fuzzy sets to which the resolution methodology could apply. Since then, hundreds of papers have been written on this and related topics (see for ex. Di Nola et al. 1989, De Baets 2000, Peeva and Kyosev 2003), with important developments. Zadeh's seminal work on the application of fuzzy sets theory and approximate reasoning to natural languages (e.g. Zadeh 1975, 1978a, 1979, 1999) has been a great source of inspiration to many researchers. In particular, the inverse problem for specifying truth-qualified fuzzy propositions has been investigated in (Sanchez 1978, 1996). In this spirit, systems of semantic equivalences of a truth-qualified proposition can be related to fuzzy relation equations involving a functional relation. Because of the specificity of these equations, new properties, that do not hold in the general case, are expected to hold. It will be seen how these results suggest the concept of the decomposition of a fuzzy relation by a fuzzy set. Finally, the problem of *'decomposability'* will be addressed.

Recall of basic definitions and general resolution of fuzzy relational equations

In what follows, U, V and W are universes of discourse.

Composition of fuzzy relations (Zadeh 1965)

Let Q be a fuzzy relation from U to V, R a fuzzy relation from V to W, then the SUP-min composition T = RoQ, T fuzzy relation from U to W, is defined as: $\forall(u,w) \in U \times W$,

$$\mu_{RoQ}(u,w) = SUP_{v \in V} [\mu_Q(u,v) \wedge \mu_R(v,w)] \quad (1)$$

(a t-norm (Schweizer and Sklar 1963) can be used instead of the min operation ($\wedge$) in the SUP-min composition.)

Note that when R and Q are crisp relations, i.e. when their membership functions can only take values 0 and 1, the following property is derived from the definition of the SUP-min composition : $(u,w) \in RoQ$, iff, $\exists\, v \in V$ such that $(u,v) \in Q$ and $(v,w) \in R$, which corresponds to a usual notation and definition of the composition of two (non-fuzzy) relations. Moreover, when R and Q are functions, $(u,w) \in RoQ$ is denoted $w = (RoQ)(u)$ or $w = R(Q(u))$.

These remarks explain the choice for notation RoQ (as in the Zadeh's original paper Zadeh 1965), instead of QoR used by some researchers, in the definition of the SUP-min composition. Notation QoR is more in the spirit of the notation for matrix multiplication.

As a particular case of composition, if A is a fuzzy subset of U and R a fuzzy relation from U to V, B = RoA, fuzzy subset of V, is defined as

$$\forall v \in V,\ \mu_{RoA}(v) = SUP_{u \in U} [\mu_A(u) \wedge \mu_R(u,v)] \quad (2)$$

Zadeh's extension principle (Zadeh 1975)

Let A be a fuzzy subset of U, and let f be a mapping from U to V. The extension principle asserts that the image of A under f is a fuzzy subset of V, defined as

$$\forall v \in V,\ \mu_{f(A)}(v) = SUP_{u \in f^{-1}(v)} \mu_A(u) \quad \text{for } f^{-1}(v) \neq \varnothing \quad (3)$$
$$= 0 \quad \text{elsewhere.}$$

A usual mathematical convention states that $SUP_{u \in \varnothing} \mu_A(u) = 0$, so that in (3) it is not necessary to write that $\mu_{f(A)}(v) = 0$ when $f^{-1}(v) = \varnothing$.

Let us consider a two-place $*$ operation defined on a product set UxV and taking values on W. From Zadeh's extension principle, this $*$ operation can also be extended to fuzzy sets : let A be a fuzzy subset of U and B a fuzzy subset of V, C = A*B, fuzzy subset of W, is defined in the case of noninteractive variables as

$$\forall w \in W,\ \mu_{A*B}(w) = SUP_{u \in U, v \in V,\ u*v=w} [\mu_A(u) \wedge \mu_B(v)] \quad (4)$$

Functional relations

A relation Q from U to V is said to be functional if, and only if, $\forall u \in U$, $\exists! v(u) \in V$ (i.e. v(u) is unique), s.t. $\mu_Q(u,v(u)) = 1$ and

$$\mu_Q(u,v) = 0\,,\ \ \forall v \in V-\{v(u)\}.$$

In the finite case, a functional relation is represented by a matrix such that for every row, there is one and only one 1, the other elements being equal to 0.

Regarding now a mapping f from U to V, as a functional relation, i.e. $\mu_f(u,v) = 1$ if $f(u) = v$ and $\mu_f(u,v) = 0$ otherwise, one has $\mu_{foA}(v) = SUP_{u \in U}\,[\mu_A(u) \wedge \mu_f(u,v)] = SUP_{u \in f^{-1}(v)}\, \mu_A(u) = \mu_{f(A)}(v)$, so that (3) is equivalent to

$$f(A) = foA \tag{5}$$

where foA expresses the SUP-min composition of f with the fuzzy set A.

Two-place α operator in the unit interval [0,1]

It is defined as, for a and b in [0,1],

$$\begin{aligned} a\,\alpha\,b &= 1, \ \text{ if } a \le b \\ &= b, \ \text{ if } a > b \end{aligned} \tag{6}$$

Remark. We introduced this algebraic α operator in (Sanchez 1974), (see also Sanchez 1976, 1977; Kaufmann 1977; Di Nola et al. 1989), in the framework of fuzzy set theory, just to solve a problem that had nothing to do with logic : the resolution of SUP-min composite fuzzy relation equations. Since then, abundant literature has presented prolific and important developments (although sometimes with some confusion) in many areas related to fuzzy set theory or fuzzy logic. In a more general setting, when a lattice L replaces the [0,1] interval as the set of values of membership functions, the α operator characterizes Brouwerian lattices. We recall that a Brouwerian lattice (Birkhoff 1967) is a lattice L in which, for any given elements a and b, the set of all $u \subset L$, such that $a \wedge u \le b$, contains a greatest element (cf. $a\,\alpha\,b$), called the relative pseudo-complement of a in b. In logic, the α operator corresponds to the so-called Gödel's implication.

The ⓐ, or INF-α, composition (Sanchez 1974, 1976) of fuzzy relations

Let Q be a fuzzy relation from U to V, and R a fuzzy relation from V to W, then the INF-α composition T = Q ⓐR, T fuzzy relation from U to W, is defined as

$$\forall (u,w) \in UxW,\ \mu_T(u,w) = INF_{v \in V}\,[\mu_Q(u,v)\ \alpha\ \mu_R(v,w)] \tag{7}$$

Transpose of a fuzzy relation

Let Q be a fuzzy relation from U to V, Q^{-1} the transpose of Q, fuzzy relation from V to U, is defined as

$$\forall (v,u) \in VxU,\ \ \mu_{Q^{-1}}(v,u) = \mu_Q(u,v) \tag{8}$$

Two-place γ (equality) operator in the unit interval [0,1]
It is defined as, for a and b in [0,1],

$$a \gamma b = 1, \text{ if } a = b$$
$$= 0, \text{ if } a \neq b \quad (9)$$

Note that a functional relation Q from U to V, can then be defined as : $\forall u \in U$, $\exists! v(u) \in V$, such that $\mu_Q(u,v) = v \gamma v(u)$.

The γ operator is commutative but not associative: for ex. $(0.5 \gamma 0.5) \gamma 1 = 1$ and $0.5 \gamma (0.5 \gamma 1) = 0$, and $\forall a \in [0,1]$, $a \alpha 0 = a \gamma 0$.

The γ operation can be extended to fuzzy sets via the extension principle as follows. Let A and B be fuzzy subsets of U, $C = A\gamma B$, fuzzy subset of {0,1} (recall that γ is a two-place operation with a result in {0,1}), is defined as

$$\forall z \in \{0,1\}, \mu_{A\gamma B}(z) = \text{SUP}_{u \in U, v \in U,\, u\gamma v = z} [\mu_A(u) \wedge \mu_B(v)] \quad (10)$$

For z = 1, (10) becomes a possibility measure (Zadeh 1978b) :

$$\mu_{A\gamma B}(1) = \text{SUP}_{u=v} [\mu_A(u) \wedge \mu_B(v)]$$
$$= \text{SUP}_{u \in U} [\mu_A(u) \wedge \mu_B(u)]$$
$$= \text{SUP}_{u \in U} [\mu_{A \cap B}(u)]$$
$$= \text{SUP}(A \cap B) = \Pi(A \mid B) \quad (11)$$

The ⓖ, or INF-γ, composition of fuzzy relations.
Let Q be a fuzzy relation from U to V, and R a fuzzy relation from V to W, then the INF-γ composition T = Q ⓖ R, T fuzzy relation from U to W is defined as

$$\forall (u,w) \in U \text{x} W, \mu_T(u,w) = \text{INF}_{v \in V} [\mu_Q(u,v) \gamma \mu_R(v,w)] \quad (12)$$

This ⓖ composition can be useful in the study of fuzzy relation equations (Sanchez 1974).

The following results were proved in (Sanchez 1974, 1976).

Lemma 1. For every pair of fuzzy relations Q from U to V and R from V to W, one has

$$R \subseteq Q^{-1} ⓐ (RoQ) \quad (13)$$

Lemma 2. For every pair of fuzzy relations Q from U to V and R from V to W, one has

$$Q \subseteq (R ⓐ (RoQ)^{-1})^{-1} \quad (14)$$

Lemma 3. For every pair of fuzzy relations Q from U to V and T from U to W, one has

$$(Q^{-1} @ T) o Q \subseteq T \tag{15}$$

Lemma 4. For every pair of fuzzy relations R from V to W and T from U to W, one has

$$Ro(R @ T^{-1})^{-1} \subseteq T \tag{16}$$

Here are now two fundamental theorems for the resolution of fuzzy relational equations.

Theorem 5. Let Q from U to V and T from U to W, be two given fuzzy relations, denote **R** the set of fuzzy relations R from V to W, such that RoQ = T; then

$$\mathbf{R} = \{R \mid RoQ = T\} \neq \varnothing$$

iff

$$Q^{-1} @ T \in \mathbf{R}; \tag{17}$$

then it is the greatest element in **R**.

Theorem 6. Let R from V to W and T from U to W, be two given fuzzy relations, denote **Q** the set of fuzzy relations Q from U to V, such that RoQ = T; then

$$\mathbf{Q} = \{Q \mid RoQ = T\} \neq \varnothing$$

iff

$$(R @ T^{-1})^{-1} \in \mathbf{Q}; \tag{18}$$

then it is the greatest element in **Q**.

Example 1. Let $U = \{u_1, u_2, u_3\}$, $V = \{v_1, v_2, v_3, v_4\}$, $W = \{w_1, w_2, w_3\}$ and let us consider two fuzzy relations Q from U to V and R from V to W :

$$Q = \begin{bmatrix} 0.2 & 0 & 0.8 & 1 \\ 0.4 & 0.3 & 0 & 0.7 \\ 0.5 & 0.9 & 0.2 & 0 \end{bmatrix} \qquad R = \begin{bmatrix} 0.3 & 0.5 & 0.2 \\ 0.8 & 1 & 0 \\ 0.7 & 0 & 0.5 \\ 0.6 & 0.3 & 1 \end{bmatrix}$$

From the o-composition one derives T = RoQ, T fuzzy relation from U to W :

$$T = \begin{bmatrix} 0.7 & 0.3 & 1 \\ 0.6 & 0.4 & 0.7 \\ 0.8 & 0.9 & 0.2 \end{bmatrix}$$

Let us now assume that Q and T are two given fuzzy relations. We can ask whether $\mathbf{R} \neq \varnothing$. We already know the answer, but the purpose is to apply (17). From (8) and (7) one derives

$$Q^{-1} = \begin{bmatrix} 0.2 & 0.4 & 0.5 \\ 0 & 0.3 & 0.9 \\ 0.8 & 0 & 0.2 \\ 1 & 0.7 & 0 \end{bmatrix} \quad Q^{-1} @ T = \begin{bmatrix} 1 & 1 & 0.2 \\ 0.8 & 1 & 0.2 \\ 0.7 & 0.3 & 1 \\ 0.6 & 0.3 & 1 \end{bmatrix}$$

And, as expected,

$$(Q^{-1} @ T)\ oQ = \begin{bmatrix} 0.7 & 0.3 & 1 \\ 0.6 & 0.4 & 0.7 \\ 0.8 & 0.9 & 0.2 \end{bmatrix} = T$$

and $R \subseteq (Q^{-1} @ T)$.

Example 2. Let U, V and W be the set of positive real numbers, and let us define two fuzzy relations Q from U to V and R from V to W, as follows.

$$\forall (u,v) \in UxV,\ \mu_Q(u,v) = \exp[-k(u-v)^2]$$

and

$$\forall (v,w) \in VxW,\ \mu_R(v,w) = \exp[-k(v-w)^2],$$

where $k \geq 1$, and both Q and R may be interpreted as "is close to."

The o-compostion of R and Q gives T = RoQ, T fuzzy relation from U to W, defined as :

$$\forall (u,w) \in UxW,\ \mu_T(u,w) = \exp[-K(u-w)^2], \qquad \text{where } K = k/4.$$

Suppose now that Q and T are given. From (8) and (7) one derives

$$\mu_{Q^{-1} @ T}(v,w) = \exp[-kw^2/4] \qquad \text{if } v \leq w/2$$

$$= \exp[-k(v-w)^2] \qquad \text{if } v \geq w/2.$$

And, as expected, one can check that

$$(Q^{-1} @ T)\ oQ = T \text{ and } R \subseteq (Q^{-1} @ T).$$

When the MAX-min composition (finite case) stands for the usual Boolean matrix product (Boolean sum-product), equation RoQ = T (where R is unknown) can be solved (Sanchez 1972) by simply checking if equality $\grave{R}\,oQ = T$ holds, with $\grave{R} = (Q^{-1}oT')'$, where prime (') stands for the Boolean complement. If $\grave{R}$ is a solution, then it is the greatest one. When $\grave{R}\,oQ \neq T$, there is no solution. The dual equation RoQ = T (where Q is unknown) can be solved, with similar results, by checking whether $Ro\grave{Q} = T$, where $\grave{Q} = (T'oR^{-1})'$.

Systems of equations, truth-qualification and functional relations

Let us first consider the following system (I), of semantic equivalences

$$
\text{(I)}\quad
\begin{array}{lcl}
(X \text{ is } F) \text{ is } \tau_1 & \leftrightarrow & X \text{ is } G_1 \\
(X \text{ is } F) \text{ is } \tau_2 & \leftrightarrow & X \text{ is } G_2 \\
\dots & & \\
(X \text{ is } F) \text{ is } \tau_w & \leftrightarrow & X \text{ is } G_w \\
\dots & & \\
(X \text{ is } F) \text{ is } \tau_N & \leftrightarrow & X \text{ is } G_N
\end{array}
$$

These equivalences are related to truth-qualification (Zadeh 1978a, 1979), through the τ_w's, of a given fuzzy proposition "X is F", like "temperature is cool is rather true". Here, F and the G_w's are given fuzzy subsets of a reference set U, and the τ_w's, which represent linguistic truth-values, are fuzzy subsets of [0,1] to be solved.

System (I) is equivalent to the following system (II), of fuzzy equations, in which the τ_w's are the unknown.

$$\text{For } w = 1,\dots, N,\ \forall u \in U,\ \mu_{G_w}(u) = \mu_{\tau_w}(\mu_F(u)). \qquad \text{(II)}$$

Remark. In the same spirit as above, instead of truth-values represented as fuzzy sets, cf. the τ_w's, one might consider (non-fuzzy) truth-relations R_τ's from [0,1] to [0,1], such that $(\mu_F(u),\mu_G(u)) \in R_\tau$, so that F and G are point-wise truth-related. Then, when the truth relation R_τ is a function (call it μ_t) one has $\mu_G(u) = \mu_t(\mu_F(u))$.

Now, the membership function μ_F can be regarded as a functional relation Φ from U to V, i.e. $\mu_\Phi(u,v) = 1$ if $\mu_F(u) = v$ and $\mu_\Phi(u,v) = 0$, otherwise. The SUP-min composition of the fuzzy sets τ_w with Φ yields the following.

$$
\begin{aligned}
\text{For } w = 1,\dots,N,\ \forall u \in U,\ \mu_{\tau_w o \Phi}(u) &= \text{SUP}_{v\in[0,1]}\,[\,\mu_\Phi(u,v) \wedge \mu_{\tau_w}(v)\,] \\
&= \mu_{\tau_w}(\mu_F(u)).
\end{aligned}
$$

In other terms (cf. system (II)), for w = 1, …,N,

$$\tau_w o \Phi = G_w \quad \Leftrightarrow \quad \mu_{\tau_w}(\mu_F(u)) = \mu_{G_w}(u),\ \forall u \in U \qquad (19)$$

Let W = {1,2,…,N}, and let us define the fuzzy relations τ and G from [0,1] to W, and from U to W, respectively, by:

$$\forall (v,w) \in [0,1]\text{x}W,\ \mu_\tau(v,w) = \mu_{\tau_w}(v)$$

and
$\forall(u,w) \in UxW, \mu_G(u,w) = \mu_{G_w}(u)$.

Now, $\forall(u,w) \in UxW, \mu_{\tau o \Phi}(u,w) = SUP_{v\in[0,1]} [\, \mu_\Phi(u,v) \wedge \mu_\tau(v,w) \,]$
$= \mu_\tau(\mu_F(u),w))$.
$= \mu_{\tau_w}(\mu_F(u))$,

so that, $\tau o \Phi = G \Leftrightarrow \forall(u,w) \in UxW, \mu_G(u,w) = \mu_{\tau_w}(\mu_F(u))$
$\Leftrightarrow \forall(u,w) \in UxW, \mu_{G_w}(u) = \mu_{\tau_w}(\mu_F(u))$.

Finally, system (II) is equivalent to the following fuzzy relational equation :

$$\tau o \Phi = G, \tag{20}$$

where Φ (hence F) and G are given, τ is unknown,
and Φ is the functional relation associated with μ_F.

Functional relations and compositions

Due to the specificity of a functional relation Q, when composing fuzzy relations, equations of the form RoQ = T will have interesting properties, which do not hold with general fuzzy relations. In what follows, prime (') denotes the usual fuzzy complementation.

Theorem 7. If R is a fuzzy relation from V to W and Q a functional relation from U to V, one has

$$(RoQ)' = R'oQ \tag{21}$$

Proof is straightforward, using the de Morgan laws and the definition of a functional relation.

Corollary 8. If the functional relation Q from U to V is a solution to RoQ = T (R and T are two given fuzzy relations from V to W and from U to W, respectively), then Q is also a solution to R'oQ = T', and reciprocally.

When Q is a functional relation from U to V and T a fuzzy relation from U to W, let us call **R** the set of fuzzy relations R from V to W, such that RoQ = T. In other words, **R** is the subset of **R** (cf. theorem 5) in which Q is functional.

Theorem 9. If **R** $\neq \varnothing$ then (**R**, $\subseteq$) is a lattice with a greatest element Q^{-1} @T and with a smallest element $(Q^{-1}$ @T')'.

Proof. If R and S are elements of **R**, then $(R \cup S) o Q = (RoQ) \cup (SoQ)$ (which always holds), hence $(R \cup S) o Q = T \cup T = T$ and $R \cup S \in$ **R**. Now, $(R \cap S) o Q = (RoQ) \cap (SoQ)$ holds only because Q is functional here (it is easy to show), hence $R \cap S \in$ **R**. So, (**R**, $\subseteq$) is a lattice. The fact that $Q^{-1} @ T$ is a greatest element comes from theorem 5. Let us show that $(Q^{-1} @ T')'$ is the smallest element. **R** $\neq \varnothing$, then $\forall R \in$ **R**, $RoQ = T$; Q being functional, from (21) one has $R'oQ = T'$, and from lemma 1 applied to Q and R', one has $R' \subseteq Q^{-1} @ (R'oQ)$, i.e. $R' \subseteq Q^{-1} @ T'$, equivalent to $(Q^{-1} @ T')' \subseteq R$, which completes the proof.

Recall that in theorem 6, **Q** was defined as the set of fuzzy relations Q from U to V, such that RoQ = T. Let us denote by **Q*** the subset of **Q** consisting of functional relations Q.

Theorem 10. $\quad$ **Q*** $= \{Q \mid Q$ functional and $RoQ = T\} \neq \varnothing$
iff
$\forall u \in U, \exists v \in V$, such that $\forall w \in W$, $\mu_T(u,w) = \mu_R(v,w)$

Proof. Suppose that $\forall u \in U, \exists v \in V$, such that $\forall w \in W$, $\mu_T(u,w) = \mu_R(v,w)$.
So, let $u \in U$ and one element $v(u) \in V$, taken from the ones such that $\forall w \in W$, $\mu_T(u,w) = \mu_R(v(u),w)$. Define a relation Q from U to V, as follows.

$$\forall u \in U,\ \mu_Q(u,v(u)) = 1 \text{ and } \mu_Q(u,v) = 0,\ \ \forall v \in V\text{-}\{v(u)\}.$$

Q is a functional relation (in general it is not a unique one), and $\forall (u,w) \in U \times W$, $\mu_{RoQ}(u,w) = SUP_{v \in V}[\mu_Q(u,v) \wedge \mu_R(v,w)] = \mu_R(v(u),w) = \mu_T(u,w)$, so that RoQ = T.

Now, suppose that **Q*** $\neq \varnothing$, and so let $Q \in$ **Q***, Q is functional so, $\forall u \in U$, $\exists! v(u) \in V$, s.t. $\mu_Q(u,v(u)) = 1$ and $\mu_Q(u,v) = 0$, $\forall v \in V\text{-}\{v(u)\}$. But also RoQ = T, so that one derives, $\forall w \in W$, $\mu_T(u,w) = \mu_R(v(u),w)$, which completes the proof.

Corollary 11. **Q*** $= \{Q\}$ $\quad$ (**Q*** has a single element)
iff
$\forall u \in U, \exists! v(u) \in V$, such that $\forall w \in W$, $\mu_T(u,w) = \mu_R(v(u),w)$

Proof. From theorem 10, it is sufficient to show that if $\forall u \in U, \exists! v(u) \in V$, such that $\forall w \in W$, $\mu_T(u,w) = \mu_R(v(u),w)$, then **Q*** $= \{Q\}$ (it is already known that **Q*** $\neq \varnothing$). So, assume that there exists Q_1 and Q_2, Q_1 and Q_2 functional, $Q_1 \neq Q_2$, with $RoQ_1 = T$ and $RoQ_2 = T$, hence :

$$\exists\, u_0 \in U \text{ s.t. } \exists\, v_1(u_0) \text{ and } \exists\, v_2(u_0) \in V,\ v_1(u_0) \neq v_2(u_0), \text{ with}$$
$$\mu_{Q_1}(u_0, v_1(u_0)) = 1 \text{ and } \mu_{Q_1}(u_0,v) = 0,\ \forall v \in V\text{-}\{v_1(u_0)\}$$

and

$\mu_{Q_2}(u_0, v_2(u_0)) = 1$ and $\mu_{Q_2}(u_0,v) = 0, \forall v \in V\text{-}\{ v_2(u_0)\}$.

Moreover, $RoQ_1 = RoQ_2 = T$ yields : $\forall w \in W, \mu_T(u_0,w) = \mu_R(v_1(u_0),w) = \mu_R(v_2(u_0),w)$. But $v_1(u_0) \neq v_2(u_0)$ contradicts the hypothesis of the corollary. Finally, **Q*** has a unique element.

Theorem 12. $\forall Q \in$ **Q*** one has $Q \subseteq (R \,\textcircled{\gamma}\, T^{-1})^{-1}$,

where the $\textcircled{\gamma}$ composition is defined in (12).

Proof. $\forall (u,v) \in U\text{X}V, \mu_{(R \textcircled{\gamma} T^{-1})^{-1}}(u,v) = \mu_{R \textcircled{\gamma} T^{-1}}(v,u)$

$= \text{INF}_{w \in W} [\mu_R(v,w) \,\gamma\, \mu_T(u,w)]$.

But Q is functional, so given $u \in U, \exists! v(u) \in V$, s.t. $\mu_Q(u,v(u)) = 1$ and $\mu_Q(u,v) = 0, \forall v \in V\text{-}\{ v(u)\}$; so

- if $v \neq v(u), \mu_Q(u,v) = 0 \leq \mu_{(R \textcircled{\gamma} T^{-1})^{-1}}(u,v)$
- if $v = v(u), \mu_Q(u,v) = 1$, but $T = RoQ$ and Q is functional, so $\mu_T(u,w) = \mu_R(v(u),w) \; \forall w \in W$, hence $\mu_R(v(u),w) \,\gamma\, \mu_T(u,w) = 1 \; \forall w \in W$, and finally :

  $\mu_{R \textcircled{\gamma} T^{-1}}(v(u),u) = 1 = \mu_Q(u,v(u))$ and $\mu_Q(u,v(u)) = \mu_{(R \textcircled{\gamma} T^{-1})^{-1}}(u,v(u))$.

Theorem 13. If **Q*** $= \{Q\}$, then $Q = (R \,\textcircled{\gamma}\, T^{-1})^{-1}$.

Proof. $\forall (u,v) \in U\text{X}V, \mu_{(R \textcircled{\gamma} T^{-1})^{-1}}(u,v) = \text{INF}_{w \in W} [\mu_R(v,w) \,\gamma\, \mu_T(u,w)] = 1$ if, and only if, $\forall w \in W, \mu_R(v,w) = \mu_T(u,w)$; but **Q*** $= \{Q\}$, so, from Corollary 11, $\forall u \in U, \exists! v(u) \in V$, such that $\forall w \in W, \mu_T(u,w) = \mu_R(v(u),w)$; so given $u \in U, \exists! v(u) \in V$, such that $\mu_{(R \textcircled{\gamma} T^{-1})^{-1}}(u,v(u)) = 1$. If $v \in V, v \neq v(u), \exists w \in W$ s.t. $\mu_T(u,w) \neq \mu_R(v(u),w)$, hence $\mu_{(R \textcircled{\gamma} T^{-1})^{-1}}(u,v) = 0$, and finally, relation $(R \,\textcircled{\gamma}\, T^{-1})^{-1}$ is functional.

Now, as $Q \in$ **Q***, from theorem 12, $Q \subseteq (R \,\textcircled{\gamma}\, T^{-1})^{-1}$, but Q and $(R \,\textcircled{\gamma}\, T^{-1})^{-1}$ are both functional, so that this inclusion implies the equality : $Q = (R \,\textcircled{\gamma}\, T^{-1})^{-1}$, by definition of functional relations.

The above results on compositions of fuzzy relations involving functional relations (cf. Q's) suggest the following concept of the decompostion of a fuzzy relation by a fuzzy set, which will be studied in next section.

Decomposition of a fuzzy relation by a fuzzy set

Definition. A fuzzy relation T from U to W is said to be decomposable by a function f, from U to V, if, and only if, there exists a fuzzy relation R from V to W, such that

$$\forall (u,w) \in U \times W, \; \mu_T(u,w) = \mu_R(f(u),w) \qquad (22)$$

In this case T will be said to be *f-decomposable* and one has the following diagram.

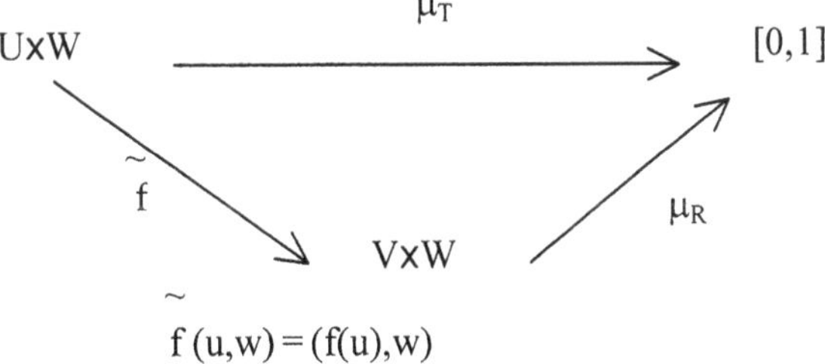

$\tilde{f}(u,w) = (f(u),w)$

With the notations of above, the following holds.

Theorem 14. A fuzzy relation T, from U to W, is *f-decomposable* (i.e. (22) holds) if, and only if, T = RoQ, where Q is the functional relation, from U to V, associated with f, i.e. $\mu_Q(u,f(u)) = 1$ and $\mu_Q(u,v) = 0$, $\forall v \neq f(u)$.

Proof. If T = RoQ, $\forall (u,w) \in U \times W$, $\mu_T(u,w) = SUP_{v \in V} [\mu_Q(u,v) \wedge \mu_R(v,w)]$

$= (\mu_Q(u,f(u)) \wedge \mu_R(f(u),w)) \vee SUP_{v \neq f(u)} [\mu_Q(u,v) \wedge \mu_R(v,w)]$

$= (1 \wedge \mu_R(f(u),w)) \vee SUP_{v \neq f(u)} [0 \wedge \mu_R(v,w)]$

$= \mu_R(f(u),w)$.

Conversely, $SUP_{v \in V} [\mu_Q(u,v) \wedge \mu_R(v,w)] = \mu_R(f(u),w) = \mu_T(u,w)$, hence T = RoQ, which ends the proof.

Returning to equation (20), where Φ is functional, one has : $\forall (u,w) \in U \times W$,

$\mu_G(u,w) = \mu_{\tau o \Phi}(u,w)$

$$= SUP_{v \in [0,1]} [\, \mu_\Phi(u,v) \wedge \mu_\tau(v,w) \,] \qquad (23)$$

From the definition of Φ (recall that Φ is the functional relation associated with

μ_F), equation (23) becomes : $\forall(u,w) \in U\times W$,

$$\mu_G(u,w) = \mu_\tau(\mu_F(u),w) \tag{24}$$

In this form, equation (24) appears as the decomposition of the fuzzy relation, G, by the membership function μ_F , as illustrated in the following diagram :

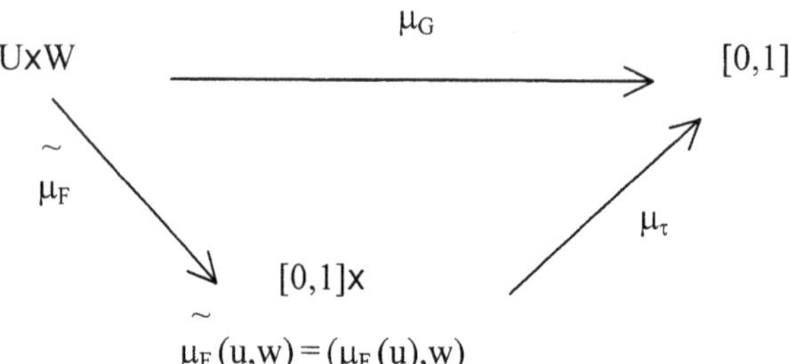

In this case, the fuzzy relation G will be said to be *F-decomposable.*

Problem. Given G and F as above, is the fuzzy relation G *F-decomposable* ? In other terms, does a fuzzy relation τ satisfying (24) exist ?

As it has been seen, this problem is equivalent to solving equation (20) : $\tau o \Phi = G$, in which the fuzzy relation G and the functional relation Φ (Φ is associated with μ_F) are given.

Theorem 9 provides an answer as follows.

Let us denote $\tau_a = \Phi^{-1} @ G$ and $\tau_b = (\Phi^{-1} @ G')'$. To solve equation $\tau o \Phi = G$, one simply needs to check whether $\tau_a o \Phi$ is equal to G or whether $\tau_b o \Phi$ is equal to G, where $\tau_b \subseteq \tau_a$.

If $\tau_a o \Phi \neq G$ (or $\tau_b o \Phi \neq G$), there is no solution to the problem. In this case, either $\tau_a \subseteq \tau_b$, or τ_a is not comparable with τ_b (in terms of $\subseteq$). In the general case, the following result always holds. Notations are given just above.

Theorem 15. $\tau_a o \Phi \subseteq G \subseteq \tau_b o \Phi$ and inclusions become equalities when $\tau_b \subseteq \tau_a$.

Proof. From (15) in lemma 3 one has $\tau_a o \Phi \subseteq G$.

Now, lemma 3 applied to Φ and to G' gives :

$$(\Phi^{-1} @ G') o \Phi \subseteq G',$$

which is equivalent to $G \subseteq ((\Phi^{-1} @ G') o \Phi)'$, or from (21), $G \subseteq (\Phi^{-1} @ G')' o \Phi$, because Φ is functional, i.e. $G \subseteq \tau_b o \Phi$.

When $\tau_b \subseteq \tau_a$, one has $\tau_b o \Phi \subseteq \tau_a o \Phi$, hence $G = \tau_a o \Phi = \tau_b o \Phi$.

On the resolution of equations

Note. In this section, truth values (τ's) are classical fuzzy subsets of the unit interval : they are not considered as fuzzy relations.

Based on the assumption that the truth-value τ, of a proposition p, is defined as the *compatibility* of a reference proposition r, with p, by using the extension principle, Zadeh (1978a, 1979) has proposed that

$$\tau = \mathit{Comp}\ (\text{"X is G" with "X is F"}) = \mu_F(G), \tag{25}$$

where τ is the truth-value of the proposition "X is F" relative to the reference proposition "X is G." When μ_F is a bijection, $\mu_F(G)$ in (25) can be computed from Zadeh's extension principle yielding, for all v in [0,1] :

$$\mu_\tau(v) = \mu_G(\mu_F^{-1}(v)) \tag{26}$$

and τ is uniquely determined from the knowledge of F and G, just in this strong particular case.

In the general case, how to truth-qualify a given proposition "X is F" to induce another given proposition "X is G", is the problem that has been investigated in (Sanchez 1978, 1996). It was searched for truth-values τ's, such that :

$$(\text{X is F}) \text{ is } \tau \leftrightarrow \text{X is G} \tag{27}$$

where "$\leftrightarrow$" denotes a semantical equivalence, i.e.

$$\forall u \in U,\ \mu_\tau(\mu_F(u)) \equiv \mu_G(u)\ , \tag{28}$$

cf. in system (II) of equations with a family of couples $(G_w,\tau_w)_{w=1,\dots,N}$, instead of a single couple, (G,τ) here.

According to the extension principle applied to the mapping μ_F, it is defined two fuzzy subsets τ_0 and τ_1, of the unit interval, as follows.

$$\tau_0 = \mu_F\ (G) \tag{29}$$

and

$$\tau_1 = (\mu_F\ (G'))' \tag{30}$$

Let us remark that τ_0 is a redefinition of (25), τ_1 was also proposed in (Prade 1980), and in general, τ_0 and τ_1 are not comparable in terms of inclusion of fuzzy sets. Moreover, definition of τ_1 can be related to the notion of 'guaranteed possibility' (Dubois and Prade 1992) as the set-valued function : $\Delta(A) = \inf_{u\in A} \pi_X(u)$, where A is a subset of U and π_X a possibility distribution function.

$\mathbf{T}$ being defined as the set of the fuzzy subsets τ of the unit interval that satisfy (28), we recall the following results.

Theorem 16. If $\mathbf{T} \neq \varnothing$, then $\tau_0 \in \mathbf{T}$ and $\tau_1 \in \mathbf{T}$.

When $\mathbf{T}$ is void, i.e. when no fuzzy susbet τ satisfying (28) existsq, the following basic theorem is very useful for practical applications, for it provides best lower and upper approximations for truth-qualification.

Theorem 17. $\forall u \in U,\ \mu_{\tau_1}(\mu_F(u)) \leq \mu_G(u) \leq \mu_{\tau_0}(\mu_F(u))$. (31)

Moreover, τ_1 is the greatest τ such that $\mu_\tau(\mu_F(u)) \leq \mu_G(u)$, τ_0 is the smallest τ such that $\mu_G(u) \leq \mu_\tau(\mu_F(u))$.

Remark. Expression (31) is the interpretation of "X is F is τ_1 $\rightarrow$ X is G $\rightarrow$ X is F is τ_0", where the symbol "$\rightarrow$" means "semantically entails" (Zadeh 1979).

Still in (Sanchez 1996), the above problem of truth-qualification was solved by using fuzzy relational equations. Let us simply recall how. From (19) where the indices (w's) have been dropped, the general problem reduces to the resolution of a fuzzy relational equation $\tau o \Phi = G$, in which τ is unknown. As previously, $\mathbf{T}$ denotes the set of the fuzzy subsets τ of the unit interval, such that $\tau o \Phi = G$, where Φ is the functional relation associated with μ_F, and F and G are given fuzzy subsets of a universe of discourse U.

Then, the following theorems hold.

Theorem 18. 1°) $\mathbf{T} \neq \varnothing$ iff $\Phi^{-1} @ G \in \mathbf{T}$
and
2°) $\Phi^{-1} @ G = (\Phi o G')' = \tau_1$.

Theorem 19. If $\mathbf{T} \neq \varnothing$, then the set of solutions is a lattice (for fuzzy set inclusion) with a smallest element, $\Phi o G$, and a greatest element, $(\Phi o G')'$, in $\mathbf{T}$. In such a case, every fuzzy subset τ of [0,1] such that $\Phi o G \subseteq \tau \subseteq (\Phi o G')'$, is a member of $\mathbf{T}$.

Concluding remarks

"What is amazing is that the more we learn about the basically simple concept of a relation, the more aware we become of its fundamental importance and wide ranging ramifications"

L.A. Zadeh, in Foreword of (Di Nola et al. 1989)

This paper has presented new results in the field of fuzzy relational equations. It is hoped that it will stimulate further theoretical developments and applications, for example in the analysis and design of intelligent systems, of fuzzy logic controllers or of systems for diagnosis assistance. We plan to contribute to the advancement of this field in forthcoming papers.

References

Birkhoff G (1967) Lattice Theory, American Mathematical Society Colloquium Publications, Vol.XXV, Providence, R.I.

De Baets B (2000) Analytical solution methods for fuzzy relational equations, In: Fundamentals of Fuzzy Sets, The Handbooks of Fuzzy Sets Series (Dubois D and Prade H, Eds.), Vol.1, Kluwer Academic Publishers, 291-340.

Di Nola A, Pedrycz W, Sanchez E, Sessa S (1989) Fuzzy Relation Equations and their Applications to Knowledge Engineering, Kluwer Academic Publishers.

Dubois D, Prade H, (1992) Possibility theory as a basis for preference propagation in automated reasoning, Proceedings of FUZZ-IEEE, San Diego, CA, 821-832.

Kaufmann A (1977) Introduction à la Théorie des Sous-Ensembles Flous, Vol.4, Compléments et Nouvelles Applications, Masson, Paris.

Peeva KG, Kyosev YK (to appear in 2003) Fuzzy Relational Calculus - Theory, Applications and Software, World Scientific Pub Co.

Prade H (automne 1980) Compatibilité, qualification, modification, niveau de précision, Busefal N°4, 71-78.

Sanchez E (1972) Matrices et Fonctions en Logique Symbolique, Ph.D. Thesis in Mathematics, Faculty of Science of Marseille, France.

Sanchez E (1974) Equations de Relations Floues, Ph.D. Thesis in Human Biology, Faculty of Medicine of Marseille, France.

Sanchez E (1976) Resolution of Composite Fuzzy Relation Equations, Information and Control, 30, 1, 38-48.

Sanchez E (1977) Solutions in Composite Fuzzy Relation Equations: Application to Medical Diagnosis in Brouwerian Logic, In: Fuzzy Automata and Decision Processes (Gupta MM, Ed.), North-Holland, 221-234. Also reproduced in (1993) Readings in Fuzzy Sets for Intelligent Systems (Dubois D, Prade H, Yager RR, Eds.), Morgan Kaufmann Pub., 159-165.

Sanchez E (1978) On Truth-Qualification in Natural Languages, Proceedings of the International Conference on Cybernetics and Society (Tokyo-Kyoto, Japan) 1233-1236.

Sanchez E (1996) Truth-qualification and fuzzy relations in natural languages, Fuzzy Sets and Systems 84, 155-167.

Schweizer B, Sklar A (1963) Associative functions and abstract semigroups, Publ. Math. Debrecen, 10, 69-81.

Zadeh LA (1965) Fuzzy Sets, Information and Control, 8, 338-353.

Zadeh LA (1975) The concept of a linguistic variable and its application to approximate reasoning, Part I, Information Sciences, 8, 199-249; (1975) Part II, Information Sciences, 8, 301-357; (1975) Part III, Information Sciences, 9, 43-80.

Zadeh LA (1978a) PRUF - A Meaning Representation Language for Natural Languages, Int. J. of Man-Machine Studies, 395-460.
Zadeh LA (1978b) Fuzzy Sets as a Basis for a Theory of Possibility, Fuzzy Sets and Systems, 1, 3-28.
Zadeh LA (1979) A Theory of Approximate Reasoning (AR), In: Machine Intelligence, Vol.9 (Hayes JE, Mitchie, Mikulich LI, Eds), J. Wiley, New York, 149-194.
Zadeh LA (1999) From Computing with Numbers to Computing with Words – From Manipulation of Measurements to Manipulation of Perceptions, IEEE Transactions on Circuits and Systems, 45, 1, 105-119.

Introduction to Modeling of Hydrogeologic Systems Using Fuzzy Differential Equations

Boris A. Faybishenko

Earth Sciences Division, Lawrence Berkeley National Laboratory, Berkeley, CA 94720, bafaybishenko@lbl.gov

Abstract

The many simultaneously occurring processes in unsaturated-saturated heterogeneous soils and fractured rocks can cause field observations to become imprecise and incomplete. As a result, field observations can become inconsistent with deterministic and stochastic mathematical models used for predictions. The performance of a system that includes soil, rock, and a monitoring network can become uncertain because of vagueness or "fuzziness," which is inherent in the system behavior, rather than being purely random. Fuzzy systems modeling, already widely used in such fields as engineering, physics, chemistry, and biology, is expected to become useful in simulating hydrogeologic system behavior. After presenting a hydrogeologic system as a fuzzy system, we derive a fuzzy-logic form of Darcy's equation. Based on this equation, a fuzzy logic form of the parabolic-type partial differential equations is derived. The elliptic-type (Laplace equation) and the parabolic-type (Richards equation) partial differential equations were approximated using fuzzy variables and solved using the basic principles of fuzzy arithmetic. The results of fuzzy systems modeling are then compared with those obtained using deterministic models. The application of fuzzy ordinary and partial differential equations to various earth sciences problems, such as flow and transport in the subsurface, is an emerging area of research.

1. Introduction

Since its introduction in 1965 by Lotfi Zadeh (Zadeh, 1965), fuzzy logic has been used to formalize "approximate knowledge" and approximate reasoning. Fuzzy logic is based on fuzzy set theory, which is an extension of classical set theory. Motivations for using the fuzzy set theory and fuzzy logic fall, in general, into two categories, theoretical (scientific) and practical. Scientific motivations for fuzzy logic are based on the fact that conventional Boolean models (based on the consideration of only two values: 0, indicating a complete non-truth, and 1, indicating an absolute truth) are inadequate for describing real practical problems with fuzzy boundaries between the system elements. A fuzzy variable has a *universe of discourse*, as defined by its minimum and maximum values, and a fuzzy membership function (FMF) that varies from 0 to 1 and indicates the degree of a membership of an element in a fuzzy set. The practical motivations for applying fuzzy logic are growing rapidly in a number of industries, especially in consumer electronics, industrial systems, and control systems (from household appliances to medicine and satellites). Several papers have recently been written on the application of fuzzy logic to soil physics and water resources, such as the application of fuzzy regression in hydrology (Bardossy and Duckstein, 1990), soil mapping (De Gruijter et al., 1997; Franssen et al., 1997), and prediction of infiltration (Bardossy and Disse, 1993; Bardossy and Duckstein, 1995). Moreover, significant progress has been made in using fuzzy models in reservoir simulations for oil exploration (Nikravesh and Aminzadeh, 2001).

Fuzzy logic is generally considered an extension of a conventional statistical approach, rather than a competitive technique (Zadeh, 1995). However, stochastic methods, based on the application of probability theory, prevent the use of perception-based information, which are fundamental to fuzzy logic (Zadeh, 2002). Such information is often gathered to solve highly complex problems arising from estimating the probabilities of imprecisely defined events and variables. These events and variables are characterized by a combination of measurement-based (which can be analyzed using statistical methods) and perception-based (using rules that can be understood by common sense) types of information. A basic difference between perceptions and measurements is that, in general, measurements could be crisp (i.e., precise) whereas perceptions are always fuzzy. Contrary to methods of probability theory, perception-based information is usually described in a natural language, using (for example) a set of *if-then* rules (Zadeh, 1978). In general, both deterministic and stochastic analyses should be considered as complementary to fuzzy logic.

To study subsurface flow and transport processes, we collect information from many sources, including objective (mostly quantitative) information using measurements, and subjective (often only qualitative) information using expert knowledge (from statistically processed past experience). Although environmental data sets contain a vast amount of information, these data are often uncertain because of the inconsistency between the real physical processes and the physics of the measurements. Examples of other causes of uncertainty in investigating flow systems are as follows:

1. Real flow processes are often deterministic-chaotic (Faybishenko, 2002), which are, however, characterized using discretely measured, random variables,
2. Data are incomplete or vague, or measurements are inaccurate,
3. Approximate estimations (volume-averaging and scaling) are used instead of direct measurements, providing qualitative information.

Thus, the main motivation for using fuzzy logic in soil sciences and hydrology arises, first of all, from the fact that imprecise and incomplete data are usually gathered in field conditions. In most practical situations, finding the best solution to a sufficiently complex problem using direct measurements could take too much time and money, or is practically impossible. As an alternative, fuzzy logic provides an intuitive approach to system definition, allowing use of a natural language to define fuzzy variables characterizing the system. The fuzzy-systems approach assumes the continuity of the medium with imprecise boundaries between different classes of the media (Franssen et al., 1997), and it is capable of handling both linear and nonlinear problems. A fuzzy-systems modeling involves the solution of mathematical equations written using fuzzy numbers (Kauffmann and Gupta, 1985).

The main goal of this paper is to introduce an approach of fuzzy-systems modeling of flow in unsaturated-saturated subsurface media, using fuzzy differential equations. To accomplish this goal, I will:

1. Develop a rationale for representing a hydrogeological system as a fuzzy system.

2. Derive a fuzzy-form of Darcy's equation for unsaturated-saturated flow in subsurface media and use this equation to derive fuzzy-type equations for flow and transport in unsaturated-saturated media (parabolic-type, a fuzzy-form of the Richards equation).

3. Present examples of fuzzy-systems modeling for solving (a) the one-dimensional fuzzy Darcy's equation for water travel time, (b) the second-order elliptic partial differential equation (Laplace equation) for steady-state, two-dimensional flow, and (c) the second-order parabolic partial differential equation for transient, one-dimensional flow.

2. Basic Concepts of Fuzzy-Systems Modeling

The basic concepts of fuzzy logic, which we can be used in fuzzy-systems modeling, are fuzziness, granularity, defuzzification, and operations on fuzzy numbers.

Fuzziness identifies a class (set) of objects with unsharp (fuzzy) boundaries. Fuzzification implies replacing a set of crisp (i.e., precise) numbers with a set of fuzzy numbers, using fuzzy membership functions based on the results of measurements and perception-based information. Several types of membership functions are used for fuzzy systems modeling, such as: triangular, trapezoid, sigmoid, gaussian, bell-curve, Pi-, *S*-, and *Z*-shaped curves. For fuzzy-systems modeling, it is important to use continuous membership functions, for which changes in input do not produce large changes in output, and with the continuous first and second derivatives. As an illustration, Figure 1 shows several membership functions for a fuzzy number

5. Note that the FMF ranges from 0 to 1 with the value of FMF=0 corresponding to complete nonmembership in a set, and the value of FMF=1 corresponding to a complete membership in a set.

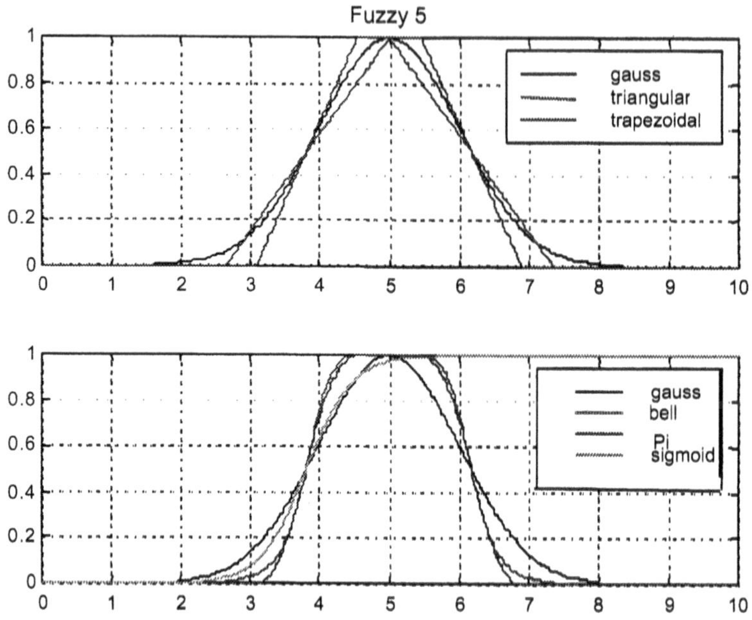

Figure 1. Presentation of fuzzy number 5 using different membership functions. A Gaussian membership function was converted into the triangular, trapezoid, bell, Pi-, and sigmoid-shape curves using Matlab.

Granularity indicates a clumpiness of structure. A granule is a clump of objects, pulled together by their indistinguishability, similarity, proximity, or functionality (Zadeh, 1997). Granules may be crisp or fuzzy. For example, granules of fractured rock could be matrix and fractures. Granulation implies a decomposition (partitioning) of a whole object into parts or number of granules. Fuzzy granulation can be considered in two senses: as part of the systems analysis or as a means of data compression or summarizing. Each fuzzy granule is characterized by a set of fuzzy attributes (e.g., in the case of the fuzzy granule *soil*, the fuzzy attributes are color, structure, texture, particle size distribution, hydraulic conductivity, etc). Each of the fuzzy attributes is characterized by a set of fuzzy values (e.g., in the case of fuzzy attribute hydraulic conductivity of soils, the fuzzy values could be very high, high, average, small, very small, etc.). A concept of granule fuzziness (including granule attributes and granule-attribute values) represents the natural human cognitive ability to form a rational decision about a system under conditions of imprecision, uncertainty, and partial truth. For numerical modeling using difference or finite-difference equations, each node (block) of the model presents a separate granule.

Defuzzification means the calculation of the crisp output from the results of fuzzy modeling, using such methods as center of gravity, largest of maximum, middle of maximum, or bisector of the area (formulae are summarized in Table 1).

Table 1. Types of defuzzification methods used in calculations

Defuzzification method	**Calculations**
Centroid (center of area or center of gravity)	$U^* = \dfrac{\int_U U\mu_U(U)dU}{\int_U \mu_U(U)dU}$
Largest of maximum (LOM)	$U^* = \max_{U^* \in U} \mu_U(U)$
Middle of maximum (MOM)	Calculated as the average of the smallest and largest maximum values, if there is more than one value of *U* with the maximum membership
Bisector of area: U^* divides the area below the FMF into two equal parts	$\int_{U<U^*} U\mu_U(U^*)dU = \int_{U>U^*} U\mu_U(U^*)dU$

Operations on fuzzy numbers. Algebraic operations on fuzzy numbers are performed based on the Zadeh's extension principle (Kauffmann and Gupta, 1985). The extension principle implies that fuzzy subsets and logic are generalizations of classical set theory and (Boolean) logic. The main concept of fuzzy algebraic operations is as follows:

$$\mu_{I^f * J^f}(z) = \bigvee_{x*y=z} \left(\mu_{I^f}(x) \wedge \mu_{J^f}(y)\right) \tag{1}$$

where I^f and J^f are fuzzy numbers defined on real lines *X* and *Y*, respectively; the symbol * denotes a fuzzy arithmetic operation (+), (-), (·), or (:). An arithmetic operation (mapping) of two fuzzy numbers denoted as $I^f * J^f$ will be defined on universe *Z*, and μ denotes a fuzzy membership function. Equation (1) can also be written in the following form:

$$\mu_{I^f * J^f}(z) = \max(\min(\mu_{I^f}, \mu_{J^f})) \qquad (2)$$

The examples of the fuzzy operations of two fuzzy numbers 2 and 3, presented as triangular membership functions in the upper panel of Figure 2, demonstrate that the summation and subtraction (middle panel) do not change the shape of the resulting membership functions. Multiplication and subdivision of fuzzy numbers (lower panel) slightly change the original shape of membership functions. The important features of fuzzy arithmetic are: addition and multiplication of fuzzy numbers is commutative and associative, but subtraction and division are neither commutative nor associative; the image and the inverse of a fuzzy number are not symmetric, as $I^f(\cdot)I^{f^{-1}} = I^{f^{-1}}(\cdot)I^f \neq 1$; and an operation to determine an unknown fuzzy number J^f from an equation $I^f(+)J^f = C^f$ or $I^f(\cdot)J^f = C^f$ is called a deconvolution of fuzzy numbers (Kaufmann and Gupta, 1991).

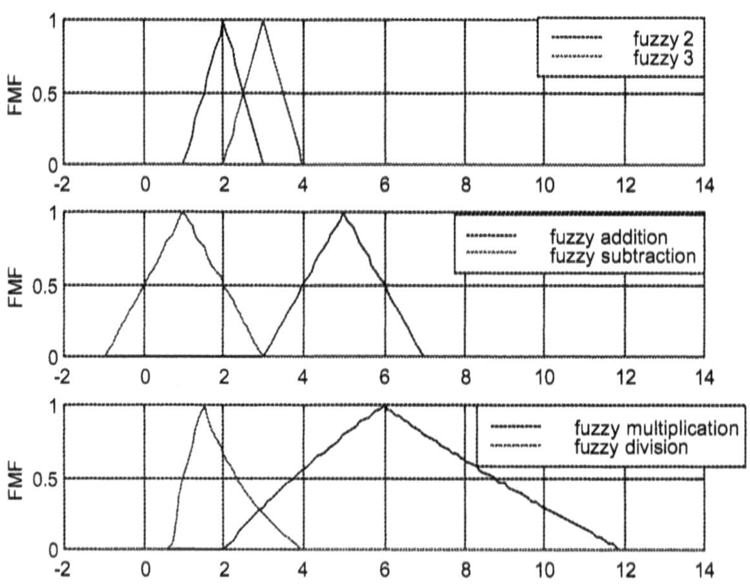

Figure 2. Examples of addition, subtraction, multiplication, and division of fuzzy numbers 2 and 3

To present a hydrogeologic system as a fuzzy system, we first provide a schematic illustration of a fuzzy system with fuzzy input, output, and state parameters, which range over fuzzy sets (Figure 3a). The relationship among system variables can be presented using the concept of a fuzzy relationship (Figure 3b).

Flow and transport in subsurface media are greatly influenced by the spatial heterogeneity of soils or rocks. The heterogeneity of soils and rocks is caused mainly by the variability of physical and hydraulic properties, such as bulk density, water content, total porosity, macroporosity, hydraulic conductivity, organic carbon content, sorption coefficients, and degradation

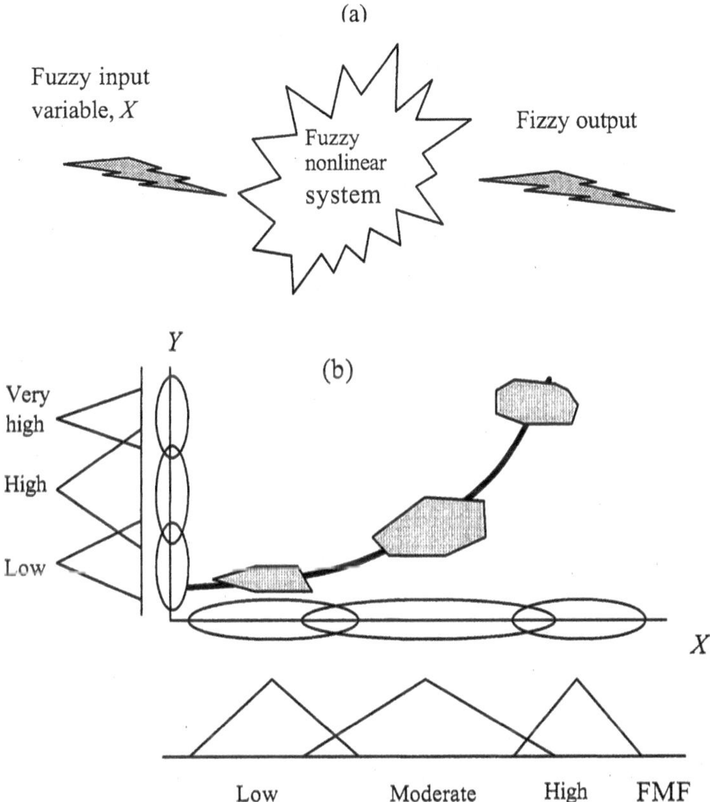

Figure 3. (a) Schematic of a fuzzy physical system with observed fuzzy inputs and fuzzy outputs, and (b) a fuzzy nonlinear relation patches (characterizing actual data), relating an input variable X to the output variable Y (Ross, 1995), and corresponding fuzzy membership functions (FMF) for X and Y.

rates, over the area and depth of the field (Wagenet and Rao, 1985). Spatial variability of subsurface properties is controlled by both extrinsic and intrinsic factors. Extrinsic factors include ambient conditions (i.e., precipitation,

atmospheric temperature and pressure) and anthropogenic conditions (field management treatments, e.g., tillage, irrigation, drainage, waste disposal). Intrinsic variability implies variations in physical, chemical, and biological properties of soils. Despite the continuous variability of soil properties, our understanding about soil processes is obtained from discrete measurements. Assuming that we have a stationary random flow field and that the relevant statistics can be derived from experimental data, we usually treat hydraulic conductivity as a log-normally distributed parameter and porosity as a normally distributed parameter. The probability distribution functions (PDFs) characterizing the spatial distribution of soil properties can be used to construct fuzzy membership functions for these properties. For example, to construct a FMF, we can normalize the PDF function to the maximum value of the PDF.

Flow processes in subsurface heterogeneous media are transient and likely to become unstable or even chaotic from small (Faybishenko, 2002) to large scales (Rodriguez-Iturbe et al., 1991). The cause of uncertainty, however, often remains unknown to the observer. For example, in fractured rocks, fractures are often infilled with sedimentary material, and fracture walls are coated with mineralized precipitates, which makes the fracture-matrix interface more fuzzy than sharp. Moreover, we do not measure the properties of soils or rocks directly, but rather the properties of the system *soils/rocks + instrumentation*, from which the properties of soils/rocks can be determined using the deconvolution of data. Using the same model and refining the parameters may cause the scaling effect, so that another granule is characterized at each level. Existing methods of direct measurement (often with limited resolution and an insufficient number of measurements) are frequently inconsistent with the physics of flow processes, so that measurements may not describe real flow processes. Moreover, not all parameters needed to describe flow processes are measurable using existing field-monitoring methods. Hydrologic quantities, which are affected by a combination of deterministic, stochastic, chaotic, and noisy components, are difficult, if not impossible, to distinguish. Measurements made with field instrumentation under field conditions are generally imprecise, ambiguous, or vague, and, therefore, uncertain. In the case of uncertainty of data, a variable is fuzzy and can be represented using fuzzy logic, which enables the hydrogeological system design to be based on human experience. Fuzzy logic is based on the use of limited data sets and the researcher's intuition.

A fuzzy approach could become particularly useful in handling uncertain or imprecise geologic, hydrogeologic, and environmental data, describing dynamically changing environments. These environments can be defined as fuzzy sets with indistinct boundaries, which might better reflect the continuous nature of a given subsurface system. For example, the conventional hydrogeological paradigm of a Representative Elementary Volume (REV) cannot be considered as a finite entity with sharp boundaries, but rather as an element (granule) with fuzzy boundaries: each element, far from being independent, interacts with others in a complex manner. Therefore, any component of the hydrogeologic system can be considered a *fuzzy granule*. Further, a hydrogeologic system that consists of heterogeneous soils or rocks can be considered a fuzzy system, with no sharp boundaries between the system elements (parts) and no deterministically (or even stochastically) defined parameters. The complexity and insufficient characterization of intrinsic flow and transport properties create the source of imprecision or vagueness of data, which could be characterized using fuzzy variables and a perception-based fuzzy-logic approach. For example, we could characterize input parameters, such as precipitation, using the terms: *heavy rain, sprinkle, drizzle*, etc. The output parameter, such as a pumping rate, could be characterized using the terms *small, moderate, high*, etc. Changes in the type of soils or rocks could be characterized as *gradual*, *moderate*, or *sharp*, with no particular crisp boundaries between the types of media.

By representing a hydrogeologic system as a fuzzy system, hydrogeological parameters such as hydraulic head, water flux, hydraulic conductivity, hydraulic gradient, and porosity etc., could be presented as fuzzy variables. This approach provides background for the development of fuzzy partial differential equations as the basis for modeling hydrogeologic systems.

4. Fuzzy Darcy's Equation

A conventional deterministic form of Darcy's law for flow in unsaturated-saturated media, which expresses the relationship between the flow rate, hydraulic gradient, and hydraulic conductivity, is given (using crisp numbers) by

$$q = - k \text{ grad } H \tag{3}$$

or for unsaturated media only

$$q = - D \text{ grad } \Theta \tag{4}$$

where q is the water flux through a unit area per unit of time, k is hydraulic conductivity, H is the hydraulic head, $H = P/\rho g + z$, P is the water pressure, z is the vertical coordinate, ρ is the water density, g is the gravity acceleration,

Θ is the moisture content, and D is the hydraulic diffusivity. Note that P, k, and D are nonlinear functions of Θ.

The past four decades have witnessed a growth in the use of stochastic methods applied to a variety of hydrogeologic problems (e.g., Dagan and Neuman, 1997). They include predictions and risk assessment under conditions of uncertainty and incompleteness of information. In a stochastic continuum approach, a random variable is represented by the sum of its mean and a spatially correlated random fluctuation. However, the stochastically averaged flux is generally nonlocal and non-Darcian.

Using fuzzy variables, a fuzzy form of Darcy's law can be presented as a fuzzy product:

$$q^{f} = k^{f}(\cdot)\,(\mathrm{grad}H^{f})^{f} \tag{5}$$

$$q^{f} = D^{f}(\cdot)\,(\mathrm{grad}\,\Theta^{f})^{f} \tag{6}$$

where q^f is the fuzzy water flux, k^f is the fuzzy hydraulic conductivity, and D^f is the fuzzy hydraulic diffusivity, H^f is the fuzzy hydraulic head, and Θ^f is the fuzzy saturation. Equation (5) takes into account the fuzzy gradient of the fuzzy hydraulic head, and Equation (6) the fuzzy gradient of fuzzy saturation. The fuzzy membership function for q^f is expressed by

$$\mu_{q^f}(X,Y) = \min\,[\mu_{k^f}(X), \mu_{(gradH^f)^f}(Y)] \tag{7}$$

where $\mu_k^f(X)$ is the fuzzy membership function for k^f, and $\mu_{(\mathrm{grad}H^f)}^f$ is the fuzzy membership function for the fuzzy hydraulic gradient. The types of membership functions for each fuzzy set can be determined using the probability distribution function for k and grad$H$. Equations (5) and (6) are a general form of fuzzy Darcy's law, from which follow that deterministic Equations (3) and (4) are particular cases (for FMF=1) of the fuzzy Darcy's law.

Using the fuzzy form of the mass conservation equation

$$\frac{\partial \Theta^f}{\partial t'} = -div q^f \tag{8}$$

and substituting (5) and (6) into (8), gives a second-order, parabolic-type, partial differential equation, given for one-dimensional flow by

$$\frac{\partial \Theta^f}{\partial t'} = \frac{\partial}{\partial x}\left[k^f(\cdot)(gradH^f)^f\right] \tag{9}$$

or

$$\frac{\partial \Theta^f}{\partial t'} = \frac{\partial}{\partial x}\left[D^f(\cdot)\left(grad\Theta^f\right)^f\right] \tag{10}$$

Fuzzy partial parabolic-type differential equations (9) and (10) can be called the fuzzy Richards equations, as the analogous deterministic equations, which describe fluid flow in unsaturated media, are called Richards equations.

5. Examples of Calculations

5.1. Calculation of the water travel time based on fuzzy Darcy's law

Based on the fuzzy Darcy's equation (Equation 5), and assuming the actual water velocity is given by

$$q^f = n^f(\cdot)\,[(dz)^f(:)\,(dt)^f) \tag{11}$$

where dz^f is the fuzzy depth to the water table, n^f is the fuzzy porosity, and dt^f is the fuzzy water travel time. Using (11), Equation (5) can be given by

$$n^f(\cdot)\,[(dz)^f(:)\,(dt)^f)] = k^f(\cdot)\,(\mathrm{grad}H^f)^f \tag{12}$$

from which the water travel tome can be determined using fuzzy operations. As an example, we consider one-dimensional, vertical flow in the vadose zone at a depth of approximately 60 m (a Gaussian FMF with a standard deviation σ=5 m, is shown in Figure 4a), with hydraulic conductivity of 1×10^{-4} m/day, or 0.365 m/year (a lognormal-shaped FMF of hydraulic conductivity with $\sigma(\ln K)=1$ is shown in Figure 4b), a unity hydraulic gradient (a Gaussian FMF with σ=0.1 is shown in Figure 4c), and a porosity of 0.1 (a Gaussian function with σ=0.02 is shown in Figure 4d). The calculated fuzzy water travel time from Equation 12 is presented in Figure 4e. The resulting asymmetric shape with a long tail of the water travel time curve, shown in Figure 4e, indicates that instead of the water travel curve of 16.44 years (which corresponds to FMF=1), which can be calculated from a deterministic analog of Equation (12), we obtain a range of possible values of the water travel time. For example, with the possibility of 80% (FMF=0.8), the water travel time would range from 11.13 years to 24.07 years, and with the possibility of 50% (FMF=0.5), it would range from 8.23 to 31.99 years. According to the possibility/probability principle (Zadeh, 1978), a *possible event* may not be *probable*, while an *impossible* event is also *improbable*.

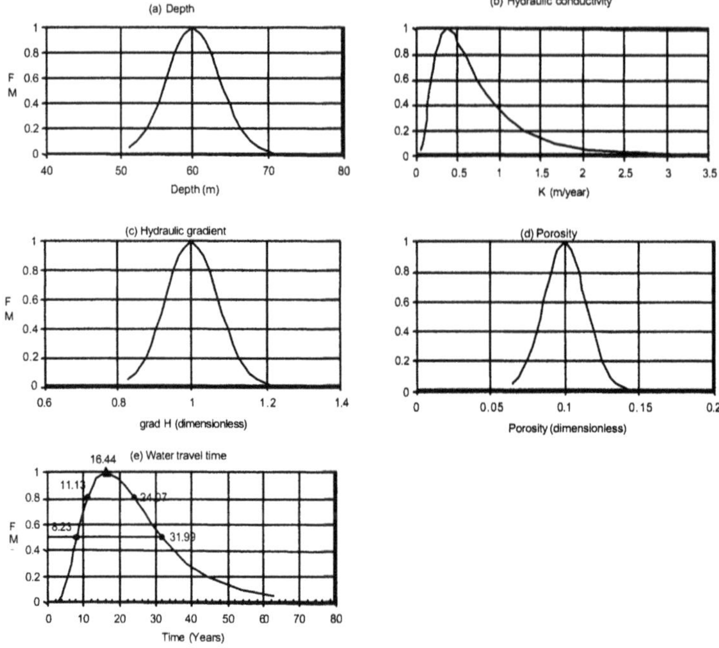

Figure 4. Fuzzy membership functions for a depth (a), hydraulic conductivity (b), hydraulic gradient (c), porosity (d), and calculated water travel time (e).

5.2. Solution of the Laplacian (elliptic) equation

Elliptical partial-differential equations describe a number of steady-state, boundary-value processes, arising in different engineering applications, such as heat and chemical transport and fluid flow. The most widely used form of elliptical equations is the Laplace equation given for a two-dimensional field by (Chapra and Canale, 1998)

$$\frac{\partial^2 U}{\partial x^2} + \frac{\partial^2 U}{\partial y^2} = 0 \qquad (13)$$

where U is either hydraulic head, temperature, or concentration, and x and y are coordinates. The derivation of Equation (13) is based on the balance of U around a discrete element (i, j) of media (Figure 5). To obtain a fuzzy solution of Equation (13), we will first transform this equation, using a so-called central difference method, into the finite-difference (algebraic) equation (the Laplacian finite-difference equation) given by

$$(U_{i+1\,j} - 2U_{i,j} + U_{i-1,j})/?x^2 + (U_{i,j+1} - 2U_{i,j} + U_{i,j-1})/?y^2 = 0 \qquad (14)$$

If $?x = ?y$, an algebraic equation, Equation (14), becomes

$$U_{i+1,j} + U_{i-1,j} + U_{i,j+1} + U_{i,j-1} = 4U_{i,j} \qquad (15)$$

To obtain a unique solution of Equation (15), a set of boundary conditions should be assigned. Then, using the notion of fuzzy numbers, Equation (15) can be written to represent a fuzzy balance equation as follows:

$$U^f_{i+1,j} \, (+) \, U^f_{i-1,j} \, (+) \, U^f_{i,j+1} \, (+) \, U^f_{i,j-1} = 4U^f_{i,j} \qquad (16)$$

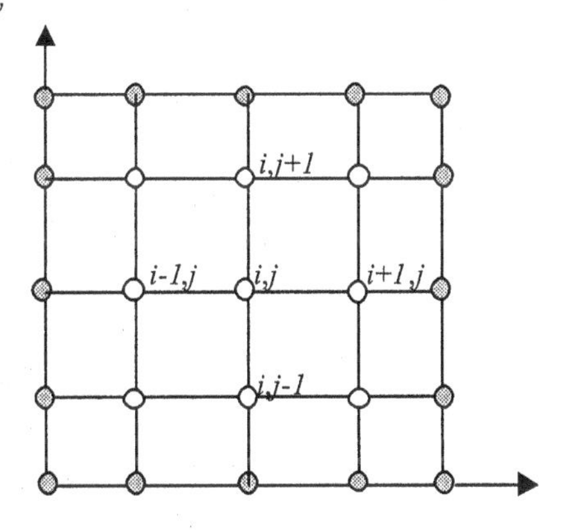

Figure 5. A two-dimensional grid used for modeling of the Laplacian finite-difference equation. Filled circles are boundary nodes. Open circles are inner nodes.

where each fuzzy number is characterized by a fuzzy membership function. Equation (15) can be called a Laplacian fuzzy-difference equation.

As an example, we solved Equation (15) using a 9×9-element (node) two-dimensional space with boundary conditions (which were also assigned using fuzzy numbers), using the Liebmann method of iterations. The solution of Equation (15), using fuzzy operations, generated a series of fuzzy membership functions for each node, shown in Figure 6. As the Gaussian-type membership functions were used, we defuzzificated the calculated values of U^f using a centroid method. (The use of the centroid method is justified for the solution of the fuzzy Laplacian equation, as the resulting membership functions are practically symmetric). The comparison of defuzzified values of U with those from the solution of the deterministic Laplace equation (Equation 12) given by Chapra and Canale (1998) showed that the average error of calculations for 9 nodes is 0.31% (after 13 iterations). Figure 6 shows the comparison of the solution for $\mu(U)$ after 5 and 13 iterations, illustrating that the solution would not change after 5 iterations.

5.3. Solution of a Parabolic Equation

A second-order partial differential parabolic equation describes the problems related to the heat-conduction, chemical diffusivity, or moisture (saturation) diffusion processes. Assuming that k and D are independent of H and Θ, the equation is given by

$$\frac{\partial U}{\partial t} = k\frac{\partial^2 U}{\partial x} \qquad (17)$$

A numerical solution of Equation (17) requires the second-order approximation along an x-coordinate and the first-order approximation for the time. Using a centeral finite difference method, an explicit form of Equation (16) for all nodes along the x-coordinate is given by:

$$\frac{U_i^{\tau+1} - U_i^{\tau}}{(\Delta t)} = k\frac{U_{i+1}^{\tau} - 2U_i^{\tau} + U_{i-1}^{\tau}}{(\Delta x)^2} \qquad (18)$$

where τ is the current time and $(\tau+1)$ is the time one step ahead, for which we seek a $U(x)$ distribution. For $\lambda = k\Delta t/(\Delta x)^2$, and using fuzzy variables, Equation (18) is given in a fuzzy form by

$$U_i^{f,\tau+1} = U_i^{f,\tau} (+) \lambda^f (\cdot) \left[U_{i+1}^{f,\tau} (-) 2U_i^{f,\tau} (+) U_{i-1}^{f,\tau} \right] \qquad (19)$$

As an example, Figure 7 presents the results of the solution of Equation (19) for a fuzzy $U^f(x,t)$ along a horizontal column of 10 cm in length, with observation points x=2, 4, 6, and 8 cm, the initial condition $U^f(x, t=0)=0$ (σ=0.1), and boundary conditions (for FMF=1) $U^f(x=0,t)=100$ (σ=0.1) and $U^f(x=10, t)=10$ (σ=0.1). Gaussian-type FMFs are used for all variables (including initial and boundary conditions).

The distributions of U (for FMF=1) along the x-axis for three moments of time, 1, 2, and 3 seconds, are shown on the upper panel of Figure 7. The distributions of predicted FMFs for U^f for four observation points are shown for three moments of time, 1, 2, and 3 seconds, in the second, third, and fourth panels of Figure 7, respectively. Figure 7 clearly shows that the width of FMFs increases with time. For example, for x=2 cm, for a possibility of 50% (FMF=0.5), at t=1 s, U^f = 18-23; for t=2 s, U^f = 38-44; and for t=3 s, U^f=62-75. Thus, as a substitute to the continuous function $U(x,t)$, we obtain a series of continuous functions $U^f = f(\mu_{Uf})$ for different nodes along the x-coordinate.

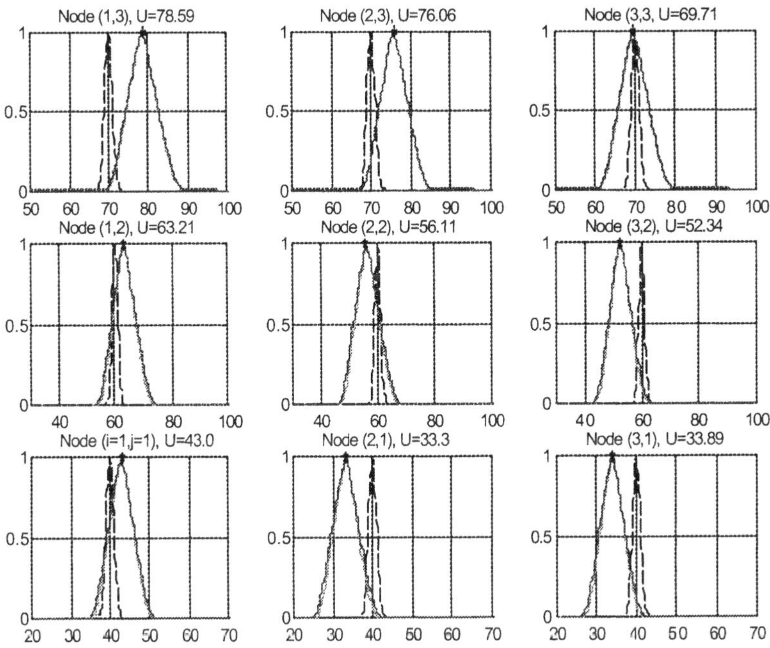

Figure 6. The solution of a fuzzy-difference Laplacian equation: FMFs of U^f for nine inner nodes of a 2D field: blue lines are initial U^f: for row j=1, $U_{\mu=1}$ =40.0, for row j=2, $U_{\mu=1}$ =60.0, and for row j=3, $U_{\mu=1}$ =70.0, and the Gaussian distribution with σ=1. Green lines – FMFs after 5 iterations, and red lines - after 13 iterations. The boundary conditions are: upper – $U_{\mu=1}$ =100.0 (σ=2), right – $U_{\mu=1}$ =50.0 (σ=2), left – $U_{\mu=1}$ =75.0 (σ=2), and lower – $U_{\mu=1}$ =0 (σ=1). U numbers for nine nodes given at the top are solutions of the deterministic Laplacian equation. x-axis - U, and y-axis - FMF of U.

6. Conclusions

Flow and transport processes in heterogeneous subsurface media, such as fluid flow, heat and chemical transport, are characterized by a combination of

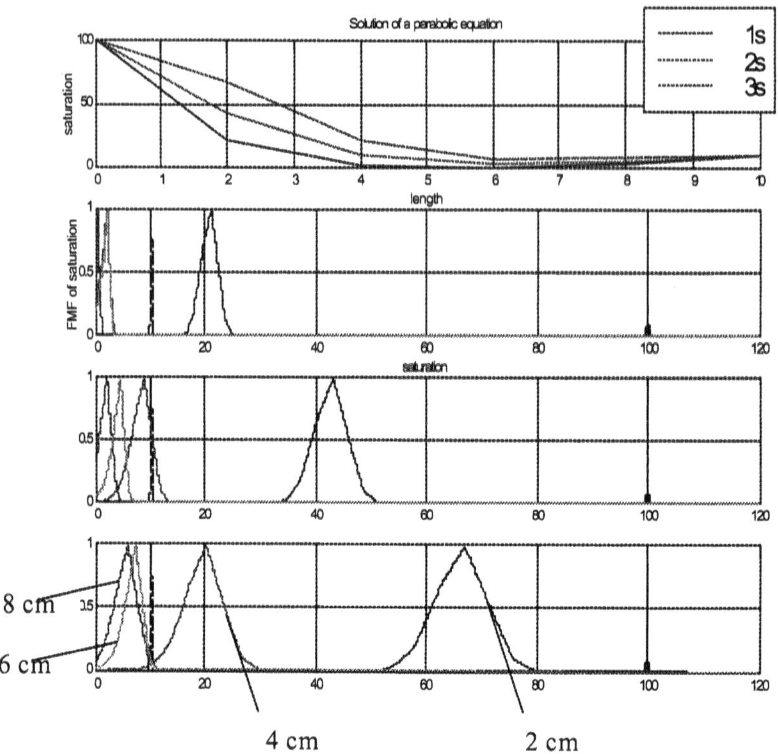

Figure 7. The results of simulations using a fuzzy-difference approximation (Equation 19) of the second-order parabolic partial differential equation. Upper panel: distribution of U (for FMF=1) along the x-axis. Second (time t=1 s, third (t=2 s), and fourth (t=3 s) panels show the distribution of FMFs for U^f, for x=2, 4, 6, and 8 cm. Dashed lines show the FMFs for boundary conditions: U^f=100 (σ=0.1), for x=0 and t>0, and U^f=10 (σ=0.1), for x=10 and t>0), and λ^f = 0.021 (σ=0.001). Initial condition $U^f(x,t=0)=0$ (σ=0.1).

deterministic, stochastic, chaotic, and noisy components. However, the amount of data collected from experiments is usually inadequate to characterize these processes in sufficient detail and to construct detailed deterministic or stochastic models, which results in imprecision, ambiguity, or vagueness (and therefore uncertainty of data). The fundamental limitations of currently used deterministic and stochastic methods to describe subsurface processes are related to their inability to use perception-based information. This information forms the basis of an alternative approach, fuzzy-systems modeling. In general, both deterministic and stochastic analyses of subsurface processes can be considered complementary to fuzzy- systems modeling.

Contrary to a stochastic approach, which is based on the hypothesis of a random field, the fuzzy-logic approach assumes the continuity of the medium, with imprecise boundaries between different parts of media (Franssen et al., 1997). Fuzzy- systems modeling does not replace conventional deterministic and stochastic simulation methods, but rather makes more sophisticated use of the information obtained using these methods. Such parameters as hydraulic head, water flux, hydraulic conductivity, hydraulic gradient, porosity, saturation, and temperature can be represented as fuzzy variables. The spatial distribution of these parameters can be used to generate fuzzy membership functions for these parameters. Fuzzy membership functions can, in turn, be used to characterize the degree of heterogeneity of a subsurface system. The relationships between fuzzy variables, for instance, hydraulic pressure and water content vs. hydraulic pressure, are given by fuzzy functions. Using the basic concepts of fuzzy logic, fuzzification and fuzzy granulation, we develop a fuzzy form of Darcy's equation and the second-order fuzzy partial differential equations. Using a fuzzy Darcy's equation, we present an example of calculations of the water travel time as a fuzzy variable.

For modeling fluid flow, chemical diffusion, and heat conduction in heterogeneous media, we approximate second-order partial differential equations (elliptic and parabolic) using fuzzy-difference equations. The solutions of the fuzzy-difference Laplacian equation (analogue of the elliptical-type equation) and the fuzzy-difference analogue of the parabolic-type equation generate a set of FMFs for each observation point (node).

Future applications of fuzzy systems modeling in earth sciences might include sensitivity analysis of fuzzy variables, simulations and predictions of flow and transport using fuzzy ordinary and partial differential equations or a fuzzy neural network approach, management of fuzzy systems, and development of fuzzy controllers to be used for measurements.

Acknowledgement: This paper is based on the presentation given at the Fuzzy Partial Differential Equation Solver (FPDES) Workshop held at BISC at the UC Berkeley in March 2002. The author appreciates very much many discussions with and a review by Lotfi Zadeh and Masoud Nikravesh of UC Berkeley, and a review by Curt Oldenburg and Dmitriy Silin of LBNL. Preparation of the paper was supported by the mini-grant program of the Earth Sciences Division of LBNL This work was supported by the Director, Office of Science, Office of Basic Energy Sciences, Division of Materials Sciences and Engineering, of the U.S. Department of Energy under Contract No. DE-AC03-76SF00098.

References

Bardossy, A., and L.Duckstein, *Fuzzy Rule-Based Modeling with Applications to Geophysical, Biological abd Engineering Systems*, CRC Press, New York, 113 p., 1995.

Bardossy, A., and M.Disse, Fuzzy rule-based models for infiltration, *Water Resour. Res.*, 29, 373-382, 1993.

Bardossy, A., and L.Duckstein, Fuzzy regression in hydrology, *Water Resour. Res.*, 26(7), 1497-1508, 1990.

Chapra, S.C. and R.P. Canale, *Numerical Methods for Engineers*, 3rd edition, The McGraw Hills Companies, Inc., Boston, 1998.

Dagan, G. and S.P. Neuman (eds.) *Subsurface Flow and Transport: A Stochastic Approach*, Paris, France, Unesco, IAHS/AISH; Cambridge; New York: Cambridge University Press, 1997.

De Gruijter, J.J., D.J.J.Walvoort, and P.F.M.Gaans, Continuous soil maps - a fuzzy set approach to bridge the gap between aggregation levels of process and distribution models, *Geoderma*, 77, 169-195, 1997.

Faybishenko, B., Chaotic Dynamics in Flow Through Unsaturated Fractured Media. *Advances in Water Resources*, 25/7, 793-816, 2002.

Franssen, Hendricks, H.J.W.M., A. C. van Eijnsbergen, and A. Stein, Use of spatial prediction techniques and fuzzy classification of mapping soil pollutants. *Geoderma* 77(2-4), 243-262, 1997.

Kaufmann, A. and M.M. Gupta. *Introduction to Fuzzy Arithmetic: Theory and Applications*, New York, N.Y.: Van Nostrand Reinhold Co., 1985.

Nikravesh M. and F. Aminzadeh, Past, present and future intelligent reservoir characterization trends. *Journal of Petroleum Science & Engineering*. 31(2-4):67-79, 2001

Rodriguez-Iturbe, I; D. Entekhabi, J.S. Lee, and R.L. Bras, Nonlinear dynamics of soil moisture at climate scales: 2. Chaotic analysis, *Water Resour. Res. 27*(8), 1907-1915, 1991.

Ross, T.J., *Fuzzy logic with engineering applications*, New York: McGraw-Hill, 1995.

Zadeh, L.A., From computing with numbers to computing with words - from manipulation of measurements to manipulation with perceptions, *Applied Mathematics and Computer Sciences*, Special Issue: *Computing with Words and Perceptions*, 307-324, 12(3), 2002.

Zadeh, L.A. Towards a theory of fuzzy information granulation and its centrality in human reasoning and fuzzy logic, *Fuzzy Sets and Systems*, 19, 111-127, 1997.

Zadeh, L. A. Probability theory and fuzzy logic are complementary rather than competitive, *Technometrics*, 37, 271-276, 1995.

Zadeh, L.A., Fuzzy sets as a basis for a theory of possibility, *Fuzzy Sets and Systems* 1, 3-28, 1978.

Zadeh, L.A., Fuzzy sets, *Information and Control*, 8, 338-353, 1965.

Wagenet, R.J. and P.S.C. Rao, Basic concepts of modeling pesticide fate in the crop root zone, *Weed Science*, 33 (suppl. 2), 25-32, 1985.

Construction of Granular Derivatives and Solution of Granular Initial Value Problem

Ildar Batyrshin[1]

Institute of Problems of Informatics, Academy of Sciences of Tatarstan, and Kazan State Technological University, K.Marx st. 68, Kazan, 420015, Russia, E-mail: batyr@emntu.kcn.ru.

Abstract. Rule based derivatives are considered. The methods of construction of granular differentials correspondent to rules are discussed. These differentials are determined by granular slope values given by rules and represented by fuzzy linear functions called directions. The method of solution of initial value problem based on the rule based derivative and on initial condition "*If X is* X_0 *then Y is* Y_0", where X_0 and Y_0 are fuzzy numbers, is proposed. This method uses a granulation of slope values, an extension of fuzzy set Y_0 in directions defined by granular slopes, a cylindrical extension of fuzzy constraints given in the left sides of rules, a concatenation and an aggregation of function values extracted from rules. The solution is represented by fuzzy relation R defined on $X\times Y$. The value $Y(X^*)$ for given fuzzy value X^* is obtained as a result of cylindrical extension of X, its intersection with R and projection of result on axes Y. The fuzzy set $Y(X^*)$ may be re-translated in linguistic form or converted into real number as a result of defuzzification procedure. The proposed approach to solution of granular initial value problem is illustrated on example. The possibilities of construction of granular derivatives from expert and from data are discussed.

Index Terms – Fuzzy relation, granular differential, initial value problem, computing with words

1 Introduction

Fuzzy rule base gives possibility of piece-wise description of functions. Such description of functions may arise as a result of human perceptions [17] about dependencies between variables and may be represented by rules, for example as follows:

[1] Research supported in part by RFBR Grant 02-01-00092 and the BISC Program of UC Berkeley.

If TEMPERATURE is LOW then DENSITY is SLOWLY DECREASING, (1)

If TEMPERATURE is HIGH then DENSITY is QUICKLY INCREASING. (2)

The fuzzy rules of such type were introduced in [4] for description of shapes of the functions. The linguistic label *SLOWLY DECREASING* in Eq. (1) may be interpreted as an evaluation of the speed of the change of the function *Y* = *DENSITY* when the input variable *X* = *TEMPERATURE* changes his values (increases) on fuzzy interval *LOW*. As a result, such rule may be considered as a linguistic evaluation of the derivative value *dY/dX* on fuzzy interval *LOW*. These linguistic rules may be rewritten in canonical form [17] as follows:

If X is LOW then dY/dX is NEGATIVE SMALL, (3)

If X is HIGH then dY/dX is POSITIVE LARGE, (4)

where *NEGATIVE SMALL* and *POSITIVE LARGE* denote the fuzzy values of derivative correspondent to linguistic evaluations *SLOWLY DECREASING* and *QUICKLY INCREASING* of the speed of the change of *DENSITY*.

A fuzzy piece-wise description of functions may be given analytically on fuzzy intervals of values of independent variable. For example, the fuzzy derivative may be given by

If X is ABOUT 20 then $dy/dx = p$, (5)

If X is BETWEEN 40 AND 50 then $dy/dx = f(x,y)$. (6)

Where p is a real value and $f(x,y)$ is a real valued function.

Generally, the rules describing derivatives may have sufficiently arbitrary form, e.g. as following:

If X is BETWEEN 60 AND 70 then Y is INCREASING and SLIGHTLY CONCAVE. (7)

This rule may be rewritten in canonical form as follows:

If X is BETWEEN 60 AND 70 then dY/dX is POSITIVE MIDDLE and d^2Y/dX^2 *is NEGATIVE SMALL.* (8)

More generally, it may be supposed that considered rules have some evaluations of certainty or confidence that add some uncertainty in rule based description of derivative.

The rules in Eqs. (1, 2, 7) are more suitable for description of shapes of the functions [3] but the rules in Eqs. (3-6, 8) are more suitable for mathematical description of fuzzy derivatives. In this paper we consider the rules similar to rules from Eqs. (1–4) but the proposed further methods of solution of granular initial value problem may be generalized on the rules of more general form.

The set of rules gives a "piece-wise" description $F(X)$ of the granular derivative dY/dX on the domain of variable X. The problem of solution of granular differential equation $dY/dX = F(X)$ with the initial condition *"If TEMPERATURE is* X_0 *then DENSITY* is Y_0", where X_0 and Y_0 are fuzzy numbers, is called a granular initial value problem. Possible solutions of this problem are considered in this paper.

The right sides of rules in Eqs. (1-4) are considered as linguistic evaluations of the slope values p in the equation of tangent line $y = px+q$. These granular slope values P define granular directions of function change on fuzzy intervals given in the left parts of rules. A granule of directions defines "a fuzzy linear function" considered as a fuzzy differential $dY = P\Delta x$ dependent on crisp values of increment Δx or differential dx. Several methods of construction of granular directions are discussed in the following sections and illustrated on examples.

The proposed approach to solution of granular initial value problem uses the general scheme of computing with words described by Zadeh [16-17]. The rule base and the initial condition given in linguistic form represent initial data set (IDS). The linguistic description of solution $Y(X^*)$, where X^* is a fuzzy value, gives the terminal data set (TDS). An inference of TDS from IDS is based on an explicitation of propositions given in natural language as result of their transformation in canonical fuzzy rule base form, a propagation of fuzzy constraints on variables during the solution of problem and a re-translation of result in linguistic form. A granular solution of initial value problem uses the basic procedures of fuzzy logic such as cylindrical extension along axes and in given directions, projection along axis etc. [14]. Several new procedures of extension in given direction, fuzzification, concatenation and aggregation of rules used in construction of fuzzy differentials and in solution of initial value problem are considered. All proposed procedures are illustrated on simulation examples.

The proposed method of solution of granular initial value problem may be considered as a generalization of Euler's solution of crisp initial value problem for ordinary differential equations. A granular solution of initial value problem is represented by a fuzzy relation R defined on $X\times Y$. The value of function $Y^*= Y(X^*)$ for given fuzzy value of X^* is obtained as a result of a cylindrical extension of X^* along the axis Y, its intersection with granular solution R and projection of the result on the axis Y. A linguistic re-translation of obtained fuzzy value Y^* may be used for linguistic description of result. A defuzzification of Y^* will give a crisp value y^* correspondent to X^*.

The granular representation of differential based on granulation of slope values and on granular extension of directions is considered in Section 2. The methods of extension of initial value in granular direction are discussed in Section 3. The steps of solution of granular initial value problem are considered in Section 4. Section 5 contains discussion of methods of construction of linguistic derivatives from data and from experts. Most of procedures used in the solution of granular initial value problem are demonstrated on examples.

2 Granular Representation of Differential

The right sides of rules in Eqs. (1, 2) contain linguistic description of speed of the change of the dependent variable Y when the input variable X is changing (increasing) within constraint given in left sides of rules. Since the speed of the function change is mathematically described by derivative, the rule

If TEMPERATURE is LOW then DENSITY is SLOWLY DECREASING,

may receive the following interpretation: if the numerical value x of input variable X= *TEMPERATURE* is increasing within the fuzzy interval correspondent to the term *LOW* then the numerical value y of the output variable Y = *DENSITY* is changing such that slope values p of lines $t= px+q$ tangent to the curve of function $y= y(x)$ on fuzzy interval *LOW* are evaluated as *SLOWLY DECREASING*. Since $dy/dx = p$ then the label *SLOWLY DECREASING* may be considered as a linguistic evaluation of fuzzy derivative value dY/dX on fuzzy interval *LOW*. From this point of view the considered rule describing the shape of function Y may be re-translated in a canonical form [16] as a rule:

$$\text{If } X \text{ is } LOW \text{ then } dY/dX \text{ is } NEGATIVE\ SMALL, \tag{9}$$

where the value of granular derivative *NEGATIVE SMALL* is correspondent to a linguistic label *SLOWLY DECREASING*.

The term *NEGATIVE SMALL* may be considered also as a granular evaluation of the slope value $P = dY/dX$. A granular direction of the change of the function defined by the tangent will be represented by a fuzzy clump of directions. From another point of view the granule of directions defines fuzzy sets of differential values dY correspondent to given crisp values of increment Δx or differential dx as follows: $dY = P\Delta x$ where P is a granular slope value defined by a rule. We will suppose that the range of crisp values of increment Δx (or differential dx) is defined by the left part of correspondent rule. As a result, the granular differential dY may be considered as a fuzzy function of crisp argument Δx. For example, the rule (9) will define a fuzzy differential as a fuzzy function $dY= P\Delta x$, where P is a fuzzy set correspondent to linguistic term *NEGATIVE SMALL* and Δx takes values in fuzzy interval defined by the term *LOW*.

Generally, granular derivatives may be given by rules

$$\text{If } X \text{ is } A \text{ then } dY/dX \text{ is } P, \tag{10}$$

where A and P are fuzzy sets defined on the domains of x and dy/dx. The rule in Eq. (10) will define fuzzy differential $dY=P\Delta x$. The slope value P will define a granular direction of the change of the function along the tangent line.

The directions [3] may be determined by means of the slope value, by the angle of the slope or by the value of the function dY correspondent to the value of Δx. The explicitation of linguistic evaluations of derivative values consists of the following steps:

1. Choose a linguistic scale of granular derivative values.
2. Choose a domain of possible slope values P.
3. Choose the method of representation of directions.
4. Choose the type of fuzzy sets representing fuzzy clump of directions.
5. Granulate the domain of slope values and define correspondence between linguistic labels and granules.

An example of linguistic scale describing a speed of function change and correspondent scale of slope values is shown in Table 1. Of course, it may be used also the scale with another number of grades. Moreover it may be used linguistic evaluations like *APPROXIMATELY 35, BETWEEN 4 and 29* etc.

Suppose, the domain of possible slope values p equals to $Dom(p)$=[-10,10]. Then for angles $\varphi=arctan(p)$ we have $Dom(\varphi)$= [-1.47, 1.47].

The following three methods of representation of directions [3] correspondent to fuzzy tangent lines may be considered. These methods use some type of fuzzy sets representing fuzzy clump of directions. It will be supposed that the central value p_i of fuzzy slope shown in Table 1 represents the "center" of fuzzy set associated with the direction l_i. This central value p_i will be considered also as an axis of the direction l_i.

Table 1. Linguistic scales of slope values

l_i	Linguistic description of the speed of function change	Linguistic value of derivative (slope)	p_i
7	*QUICKLY INCREASING*	*POSITIVE LARGE*	9
6	*INCREASING*	*POSITIVE MIDDLE*	6
5	*SLOWLY INCREASING*	*POSITIVE SMALL*	3
4	*CONSTANT*	*ZERO*	0
3	*SLOWLY DECREASING*	*NEGATIVE SMALL*	-3
2	*DECREASING*	*NEGATIVE MIDDLE*	-6
1	*QUICKLY DECREASING*	*NEGATIVE LARGE*	-9

In the following for constructing granular models we will use parametric generalized bell membership functions defined on the domain $Dom(u)$ of some variable u as follows [10]:

$$\mu_D(u) = \frac{1}{1 + \left|\frac{u-c}{a}\right|^{2b}} \tag{11}$$

where a, b, c are real valued parameters chosen in suitable way. In parametric form we will write Eq. (11) as: $\mu_D = GBMF(Dom(u);a,b,c)$. Generally, in all constructions it may be used a membership function of another type.

Below we consider three types of granular extension of directions l_i which for each real value of increment Δx of input variable x define a fuzzy set of differential dY of dependent variable y as follows: $\mu_{dY}(dy) = \mu_D(u)$, where $\mu_D(u) = GBMF(Dom(u);a_i,b_i,c_i)$ and determined by the values of Δx and u. As a variable u we use the differential dy of function y, the slope p or the angle $\varphi = arctan(p)$. Each type of variable u used in definition of a granular direction defines some fuzzy relation on $X{\times}Y$ called an extension of direction l_i.

The first method of construction of granular direction is called **a proportional extension of direction**. For each value $\Delta x > 0$ from the domain of increments $Dom(\Delta x)$ the correspondent fuzzy set dY_i of differential values dy associated with direction l_i is defined by:

$$\mu_{dY_i}^{prop}(dy) = \frac{1}{1 + \left|\frac{p - p_i}{a_i}\right|^{2b_i}}, \tag{12}$$

where $p = dy/\Delta x$. The examples of fuzzy clump of directions 3 and 4 constructed by such method are shown on Fig. 1. Fuzzy relations presented on Fig. 1 will be considered as granular differentials which for given value of increment Δx define a fuzzy set of differential values dY. The corresponding fuzzy relation may be considered as an extending fuzzy linear function representing granular differential values. A granular differential obtained by Eq. (12) will be called also as **a proportional differential.** Generally, this type of differential is defined by the extension principle of fuzzy logic from equation $dY = P_i\Delta x$ where P_i is a fuzzy set (fuzzy slope value) defined on $Dom(p)$ as follows:

$$\mu_{dY_i}^{prop}(dy) = \mu_{P_i}\left(\frac{dy}{\Delta x}\right). \tag{13}$$

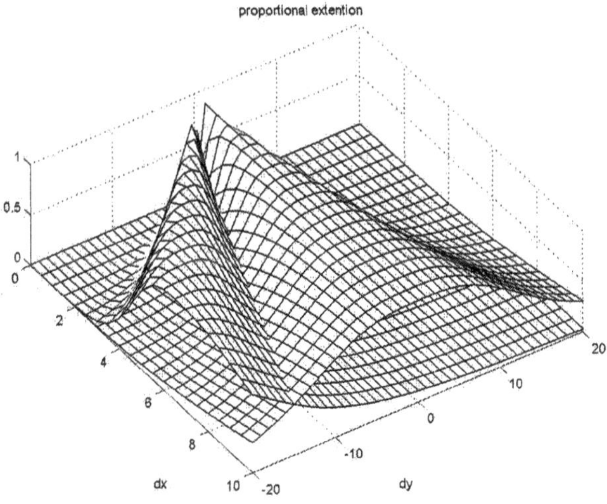

Fig. 1. Proportional extension of directions correspondent to granular slope values *ZERO* and *NEGATIVE SMALL*

Another way for calculating a proportional differential may be based on a fuzzy arithmetic. For example, if P_i is given by a trapezoidal membership function with parameters (a,b,c,d) then the proportional fuzzy differential correspondent to increment Δx will be represented by a trapezoidal membership function with parameters $(a\Delta x, b\Delta x, c\Delta x, d\Delta x)$.

Denote $\varphi_i = arctan(p_i)$. **A conical extension of direction** [3] is defined as follows:

$$\mu_{dY_i^{con}}(dy) = \frac{1}{1 + \left|\frac{\varphi - \varphi_i}{a_i}\right|^{2b_i}}, \tag{14}$$

where $\varphi = arctan(dy/\Delta x)$. Eq. (14) defines a fuzzy set of values of the differential dy in the point Δx associated with direction l_i. The examples of fuzzy clump of directions 3 and 4 constructed by such method are shown on Fig. 2. The conical extension will coincide with proportional extension if we choose the value of parameter b_i in Eq. (14) equal to the parameter b_i in Eq. (12) and the value of parameter a_i in Eq. (14) will be received by multiplication of a_i in Eq. (12) on coefficient $k=(arctan(p)-arctan(p_i))/(p-p_i)$. But, as we can see, in this case the parameter a_i in Eq. (14) should change his value with the change of p.

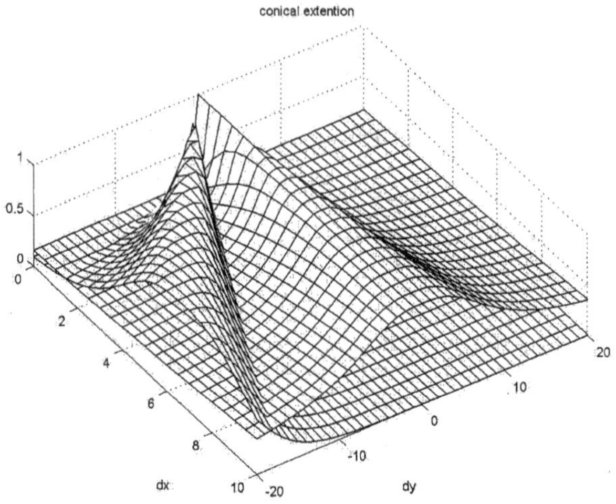

Fig. 2. Conical extension of directions correspondent to granular slope values *ZERO* and *NEGATIVE SMALL*

Since $\Delta x > 0$ the fuzzy set of differentials for increment $\Delta x = 0$ in proportional and conical extensions do not defined. Nevertheless we will define a fuzzy sets D_{i0} in the point $\Delta x = 0$ as singletons, such that $D_{i0}(dy) = 1$ for $dy = 0$ and $D_{i0}(dy) = 0$ for all other values of dy. These fuzzy sets D_{i0} defined for $\Delta x = 0$ will de called a starting sets for conical and proportional extensions of direction l_i.

A "width" of proportional and conical differentials dY is an extending value with the increasing of the increment value Δx. This property of proportional differential may be considered as reasonable if it reflects the increase of the difference between the increment ΔY and differential dY of function Y.

If the extending "width" of fuzzy differential dY is not desirable then it may be used **a cylindrical extension of direction** and correspondingly **cylindrical differential** (for all $\Delta x > 0$):

$$\mu_{dY_i}^{cyl}(dy) = \frac{1}{1 + \left| \frac{dy - dy_i}{a_i} \right|^{2b_i}}, \tag{15}$$

where $dy_i = p_i \Delta x$. The fuzzy value of cylindrical differential will have constant cross-section. The examples of fuzzy clump of directions 3 and 4 from Table 1 constructed by such method are shown on Fig. 3. For $\Delta x = 0$ we define $D_{i0}^{cyl} = GBMF(Dom(y); a_i, b_i, 0)$ which will be called a starting set for proportional extension of direction l_i. If $a_i = a$, $b_i = b$ (i=1,...,m) where a and b are some constants then all D_{i0} will coincide and will be denoted as D_0.

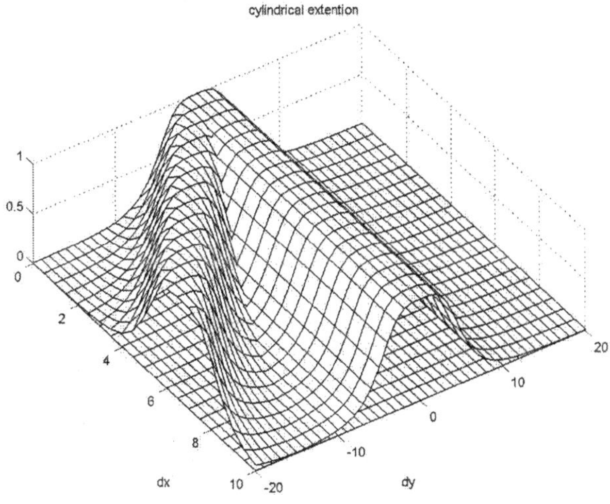

Fig. 3. Cylindrical extension of directions correspondent to granular slope values *ZERO* and *NEGATIVE SMALL*

Formally, the cylindrical differential may be considered as a result of action of generalized constraint *dY isr D* on differential defined by the rule

$$R_i\text{: If } X \text{ is } A_i \text{ then } dy/dx = p_i, \tag{16}$$

where p_i is a crisp slope value. Generally, the constraining relation D can constrain a value of differential given by Eq. (16) by different ways defined by the type of this constraint [17,18]. We will suppose that D equals to $D_{i0} = GBMF(Dom(y);a_i,b_i,0)$ in the rule R_i and fuzzify the line $dy_i = p_i\Delta x$ by Eq. (15).

Granular differential defines some fuzzy relation on $X\times Y$. If we compare the considered three types of fuzzy differential then we can see that a cylindrical extension gives possibility to preserve the width of relation along the axis of direction. The constant width of cylindrical direction may be considered as a useful property in comparison with conical and proportional directions, which define relations with expanding width. Such expansion may be not desirable if the angle of this expansion and the length of fuzzy interval A in the left part of correspondent rule are sufficiently large. In this case the width of this relation near the right side of this interval will became “too wide” and, as a result, the uncertainty in the value of differential will be large. The angle of such expansion is controlled by the value of the parameter a_i for considered bell membership functions. If we want to decrease the value of this parameter then for covering of all range of possible directions it should be chosen the scale of slope values with larger number of grades. From another point of view, the intersection of neighboring fuzzy directions in conical and proportional extensions not much depends on the value of Δx and determined by the angle of the deviation from the axis of direction. For cylindrical extension the intersection of neighboring fuzzy directions is high for small values of Δx and small for large values of Δx.

A cylindrical differential has the principal difference from proportional one: the first defines only one slope value together with a fuzzy deviation of function from correspondent tangent line defined by constraint D whereas a proportional differential defines a fuzzy set of slope values.

3 Granular Directions Induced by Initial Value

The methods of construction of granular directions, considered in Section 2, use as an initial point the element $dy = 0$. Suppose Y_0 is a given fuzzy set on $Dom(Y)$. If this fuzzy set is considered as "an initial fuzzy point" for construction of differential dY_i in given direction l_i then we should construct fuzzy direction l_i from each element z from the $Dom(dY)$ weighted by membership value $Y_0(z)$. Resulting fuzzy directions should be obtained as a result of fuzzification of fuzzy directions constructed from all elements z. We describe here the several possible methods of such fuzzification of directions.

Suppose that direction l_i is given. The cylindrical extension of direction may be considered as a fuzzification of axis line. This fuzzification is defined by a constraint D, i.e. by a starting set D_0 (or D_{i0}). In the point $\Delta x = 0$ the starting set D_0 gives a fuzzification of singleton. Consider a way of fuzzification of fuzzy set Y_0 given in the point $x = x_0$. In each point $z \in Y$ we build starting set D_z for this direction by $D_z = GBMF(Dom(dY); a_i, b_i, z)$. The membership value $Y_0(z)$ is considered as a weight of this starting set. The weighted starting set in the point z is calculated as $T(Y_0(z), D_z)$, where T is some conjunction operation. The resulting starting set (initial fuzzy s et) for granular direction is defined as follows: $D_{Y0} = Fuzz_{z \in Y}(Y_0(z), D_z)$ where $Fuzz$ is some fuzzification procedure based on a conjunction T. We can consider the following types of such fuzzification:

$$D_{Y0} = \underset{z \in Y}{\Upsilon} T(Y_0(z), D_z),$$

$$D_{Y0} = \frac{\sum\limits_{z \in Y} Y_0(z) \cdot D_z}{\sum\limits_{z \in Y} Y_0(z)} \tag{17}$$

These two types of fuzzification will be called a max-T and a weighted-average fuzzification respectively. Fig 4 shows the examples of initial fuzzy sets for cylindrical extension of direction obtained as the result of max-min, max-prod and weighted average fuzzifications of initial value Y_0 and starting set D_0.

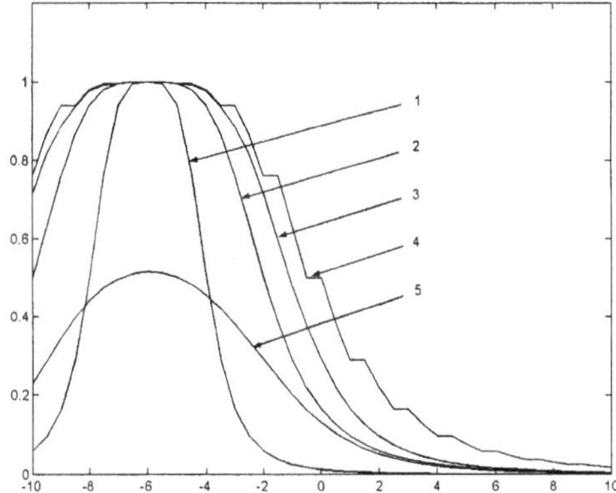

Fig. 4. Initial fuzzy sets for cylindrical extension constructed from starting set D_0 and initial value Y_0: 1 – D_0, 2 – Y_0, 3 – max-prod fuzzification, 4 – max-min fuzzification, 5– weighted-average fuzzification

In this examples *Dom(dY)* = {–10, -9.5, -9, …, 9.5, 10}, the initial value is defined as Y_0 = *GBMF*(*Dom*(*dY*); 4, 2, -6) and the starting set D_0 as D_0 = *GBMF*(*Dom*(*dY*);2,2,0). On Fig. 4 the starting set D_0 is located in the position y = -6. We see that the weighted average fuzzification does not give sufficiently good results. A max-prod fuzzification is more preferable for the use because it is smooth and not so wide as max-min fuzzification. The max-min fuzzification will also give a smooth shape if the distance between points y in *Dom*(*dY*) will be decreased.

The cylindrical extensions of D_{Y0} in the direction number 7 *(QUICKLY INCREASING)* obtained for all three methods of fuzzification of Y_0 and D_0 are shown on Fig. 5. This cylindrical extension may be constructed in each point x as the fuzzy set obtained as a bias of the fuzzy set D_{Y0} on the value $y - y_0 = p_i(x-x_0)$.

More general approach to construction of granular direction may be based on an extension of a two-dimensional fuzzy initial point (X_0, Y_0) in given direction. This type of cylindrical extension in direction was considered by Zadeh [14, 17].

The method of construction of proportional extension of direction based on a given initial value Y_0 is similar to the methods of construction of cylindrical extension. But instead of weighting and aggregating of starting sets D_0 and extension of resulting initial fuzzy set in the direction here we weigh and aggregate directly the proportional extensions. For each point z from Y we calculate the proportional extension D_{iz} of given direction. Each such proportional extension is weighted by membership value $Y_0(z)$ and the results are aggregated as follows:

$$D_{Yi} = \bigcup_{z\in Y} T(Y_0(z), D_{iz}), \qquad (18)$$

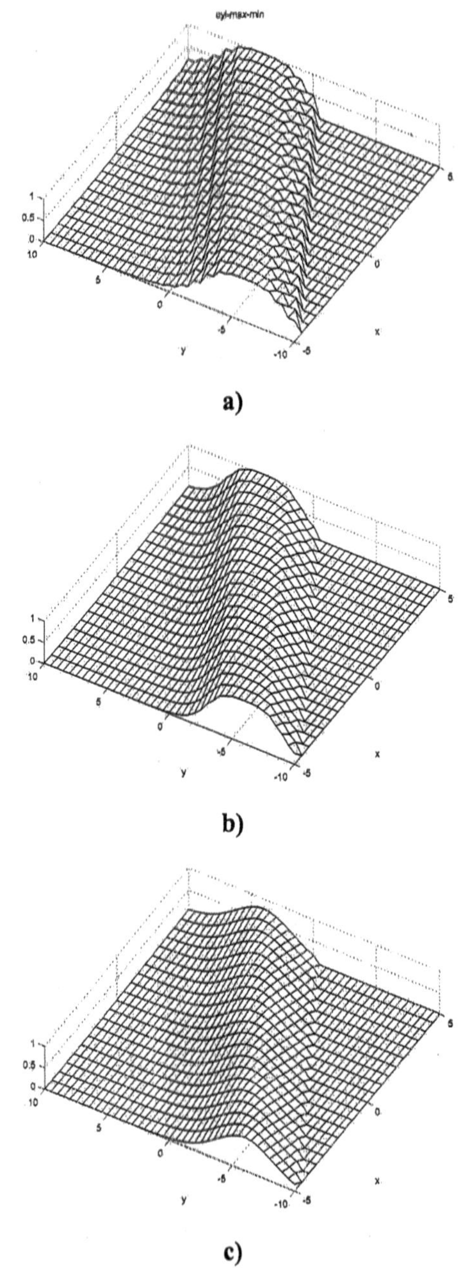

a)

b)

c)

Fig. 5. Cylindrical extensions along direction based on a) max-min, b) max-prod, c) weighted-average aggregations of initial value Y_0 and starting set D_0

The examples of max-min and max-prod fuzzifications of proportional extensions based on given fuzzy initial value Y_0 are shown on Fig. 6.

The method of construction of conical extension of direction based on a given initial value Y_0 is quite similar to the case of proportional extension. The examples of construction of such conical extensions of directions are shown on Fig. 7.

It should be noted that the method of weighted-average fuzzifications of conical and proportional extensions was not satisfactory because the fuzzy sets Y_x obtained as result of cross-section of resulting granular direction in some point have a small modal values.

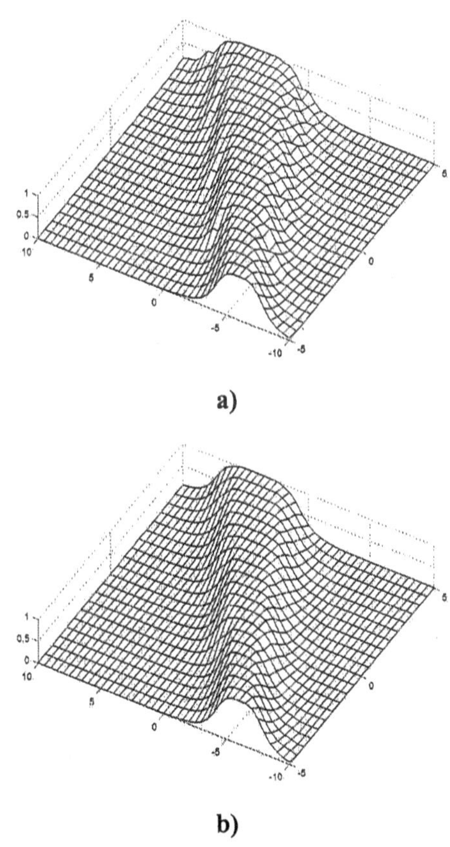

Fig. 6. Proportional extensions of initial value Y_0 along granular direction based on a) max-min, and b) max-prod aggregations

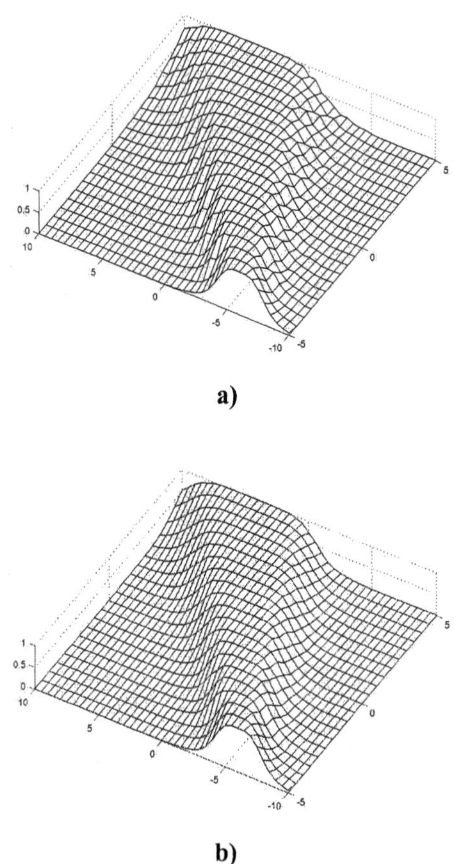

Fig. 7. Conical extensions of initial value Y_0 along granular direction based on a) max-min, and b) max-prod aggregations

4 Solution of a Granular Initial Value Problem

The set of rules with granular slopes in the right parts of rules may be considered as a granular description of a derivative $dY/dX=F(X)$ piece-wise defined on the domain of variable X. We suppose that variables X and Y are defined on some intervals $Dom(X)$ and $Dom(Y)$ of real values. Each rule defines some peace of derivative on the fuzzy interval correspondent to the value of X in the left part of rule.

Suppose we have the canonical set of rules

$$R_i\text{: If } X \text{ is } T_i \text{ then } dY/dX \text{ is } P_i, \ (i=1,\dots,m), \tag{19}$$

where T_i are linguistic terms describing some fuzzy intervals (normal and convex fuzzy sets $A_i\!:X\rightarrow[0,1]$). The right sides of rules contain linguistic terms P_i from the scale of linguistic values of slopes L considered in Section 2. The use of linguistic terms in the left parts of rules implies that it is defined the set of terms of linguistic variable X [16]. This set of terms can include the labels *VERY SMALL, SMALL, MIDDLE, LARGE, VERY LARGE, APPROXIMATELY N, BETWEEN N AND M, GREATER THAN N* etc., where N and M are some real values or fuzzy numbers. The meaning of these terms may be precisiated by semantic rule establishing the correspondence between linguistic terms T_i and fuzzy sets A_i defined on X.

Generally, for the same rule base there may be defined several precisiations of linguistic terms of X dependent on some parameter or context [3]. The role of such parameter or context may play some another variable. The precisiation of granular slopes also may depend on the value of this parameter. In this case the rule base will describe the parametric family of granular derivatives with precisiation dependent on the value of this parameter. For example, the rules in Eqs.(1, 2) may describe linguistically two pieces of derivative *dDENSITY / dTEMPERATURE* but a precisiation of this derivative may be determined by the value of the third parameter Z= *PRESSURE*.

Below, instead of the rule base in Eq. (19) we will consider the rule base

$$R_i\text{: If } X \text{ is } A_i \text{ then } dY/dX \text{ is } P_i, \ (i=1,\dots,m), \tag{20}$$

where A_i are fuzzy intervals correspondent to terms T_i, i.e. all A_i are normal and convex fuzzy sets defined on the domain $Dom(X)$ of X. Suppose that fuzzy intervals A_i (i=1,…,m) define some fuzzy partition of $Dom(X)$, i.e. the following conditions are fulfilled: $sup_x(A_j\cap A_k)(x)= s_1$, $inf_{x\in X}\ ((\bigcup_{i=1}^{m} A_i)(x)) = s_2$, where s_1 and s_2 belong to [0,1] such that $s_1 < 1$ and $s_2 > 0$. Since the cores of fuzzy intervals in fuzzy partition do not intersected then these fuzzy intervals may be linearly ordered such that $A_j < A_k$ iff $x_j < x_k$ for some points x_j and x_k from the cores of A_j and A_k respectively. We will suppose that this ordering coincide with the numeration of rules such that $A_i < A_{i+1}$ for all i= 1, …, m.

Let us consider the methods of solution of granular ordinary differential equation given by rule base from Eq. (20) which define fuzzy graph [16]

$$Q= A_1\times P_1 +\dots+ A_m\times P_m, \tag{21}$$

and satisfying the initial condition:

$$If\ X\ is\ X_0\ then\ Y\ is\ Y_0, \tag{22}$$

where Eq. (21) is the notation for the rule base with piece-wise description of the derivative of Y, and X_0, Y_0 are fuzzy intervals defined on X and Y respectively and given linguistically, e.g.

If TEMPERATURE is APPROXIMATELY 40 then DENSITY is APPROXIMATELY 5.

The problem of solution of granular differential equation in Eq. (21) with initial condition in Eq. (22) will be called a granular initial value problem.

We will suppose that the type of extension of direction is determined. Without the loss of generality we will suppose that the intersection of initial value X_0 with fuzzy interval A_1 from the first rule will be a normal fuzzy set. The solution of granular initial value problem will include the following steps.

1. Find the core $[x_{11}, x_{12}]$ of fuzzy set $X_0 \cap A_1$, $k = 1$.
2. Select a starting point x_0 in $[x_{11}, x_{12}]$.
3. Construct a fuzzy set Y_0 in x_0.
4. Fuzzify by a fuzzy set Y_0 a starting fuzzy set D_0 for direction l_i determined by slope P_1: $D_{Y0} = Fuzz_{z \in Y}(Y_0(z), D_z)$, where $Fuzz$ is one of the fuzzification operations described in the previous section.
5. Construct a granular extension D_1 of the direction l_i based on the initial fuzzy set D_{Y0}. k=2.
6. Select a starting point x_{k-1} in the interval $[x_{k1}, x_{k2}]$ maximizing the intersection of fuzzy sets A_{k-1} and A_k.
7. Cut the granular extension of direction D_{k-1} in the point x_{k-1}. The result will give a fuzzy set $D_{Yk-1}(y) = D_{k-1}(y, x_{k-1})$.
8. Construct a granular extension D_k based on a fuzzy set D_{Yk-1} and on a slope value defined by P_k. $k=k+1$.
9. Repeat Steps 6 - 10 while $k \le m$.
10. Construct a cylindrical extensions of constraints A_k ($k=1,\ldots,m$) along the axis Y: $C_Y(A_k)(x,y) = A_k(x)$.
11. Propagate the cylindrical extensions of constraint A_k ($k=1,\ldots,m$) on correspondent granular directions D_k.
12. Aggregate in overall fuzzy graph the constrained directions obtained on Step 11.

As a result of described procedure a fuzzy relation R on $X \times Y$ will be constructed which will give a solution $Y_R(X)$ of granular initial value problem. The calculation of the value of this function for given fuzzy value X^* of input variable X represented by fuzzy set A^* may be found as a result of the following steps.

13. Construct a cylindrical extension $C_Y(A^*)$ of A^* along the axis Y.

14. Calculate a granular solution $Y_R(X^*) = C_Y(A^*) \cap R$.
15. Find a projection $B^* = P_Y(Y_R(X^*))$ on axis Y.
16. Find a linguistic re-translation of fuzzy set $Y(X^*) = B^*$.
17. Find a numerical solution $y^* = Defuz(Y(X^*))$ as a result of defuzzification procedure.

The linguistic value of the function Y obtained as a result of the re-translation of fuzzy set $Y(X^*) = B^*$ may be considered as a reply on a query "*What is a value of Y if X is A*?*"

Let us discuss the steps of considered procedure.

A starting point on Step 2 may be defined, for example, as follows: $x_0 = (x_{11}+x_{12})/2$.

Instead of two fuzzy sets X_0 and Y_0 initial condition may be given as initial fuzzy relation R_0 correspondent to two-dimensional initial point. In this case an initial values X_0 and Y_0 on Step 3 may be received as result of projection [15] of this relation R_0 on axis X and Y as follows:

$$X_0(x) = P_X(R_0)(x) = max_y(R_0(x,y)),$$
$$Y_0(y) = P_Y(R_0)(y) = max_x(R_0(x,y)).$$

Steps 4, 5, 8 are described in Section 3.

If we will use fuzzy intervals A_k with strict monotonic membership functions from both sides of cores then on Step 6 each interval $[x_{k1},x_{k2}]$ will contain only one point.

Steps 11 and 12 may be realized by several methods. One method may be based on the intersection of each granular direction with correspondent cylindrical extension of fuzzy constraint A_k on Step 11 and aggregation of results obtained for each rule with union operation on Step 12:

$$R = \bigcup_{k=1}^{m} (D_k \cap C_Y(A_k)).$$

This method will be called a max-min aggregation of rules. For this method Steps 11, 12 are reduced to:

$$R(x,y) = max_{k=1,\ldots,m}(min(D_k(x,y),A_k(x))).$$

Another method may be based on the weighing of granular directions by correspondent membership values of cylindrical extensions of fuzzy constraints A_k on Step 11 and on averaging of results on Step 12:

$$R(x,y) = \frac{\sum_{k=1}^{m}(D_k(x,y) \cdot A_k(x))}{\sum_{k=1}^{m} A_k(x)}.$$

This method will be called a weighted-average aggregation.

As it follows from the last formulas, for both methods we really do not need to construct cylindrical extensions of constraints A_k.

Both methods of aggregation of rules are illustrated on Fig. 8, where it is considered an example of solution of initial value problem with the rule base

R_1: *If X is SMALL then Y is QUICKLY INCREASING,*
R_2: *If X is MIDDLE then Y is SLOWLY INCREASING,*
R_3: *If X is LARGE then Y is DECREASING,*

and with initial value

R_4: *If X is APPROXIMATELY -5 then Y is APPROXIMATELY –6.*

From the shapes of solutions it is easy to see that the weighted-average aggregation gives possibility to construct smoother overall fuzzy graph with respect to the max-min aggregation.

The overall fuzzy graph may be considered as an analogue of Euler's piece-wise linear function correspondent to the solution of initial value problem.

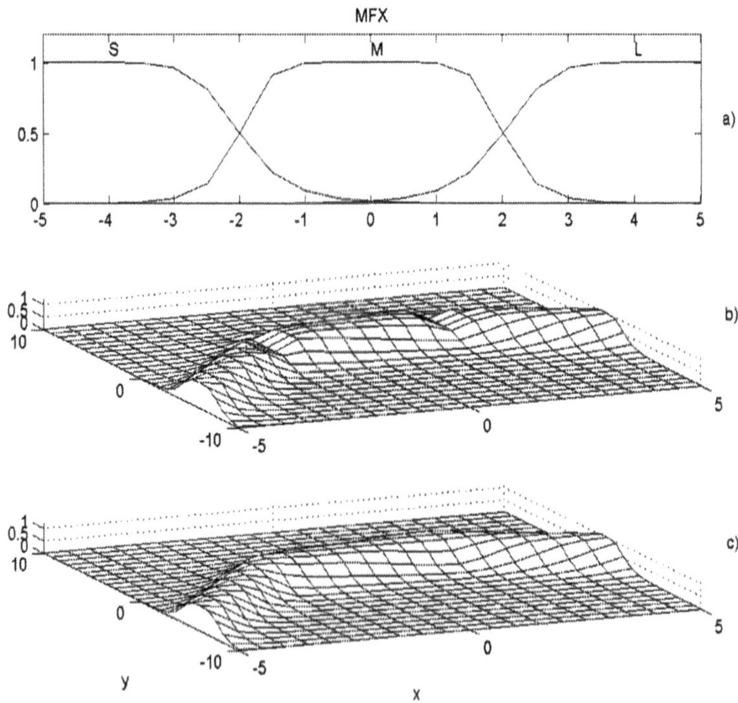

Fig. 8. Solution of example on granular initial value problem: a) fuzzy constraints A_k on X: *S - SMALL, M- MIDDLE, L - LARGE; b*) max-min aggregation of granular directions with constraints; c) weighted-average aggregation of granular directions with constraints

For calculation of Steps 13 and 14 of considered procedure it may be used the following formula:

$$Y_R(X^*)(x,y) = min(C_Y(A^*)(x,y),R(x,y)) = min(A^*(x),R(x,y)).$$

Step 15 is calculated as follows:

$$B^*(y)= max_x(min(A^*(x),R(x,y))).$$

Steps 13 – 15 are illustrated on Fig. 9.

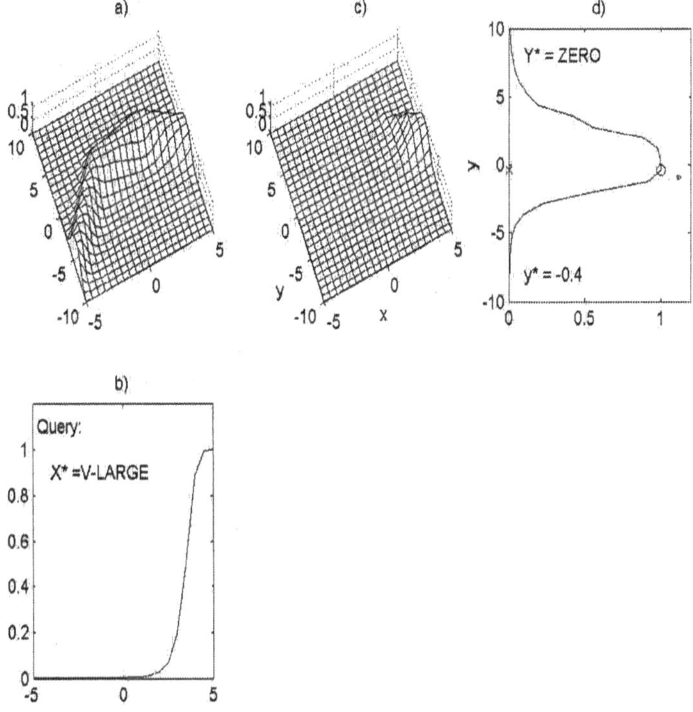

Fig. 9. Calculation of reply on the query "*What is a value of Y if X is VERY LARGE?*"; a) a granular solution of the initial value problem; b) a fuzzy value of X in the query; c) a granular reply on the query; d) a fuzzy value of Y in reply with linguistic and numeric re-translations

A linguistic re-translation of fuzzy set $Y(X^*) = B^*$ may be done by several methods [4]. One of them is the following. Suppose G is some grid of meaningful numerical values and $a\in[0,1]$. Suppose that a-level of B^* is an interval $B_a^* = [b_1,b_2]$. Let us replace this interval by $[g_1,g_2]$ where g_1 and g_2 are the knots of the grid G nearest to b_1 and b_2 respectively. As result, a linguistic interpretation of B^* may be given as *"Y is BETWEEN* g_1 *AND* g_2*"*. On the higher level of linguistic generalization such fuzzy descriptions may be re-translated in linguistic terms like *SMALL, MIDDLE* etc. In this case these terms should be defined on the set of intervals $[g_1,g_2]$ correspondent to given grid G. Linguistic values of the solution of initial value problem may be considered as the elements of the terminal data set in the process of calculation with words [17].

The fuzzy solution $Y(X^*) = B^*$ may be converted into a numerical value by suitable defuzzification procedure [10]. The real value $y^*= -0.4$ correspondent to fuzzy solution $Y(X^*)= B^*$ shown on Fig. 9 was obtained as a mean value of elements from Y maximizing the membership value B^*.

5 Construction of Linguistic Derivatives

Different approaches to qualitative description of dependencies between variables based on expert knowledge and on experimental data are developed [1, 5, 6, 8–12, 15–17] and the rules of new type may be used as additional tool for construction of qualitative models. The rules describing linguistic derivatives may be obtained from experts and from experimental data. In the first case the rules may formalize qualitative expert knowledge about dependencies between input and output parameters of processes. In the second case the rules may be obtained as a result of "linguistic approximation" of dependencies in data [3]. The last problem may be formulated as a problem of construction of optimal fuzzy partition of X and optimal approximation of data on these fuzzy intervals by linear functions. The application of re-translation procedure to obtained partition and linear functions will give required rule base. It may be used different methods for construction of such partition [1, 7, 10, 12]. Below we give a short description of the method used in [4] for construction of linguistic rules of considered type.

Suppose $D=\{(x_i,y_i)\}$, $x_i\in X$, $y_i\in Y$, $(i=1,...,n)$ to be a given set of data where X and Y are real intervals: $X=[x_l,x_r]$, $x_l<x_r$, $Y=[y_l,y_r]$, $y_l<y_r$. The representation of this set of data by the set of rules in Eq.(19) consists on the following steps:

1. Choose a sufficiently large number of fuzzy intervals m.
2. Construct a fuzzy partition of X on m fuzzy intervals $\{A_k\}$, $(k=1,\ldots,m)$.
3. On each fuzzy interval A_k find the optimal approximation of data by linear function $y_k=p_kx+q_k$.

4. Calculate a value of fitness function for obtained approximation $\{A_k, y_k\}$ $(k=1,\ldots,m)$.
5. Find the optimal partition $\{A_k\}$, $(k=1,\ldots,m)$ of X by means of genetic algorithm.
6. If some intervals A_k of optimal partition $\{A_k\}$, $(k=1,\ldots,m)$ are "not admissible" then merge one of such intervals with neighboring interval, decrease the number of intervals $m=m-1$ and go to Step 5, else go to Step 7.
7. Apply procedures of linguistic re-translation to fuzzy intervals A_k and to linear functions $y_k=p_kx+q_k$.

Another method of construction of rule base of considered type might be based on the description of expert knowledge about dependencies between parameters of process. In [5] it was discussed the expert system modeling processes in chemical reactor and based on the rules like the following:

AN INCREASING OF THE TEMPERATURE OF THE PROCESS LEADS TO AN INCREASING OF THE DENSITY OF THE PRODUCT.

The set of such rules was transformed to the rules of expert system like the following:

If DENSITY= MUST_BE_INCREASED and TEMPERATURE= CAN_BE_INCREASED then RECOMMENDATION= TEMPERATURE_OUGHT_TO_BE_INCREASED, pv= 4.

The evaluation of the rule *pv= 4* denotes a qualitative degree of influence of *TEMPERATURE* on the *DENSITY* measured in linguistic scale with 5-7 grades. Such rules may be easily transformed to the rules of type Eqs.(1, 2) considered above.

Conclusions

The methods of representation of derivatives in the form of linguistic rules have been studied. Such rules give piece-wise description of derivative and may be used for description of dependencies between variables of uncertain processes. This qualitative description may formalize expert knowledge about process or represent experimental data about dependencies of variables. Considered rules linguistically describe the speeds of change of dependent variable when independent variable is changed on given fuzzy intervals. The right sides of rules are considered as a granular evaluations of the slope value p in the equation of tangent line $y=px+q$.

The problem of solution of differential equation given by rule base with initial condition represented also by rule called a granular initial value problem. The solution of this problem may be considered as a granular generalization of Euler's method based on piece-wise linear representation of function. The method of solution of granular initial value problem uses the following steps of computing with words: an explicitation of linguistic rule base by granulation of domains of independent variable and slope values; precisiation the meaning of the words as fuzzy constraints on variables; constraints propagation during the solution of the problem and linguistic re-translation of result.

Additionally to the known procedures of processing of fuzzy relations such as cylindrical extension along axis and direction, projection etc. several new procedures have been proposed: proportional extensions of direction, concatenation, aggregation of rules etc.

The methods of extraction of linguistic derivatives from experts and from experimental data have been discussed.

This work may be considered as an initial step in the modeling and solution of granular differential equations based on the new types of rules. Considered approach may be extended on systems of differential equations, on higher order differential equations and on partial differential equations.

The main feature of proposed approach is a definition of a granular differential (represented by a granular direction) as a fuzzy linear function defined on a fuzzy interval. The set of such differentials together with fuzzy initial value determined by rule base is used for construction of solution of granular initial value problem. The proposed concept of granular differential may be generalized on the case of functions with several variables and used for formalization of fuzzy partial differential equations.

The new procedures for solution of granular initial value problem were proposed and demonstrated on simulation examples. These procedures may be used in solution of systems of differential equations, higher order differential equations and partial differential equations given by rules.

The definition of derivatives by fuzzy rule base gives possibility to formalize uncertainty in mathematical models caused by fuzziness and errors in parameters of models. The rules considered in the work may be generalized and may include right sides with analytical functions.

The proposed approach may be applied to description of the uncertain dependencies between variables of complex systems and technological processes. It would be interesting to combine proposed approach with the methods of qualitative reasoning about processes and systems based on the use of the signs of derivatives [6, 8, 9].

Acknowledgements

The author is thankful for useful discussions to Dr. Alexandra Panova and Dr. Michael Wagenknecht.

References

1. Babuska R (1998) Fuzzy Modeling for Control. Kluwer, Boston
2. Batyrshin I Fatkullina R (1995) Context-dependent fuzzy scales and context-free rules for dependent variables. In: Sixth IFSA World Congress, IFSA'95. Sao Paulo, Brasil, vol 1, pp 89-91
3. Batyrshin I, Panova A (2001) On granular description of dependencies. In: Proc. 9th Zittau Fuzzy Colloquium, Zittau, Germany, pp 1–8
4. Batyrshin I, Wagenknecht M (2002) Towards a linguistic description of dependencies in data. Int J Appl Math Comp Sci. Special Issue Computing with Words, Zadeh LA, Kacprzyk J, Rutkowska D (eds) (submitted)
5. Batyrshin I, Zakuanov R, Bikushev G (1994) Expert system based on algebra of uncertainties with memory in process optimization. In: Fuzzy Logic Intell Technol Nucl Sci. 1st FLINS Workshop. Mol, Belgium. World Scientific, pp 156-159
6. D'Ambrosio B (1989) Extending the mathematics in qualitative process theory. Int J Intell Syst, vol 4, pp 55-80
7. De Boor C (1978) A Practical Guide to Splines. Springer-Verlag, New York
8. De Kleer J, Brawn J (1984) A qualitative physics based on confluences. Artific Intelligence, vol 24, pp 7-83
9. Forbus KD (1984) Qualitative process theory. Artificial Intelligence, vol 24, pp 85-168
10. Jang JSR, Sun CT, Mizutani E. (1997) Neuro-Fuzzy and Soft Computing. A Computational Approach to Learning and Machine Intelligence. Prentice Hall, New York
11. Kivikunnas S (1999) Overview of process trend analysis methods and applications. In: Workshop Appl. Chem. Biochem. Industry, Aachen, Germany, (http://www.erudit.de/erudit/committees/Publications.htm)
12. Kosko B (1997) Fuzzy Engineering. Prentice-Hall, Upper Saddle River, NJ
13. Kunz KS (1957) Numcrical Analysis. McGraw-IIill, New York
14. Zadeh LA (1966) Shadows of fuzzy sets (in Russian), Problems of Information Transmission, vol 2, pp 37–44, 1966
15. Zadeh LA (1975) The concept of a linguistic variable and its application to approximate reasoning. Part I. Inform. Sci., vol 8, pp 199-249; Part II. Inform. Sci., vol 8, pp 301-357; Part III. Inform. Sci., vol 9, pp 43-80
16. Zadeh LA (1997) Toward a theory of fuzzy information granulation and its centrality in human reasoning and fuzzy logic. Fuzzy Sets Syst., vol 90, pp 111-127
17. Zadeh LA (1999) From computing with numbers to computing with words - from manipulation of measurements to manipulation of perceptions. IEEE Trans Circuits Systems - 1: Fundamental Theory and Applications 45:105-119

Numerical solutions of fuzzy partial differential equations and its applications in computational mechanics

Andrzej Pownuk[1]

Chair of Theoretical Mechanics, Department of Civil Engineering,
Silesian University of Technology

Abstract: Calculation of the solution of fuzzy partial differential equations is in general very difficult. We can find the exact solution only in some special cases. Fortunately, in most of engineering applications relations between the solutions and uncertain parameters are monotone (we can assume that, when the uncertainty of the parameters is sufficiently small). In this case, the exact solution can be calculated using only endpoints of given intervals. In order to improve the efficiency of calculation we can apply sensitivity analysis.

In this paper, a very efficient algorithm of solution was presented. This algorithm is based on finite element method (or any other numerical method of solution PDE like for example FEM or BEM) and sensitivity analysis. Using this method we can solve engineering problems with thousands degree of freedom. Fuzzy partial differential equations can be applied for modeling of mechanical system (structures) with uncertain parameters.

To construct the fuzzy membership function random sets can be applied. This theory contains fuzzy sets and probability theory as special cases. Using algorithms, which are described in this paper we can solve partial differential equations with random and fuzzy parameters.

1 Introduction

Fuzzy number is a fuzzy set F of the real line R, which is convex and normal. Let $F(R)$ denote the set of all fuzzy numbers, which are upper semicontinuous and have compact support. If $F \in F(R)$

[1] E-mail address: pownuk@zeus.polsl.gliwice.pl

$$\mu_F : R \ni x \to \mu_F(x) \in [0,1] \subset R \tag{1}$$

we can also write

$$\mu(.|.) : R \times F(R) \ni (x, F) \to \mu(x \mid F) \in [0,1] \subset R \tag{2}$$

2 Fuzzy equation

Let us consider the following equations with fuzzy parameter

$$\frac{du}{dx} = hx, u(0) = u_0. \tag{3}$$

An analytical solution of this equation is the following

$$u(x,h) = \frac{hx^2}{2} + u_0. \tag{4}$$

The membership function of the fuzzy solution $u_F(x) \in F(R)$ can be calculated using the extension principle

$$\mu(\xi \mid u_F(x)) = \sup_{h:\xi = \frac{hx^2}{2} + u_0} \mu_F(h). \tag{5}$$

Let us consider some partial differential equations with vector of fuzzy parameters $\mathbf{h} \in F$

$$\mathbf{H}\left(\mathbf{x}, \mathbf{h}, \mathbf{u}, \frac{\partial \mathbf{u}}{\partial \mathbf{x}}, \ldots, \frac{\partial^k \mathbf{u}}{\partial \mathbf{x}^k}\right) = 0, \quad \mathbf{u} \in V \tag{6}$$

where V is some functional space. If we know the exact solution of the problem (6) $\mathbf{u} = \mathbf{u}(\mathbf{x}, \mathbf{h})$ we can calculate the fuzzy solution using the extension principle:

$$\mu(\xi \mid \mathbf{u}_F(\mathbf{x})) = \sup_{\mathbf{h}:\xi = \mathbf{u}(\mathbf{x},\mathbf{h})} \mu_F(\mathbf{h}) \tag{7}$$

The same solution can be calculated using α-level cut method (Buckley, Qy 1990, Buckley, Feuring 2000). The algorithm is the following:

Algorithm 1
1) Calculate α-level cut of fuzzy parameters $\mathbf{h} \in F$

$$\hat{\mathbf{h}}_{\alpha} = \{\mathbf{h} : \mu_F(\mathbf{h}) \geq \alpha\}. \tag{8}$$

2) Calculate the solution of partial differential equations with interval parameters:

$$\hat{\mathbf{u}}_{\alpha}(\mathbf{x}) = \{\mathbf{u}(\mathbf{x},\mathbf{h}) : \mathbf{h} \in \hat{\mathbf{h}}_{\alpha}\}. \tag{9}$$

3) Calculate fuzzy membership function of the solution:

$$\mu(\xi \mid \mathbf{u}_F(\mathbf{x})) = sup\{\alpha : \xi \in \hat{\mathbf{u}}_{\alpha}\}. \tag{10}$$

The most difficult part of this algorithm is the step 2.

3 Random sets interpretation of fuzzy

Let us consider probability space $(\Omega, P_{\Omega}, \Sigma_{\Omega})$ and interval-valued random variable:

$$\hat{H}_{\Omega} : \Omega \ni \omega \to \hat{H}_{\Omega} \in I(R) \tag{11}$$

where $I(R)$ is a set of all intervals.
Using such random variable, we can define upper and lower probability

$$Pl(A) = P_{\Omega}\{\omega : \hat{H}_{\Omega}(\omega) \cap A \neq \varnothing\} \tag{12}$$

$$Bel(A) = P_{\Omega}\{\omega : \hat{H}_{\Omega}(\omega) \subseteq A\} \tag{13}$$

Let us consider discrete random variable, which satisfy the following condition:

$$\hat{H}_{\Omega}(\omega_1) \supseteq \hat{H}_{\Omega}(\omega_2) \supseteq \ldots \supseteq \hat{H}_{\Omega}(\omega_n) \tag{14}$$

then we can define fuzzy membership function in the following way:

$$\mu_F(h) = P_\Omega\{\omega : h \in \hat{H}_\Omega(\omega)\} \qquad (15)$$

Let us consider some mechanical system and performance function $g(\mathbf{h})$, which has the following properties:
- if $g(\mathbf{h}) \geq 0$, then the structure is safe,
- if $g(\mathbf{h}) < 0$, then the structure failed.

Upper probability of failure of the structure can be defined in the following way:

$$P_f^+ = P_\Omega\{\omega : g(\hat{H}_\Omega(\omega)) \cap (-\infty,0) \neq \varnothing\} \qquad (16)$$

If the structure has fuzzy parameters $\mathbf{h} \in F$, then we can calculate upper probability of failure in the following way:

$$P_f^+ = \sup_{\mathbf{h}:g(\mathbf{h})<0} \mu_F(\mathbf{h}) \qquad (17)$$

Upper probability of failure of the structures with random ($\mathbf{X}_\Omega : \Omega \ni \omega \to \mathbf{X}_\Omega(\omega) \in R^n$) and fuzzy ($\mu_F : R^m \in \mathbf{h} \to \mu_F(\mathbf{h}) \in R$) parameters (i.e. $y = g(\mathbf{x},\mathbf{h})$) can be calculated using the following formula

$$P_f^+ = \sum_{\mathbf{x}} \mu_{\mathbf{x}(F)}(\mathbf{x}) P_\Omega\{\mathbf{x}\} = E_\Omega\{\mu_{\mathbf{x}(F)}(\mathbf{x})\} \qquad (18)$$

where

$$\mu_{\mathbf{x}(F)}(\mathbf{x}) = \sup_{\mathbf{h}:g(\mathbf{x},\mathbf{h})<0} \mu_F(\mathbf{h}) \qquad (19)$$

4 Numerical methods of solution of partial differential equations

Many problems in engineering can be described using partial differential equations particularly:

- static and dynamic of structures,
- biomechanics,
- heat and mass transfer,
- electromagnetic fields,

- meteorology etc.

The most popular methods of solution of such equations are:

- finite element method (FEM),
- boundary element method (BEM),
- finite difference method (FDM).

Most universal and popular is finite element method (Ciarlet 1978). The FEM algorithm has the following steps:

1) Formulate boundary value problem

$$\mathbf{L}(\mathbf{x},\mathbf{u}) = \mathbf{f}(\mathbf{x}), \quad \mathbf{u} \in V \tag{20}$$

where $\mathbf{L}(\mathbf{x},.)$ is differential operator, $\mathbf{f}(\mathbf{x})$ is some function and V is a functional space (e.g. Sobolev space).

2) Formulate variational equation of the problem

$$\forall \mathbf{v} \in V, \quad a(\mathbf{u},\mathbf{v}) = l(\mathbf{v}) \tag{21}$$

3) Discretize domain Ω using finite elements $\bigcup_e \Omega^e = \Omega$ and build a solution space V_h

$$\forall \mathbf{u}_h \in V_h, \exists \mathbf{u} \in R^N, \mathbf{u}_h(\mathbf{x}) = \mathbf{N}(\mathbf{x})\mathbf{u} \tag{22}$$

where $\mathbf{N}(\mathbf{x}) = \begin{bmatrix} N_1^1(x) & \dots & N_1^N(x) \\ \dots & \dots & \dots \\ N_n^1(x) & \dots & N_n^N(x) \end{bmatrix}$ is a matrix of shape functions. An approximate solution has the following form:

$$u_i^e(\mathbf{x}) = \sum_j N_{ij}^e(\mathbf{x}) \cdot u_j^e, \quad \mathbf{x} \in \Omega^e \quad (\text{i.e.}\, \mathbf{u}_h^e(\mathbf{x}) = \mathbf{N}^e(\mathbf{x})\mathbf{u}^e) \tag{23}$$

$$u_i^h(\mathbf{x}) = \sum_j N_{ij}(\mathbf{x}) \cdot u_j, \quad \mathbf{x} \in \Omega \quad (\text{i.e.}\, \mathbf{u}_h(\mathbf{x}) = \mathbf{N}(\mathbf{x})\mathbf{u}) \tag{24}$$

4) The approximate solution satisfies the following variational equation:

$$\forall \mathbf{v}_h \in V_h, \quad a(\mathbf{u}_h, \mathbf{v}_h) = l(\mathbf{v}_h) \tag{25}$$

5) Vector of nodal solution can be calculated as a solution of the following system of linear equation

$$\mathbf{Ku} = \mathbf{Q} \tag{26}$$

where **K** (stiffness matrix) and the vector **Q** are defined in the following way:

$$K_{ij} = a(\mathbf{N}_i, \mathbf{N}_j) \tag{27}$$

$$Q_i = l(\mathbf{N}_i) \tag{28}$$

6) If we know, the solution of the system of equation (26) **u** (vector of nodal solution) we can calculate the value of approximate solution between the nodes using equations (23, 24).

Finite element method was implemented in many commercial engineering programs e.g.:

- ABACUS (http://www.hks.com/)
- ADINA (http://www.adina.com/)
- ANSYS (http://www.ansys.com/)
- etc.

5 Using this method we can solve problems with thousands or even millions degree of freedom. Numerical methods of solution of the fuzzy partial differential equations

To solution of the fuzzy partial differential equations, we can also apply algorithm of the finite element method. In this case, equation (26) has the following form:

$$\mathbf{K}(\mathbf{h})\mathbf{u} = \mathbf{Q}(\mathbf{h}), \quad \mathbf{h} \in \hat{\mathbf{h}}_\alpha \tag{29}$$

or in nonlinear problems

$$\mathbf{K}(\mathbf{h}, \mathbf{u})\mathbf{u} = \mathbf{Q}(\mathbf{h}), \quad \mathbf{h} \in \hat{\mathbf{h}}_\alpha \tag{30}$$

If we know the solution $\hat{\mathbf{u}}_\alpha$ of parameter dependent system of equation (29)

$$\hat{\mathbf{u}}_{\alpha} = hull\{\mathbf{u} : \mathbf{K}(\mathbf{h})\mathbf{u} = \mathbf{Q}(\mathbf{h}), \mathbf{h} \in \hat{\mathbf{h}}_{\alpha}\} \tag{31}$$

(the symbol $hull\ S$ denote the smallest interval, which contain the set S) we can calculate a fuzzy membership function of the fuzzy nodal solution $\mathbf{u}_F$ in the following way:

$$\mu(\mathbf{u} \mid \mathbf{u}_F) = sup\{\alpha : \mathbf{u} \in \hat{\mathbf{u}}_{\alpha}\} \tag{32}$$

The exact solution set $\{\mathbf{u} : \mathbf{K}(\mathbf{h})\mathbf{u} = \mathbf{Q}(\mathbf{h}), \mathbf{h} \in \hat{\mathbf{h}}_{\alpha}\}$ of the problem (29) is very complicated because of this in applications we use only the smallest interval which contain the exact solution i.e. $hull\{\mathbf{u} : \mathbf{K}(\mathbf{h})\mathbf{u} = \mathbf{Q}(\mathbf{h}), \mathbf{h} \in \hat{\mathbf{h}}_{\alpha}\}$ (Kulpa et all 1998).
The fuzzy solution between the nodes $\mathbf{u}_F(\mathbf{x})$ can be calculated using the following formula:

$$\mu(\mathbf{u} \mid \mathbf{u}_F(\mathbf{x})) = sup\{\alpha : \mathbf{u} \in \hat{\mathbf{u}}_{\alpha}(\mathbf{x})\} \tag{33}$$

where $\hat{\mathbf{u}}_{\alpha}(\mathbf{x})$ is defined as follows

$$\hat{\mathbf{u}}_{\alpha}(\mathbf{x}) = hull\{\mathbf{u}_h(\mathbf{x},\mathbf{h}) : \mathbf{h} \in \hat{\mathbf{h}}_{\alpha}\} \tag{34}$$

i.e.

$$\hat{\mathbf{u}}_{\alpha}(\mathbf{x}) = hull\{\mathbf{N}(\mathbf{x})\mathbf{u} : \mathbf{K}(\mathbf{h})\mathbf{u} = \mathbf{Q}(\mathbf{h}),\ \mathbf{h} \in \hat{\mathbf{h}}_{\alpha}\} \tag{35}$$

The finite difference method and the boundary element method can be also applied to calculation of numerical solution of the fuzzy partial differential equation. With all mentioned methods, we can apply the following general algorithm.

Algorithm 2

1) Calculate α-cut of fuzzy parameters $\hat{\mathbf{h}}_{\alpha} = \{\mathbf{h} : \mu_F(\mathbf{h}) \geq \alpha\}$.
2) Solve system of parameter dependent system of equation

$$\hat{\mathbf{u}}_{\alpha} = hull\{\mathbf{u} : \mathbf{K}(\mathbf{h})\mathbf{u} = \mathbf{Q}(\mathbf{h}), \mathbf{h} \in \hat{\mathbf{h}}_{\alpha}\} \tag{36}$$

3) Calculate fuzzy nodal solution $\mathbf{u}_F$

$$\mu(\mathbf{u} \mid \mathbf{u}_F) = sup\{\alpha : \mathbf{u} \in \hat{\mathbf{u}}_\alpha\} \tag{37}$$

4) Calculate fuzzy solution between the nodal points $\mathbf{u}_F(\mathbf{x})$

$$\mu(\mathbf{u} \mid \mathbf{u}_F(\mathbf{x})) = sup\{\alpha : \mathbf{u} \in \hat{\mathbf{u}}_\alpha(\mathbf{x})\} \tag{38}$$

6 Systems of algebraic equations with interval parameters

The most difficult part of second algorithm is the step 2. It can be shown that finding the solution of system of a linear interval equation is NP-hard (Kreinovich et all 1998). Because of that the interval solution $\hat{\mathbf{u}}_\alpha$, which is defined in the equation (36) can be found only in special cases.

6.1 Application of monotone functions

Many numerical examples shows, that the relation between the solution $\mathbf{u}=\mathbf{u}(\mathbf{h})$ and uncertain parameters $\mathbf{h}$ is monotone (McWilliam 2000, Noor et all 2000, Pownuk 2000).

Let us consider function $y=u(h)$ and the interval $\hat{h} = [h^-, h^+]$. If function $u(h)$ is monotone, then the extreme values of the function u over the interval $\hat{h}$ can be calculated using only the endpoints h^-, h^+.

$$\hat{y} = [y^-, y^+] = [min\{u(h^-), u(h^+)\}, max\{u(h^-), u(h^+)\}] \tag{39}$$

where

$$y^- = \inf_{h:h\in\hat{h}} u(h), \quad y^+ = \sup_{h:h\in\hat{h}} u(h). \tag{40}$$

Let's assume that the function $\mathbf{u}$ depends on m parameters $h_1, ..., h_m$ (i.e. $\mathbf{u} : R^m \ni \mathbf{h} \to \mathbf{u}(\mathbf{h}) \in R^n$), which belong to the m intervals $\hat{h}_1, ..., \hat{h}_m$ (i.e.

$\mathbf{h} \in \hat{\mathbf{h}}$). If this function is monotone then the extreme values can be found after calculation all combination of endpoints of the multidimensional interval $\hat{\mathbf{h}}$.

$$u_i^- = min\{u_i(\mathbf{h}) : \mathbf{h} \in Vertex(\hat{\mathbf{h}})\}, \tag{41}$$

$$u_i^+ = max\{u_i(\mathbf{h}) : \mathbf{h} \in Vertex(\hat{\mathbf{h}})\}. \tag{42}$$

where $Vertex(\hat{\mathbf{h}})$ is a set of all vertex of the interval $\hat{\mathbf{h}}$. Now we can write

$$\hat{\mathbf{u}} = [u_1^-, u_1^+] \times [u_2^-, u_2^+] \times ... \times [u_n^-, u_n^+] = hull\, \mathbf{u}(\hat{\mathbf{h}}) \tag{43}$$

To calculation of the vector $\hat{\mathbf{u}}$ we have to calculate the value of the function $\mathbf{u}(\mathbf{h})$ 2^m times. In the same way, we can calculate the solution of the problem (36)

$$u_i^- = min\{u_i : \mathbf{K}(\mathbf{h})\mathbf{u} = \mathbf{Q}(\mathbf{h}), \mathbf{h} \in Vertex(\hat{\mathbf{h}})\}, \tag{44}$$

$$u_i^+ = max\{u_i : \mathbf{K}(\mathbf{h})\mathbf{u} = \mathbf{Q}(\mathbf{h}), \mathbf{h} \in Vertex(\hat{\mathbf{h}})\}. \tag{45}$$

Unfortunately, this method has very high computational complexity and cannot be applied to problems that are more complicated.

6.2 Application of sensitivity analysis

Let us consider a function $u : R \ni h \to u(h) \in R$. If the derivative $\frac{\partial u}{\partial h}$ has constant sign then extreme values can be calculated using the following formulas

$$\text{If } \frac{\partial u(h_0)}{\partial h} > 0 \text{, then } u^- = u(h^-),\ u^+ = u(h^+) \tag{46}$$

$$\text{If } \frac{\partial u(h_0)}{\partial h} < 0 \text{, then } u^- = u(h^+),\ u^+ = u(h^-) \tag{47}$$

where $h_0 = \dfrac{h^- + h^+}{2} = mid(\hat{h})$.

In multidimensional case in order to calculate extreme values of function **u**=**u**(**h**) ($u_i = u_i(\mathbf{h})$) we can compute the sign vector $\mathbf{S}^i$

$$\mathbf{S}^i = \left[sign\left(\frac{\partial u_i(\mathbf{h}_0)}{\partial h_1}\right) \quad \dots \quad sign\left(\frac{\partial u_i(\mathbf{h}_0)}{\partial h_m}\right) \right]^T \tag{48}$$

$$\mathbf{S}^i = [S_1^i, \dots, S_m^i]^T \tag{49}$$

where $\mathbf{h}_0 = \left[\dfrac{h_1^- + h_1^+}{2} \quad \dots \quad \dfrac{h_m^- + h_m^+}{2} \right]^T = mid(\hat{\mathbf{h}})$.

If function $u_i = u_i(\mathbf{h})$ is monotone, then upper bound can be calculated using the following point

$$\mathbf{h}_i^{upper} = \left[{}^i h_1^{upper} \quad \dots \quad {}^i h_m^{upper} \right]^T \tag{50}$$

where

$$\text{if } S_k^i > 0 \text{ then } {}^i h_k^{upper} = h_k^+ \tag{51}$$

$$\text{if } S_k^i < 0 \text{ then } {}^i h_k^{upper} = h_k^- . \tag{52}$$

In this case $u_i^+ = u_i(\mathbf{h}_i^{upper})$. In the same way, we can construct lower bound of the function $u_i = u_i(\mathbf{h})$.

$$\mathbf{h}_i^{lower} = \left[{}^i h_1^{lower} \quad \dots \quad {}^i h_m^{lower} \right]^T \tag{53}$$

where

$$\text{if } S_k^i > 0 \text{ then } {}^i h_k^{upper} = h_k^- \tag{54}$$

$$\text{if } S_k^i < 0 \text{ then } {}^i h_k^{upper} = h_k^+ . \tag{55}$$

and finally

$$u_i^- = u_i(\mathbf{h}_i^{lower}), \quad u_i^+ = u_i(\mathbf{h}_i^{upper}), \tag{56}$$

$$\hat{\mathbf{u}} = [u_1^-, u_1^+] \times \ldots \times [u_m^-, u_m^+]. \tag{57}$$

We see that extreme values of the function $\mathbf{u} = \mathbf{u}(\mathbf{h})$ can be calculated using the sign vectors $\mathbf{S}^i$ and the endpoints of interval $\hat{\mathbf{h}}$

$$\mathbf{h}_i^{lower} = \mathbf{h}^{lower}(\mathbf{S}^i, \hat{\mathbf{h}}), \quad \mathbf{h}_i^{upper} = \mathbf{h}^{upper}(\mathbf{S}_i, \hat{\mathbf{h}}) \tag{58}$$

$$\hat{u}_i = [u_i^-, u_i^-] = [u\left(\mathbf{h}^{lower}(\mathbf{S}_i, \hat{\mathbf{h}})\right), u\left(\mathbf{h}^{upper}(\mathbf{S}^i, \hat{\mathbf{h}})\right)] \tag{59}$$

The vector $\mathbf{S}^i$ have to be calculated for each coordinate of the vector $\mathbf{u}$ i.e. n times. From the definition of the vectors $\mathbf{S}^i$ arise that

$$\mathbf{h}^{lower}(\mathbf{S}^i, \hat{\mathbf{h}}) = \mathbf{h}^{lower}((-1) \cdot \mathbf{S}^i, \hat{\mathbf{h}}) \tag{60}$$

$$\mathbf{h}^{upper}(\mathbf{S}^i, \hat{\mathbf{h}}) = \mathbf{h}^{upper}((-1) \cdot \mathbf{S}^i, \hat{\mathbf{h}}) \tag{61}$$

i.e. the sign vector $\mathbf{S}^i$ generate the same lower and upper bound as vector $(-1) \cdot \mathbf{S}^i$. From computational point of view, it is convenient to find independent sign vectors, which generate different lower and upper bounds of the solution.

$$IndS = \{\mathbf{S}^{1*}, \ldots, \mathbf{S}^{k*}\} \tag{62}$$

$$\forall \mathbf{S}^{i*}, \mathbf{S}^{j*} \in IndS, (i \neq j) \Rightarrow ((\mathbf{S}^{j*} \neq \mathbf{S}^{i*}) \wedge (\mathbf{S}^{i*} \neq (-1) \cdot \mathbf{S}^{j*})) \tag{63}$$

According to my experience, number of the sign vectors $\mathbf{S}^{j*}$ is much lower than number of the vectors $\mathbf{S}^{i}$ (Pownuk 2001).
Now we can apply sensitivity analysis method to solution of the problem (36). Whole algorithm has the following steps

Algorithm 3
1) Formulate parameter dependent system of equation with interval parameters in the form (36). And calculate $\mathbf{u}(\mathbf{h}_{\alpha}^{0})$

$$\mathbf{K}(\mathbf{h}_{\alpha}^{0})\mathbf{u}(\mathbf{h}_{\alpha}^{0}) = \mathbf{Q}(\mathbf{h}_{\alpha}^{0}), \quad \mathbf{h}_{\alpha}^{0} = mid(\hat{\mathbf{h}}_{\alpha}) \tag{64}$$

2) For i=1,...,m calculate $\dfrac{\partial \mathbf{u}(\mathbf{h}_{\alpha}^{0})}{\partial h_i}$, where

$$\mathbf{K}(\mathbf{h}_{\alpha}^{0})\frac{\partial \mathbf{u}(\mathbf{h}_{\alpha}^{0})}{\partial h_i} = \frac{\partial \mathbf{Q}(\mathbf{h}_{\alpha}^{0})}{\partial h_i} - \frac{\partial \mathbf{K}(\mathbf{h}_{\alpha}^{0})}{\partial h_i}\mathbf{u}(\mathbf{h}_{\alpha}^{0}). \tag{65}$$

3) For i=1,...,n (n – number of degree of freedom) calculate the sign vector

$$\mathbf{S}_{\alpha}^{i} \cdot = \left[sign\left(\frac{\partial u_i(\mathbf{h}_{\alpha}^{0})}{\partial h_1}\right) \ \ldots \ sign\left(\frac{\partial u_i(\mathbf{h}_{\alpha}^{0})}{\partial h_m}\right)\right] \tag{66}$$

4) Calculate independent sign vectors $IndSign_{\alpha} = \{\mathbf{S}_{\alpha}^{1*},...,\mathbf{S}_{\alpha}^{k*}\}$ using condition (63) and create vector **U** such, that

$$\mathbf{S}_{\alpha}^{i} = \mathbf{S}_{\alpha}^{j*}, \text{ where } j = U_i \tag{67}$$

5) For i=1,...,k calculate interval solution $\hat{\mathbf{u}}_{\alpha}^{i*}$

$$\hat{\mathbf{u}}_{\alpha}^{i*} = [\mathbf{u}\left(\mathbf{h}^{lower}(\mathbf{S}_{\alpha}^{i*},\hat{\mathbf{h}})\right), \mathbf{u}\left(\mathbf{h}^{upper}(\mathbf{S}_{\alpha}^{i*},\hat{\mathbf{h}})\right)] \tag{68}$$

$$IndSolutin_{\alpha} = \{\hat{\mathbf{u}}_{\alpha}^{1*},...,\hat{\mathbf{u}}_{\alpha}^{k*}\} \tag{69}$$

6) Calculate extreme interval solution $\hat{\mathbf{u}}_{\alpha}$.

For $i = 1,...,n$

$$\hat{u}_{\alpha i} = [u_{\alpha i}^{j*-}, u_{\alpha i}^{j*-}], \text{ where } j = U_i \tag{70}$$

Computational complexity of this algorithm:
- step 1 – 1 solution systems of equations,
- step 2 – m solution systems of equations,
- step 5 – $2 \cdot k$ solution systems of equations ($1 \le k \le n$).

In presented algorithm we have to calculate a system of equation between $1+m+2$ and $1+m+2\cdot n$ times.

6.3 Calculation of the solution between the nodes

Sometimes we would like to know the interval solution $\hat{\mathbf{u}}_{\alpha}(\mathbf{x})$ of the boundary value problem in the point $\mathbf{x} \in \Omega$ between the nodal points. If we assume that the function $u_i = u_i(\mathbf{x}, \mathbf{h})$ is monotone (for fixed $\mathbf{x} \in \Omega$), then to calculation of extreme values sensitivity analysis can be applied.

First, we have to calculate sensitivity vector $\mathbf{S}_{\alpha}^{\mathbf{x}}$

$$\mathbf{S}_{\alpha}^{\mathbf{x}} = \left[sign\left(\frac{\partial u_i(\mathbf{x}, \mathbf{h}_{\alpha}^0)}{\partial h_1} \right) \quad \dots \quad sign\left(\frac{\partial u_i(\mathbf{x}, \mathbf{h}_{\alpha}^0)}{\partial h_m} \right) \right]. \tag{71}$$

From the equation (24) arise that:

$$\frac{\partial u_i(\mathbf{x}, \mathbf{h}_{\alpha}^0)}{\partial h_q} = \frac{\partial N_{ij}(\mathbf{x}, \mathbf{h}_{\alpha}^0)}{\partial h_q} u_j(\mathbf{h}_{\alpha}^0) + N_{ij}(\mathbf{x}, \mathbf{h}_{\alpha}^0) \frac{\partial u_j(\mathbf{h}_{\alpha}^0)}{\partial h_q} \tag{72}$$

The vectors $\mathbf{u}(\mathbf{h}_{\alpha}^0)$, $\frac{\partial \mathbf{u}(\mathbf{h}_{\alpha}^0)}{\partial h_j}$ were calculated in the algorithm 3, because of that to calculation of $\frac{\partial u_i(\mathbf{x}, \mathbf{h}_{\alpha}^0)}{\partial h_q}$ we don't have to solve any system of linear equation.

Now we have to check if the sign vector $\mathbf{S}_{\alpha}^{\mathbf{x}}$ is unique.

We assume that the sign vector $\mathbf{S}_{\alpha}^{\mathbf{x}}$ is unique if

$$\forall \mathbf{S}_{\alpha}^{i*} \in IndSign_{\alpha},\ (\mathbf{S}_{\alpha}^{i*} \neq \mathbf{S}_{\alpha}^{\mathbf{x}}) \wedge (\mathbf{S}_{\alpha}^{i*} \neq (-1)\cdot \mathbf{S}_{\alpha}^{\mathbf{x}}) \tag{73}$$

If the sign vector $\mathbf{S}_{\alpha}^{\mathbf{x}}$ is not unique i.e.

$$\exists \mathbf{S}_{\alpha}^{p*} \in IndSign_{\alpha},\ (\mathbf{S}_{\alpha}^{p*} = \mathbf{S}_{\alpha}^{\mathbf{x}}) \wedge (\mathbf{S}_{\alpha}^{p*} = (-1)\cdot \mathbf{S}_{\alpha}^{\mathbf{x}}) \tag{74}$$

then extreme solution can be calculated using the following formulas

$$u_{\alpha i}^{-}(\mathbf{x}) = N_{ij}(\mathbf{x}, \mathbf{h}^{lower}(\mathbf{S}_{\alpha}^{p*}, \hat{\mathbf{h}}_{\alpha}))u_{\alpha j}^{p*-} \tag{75}$$

$$u_{\alpha i}^{+}(\mathbf{x}) = N_{ij}(\mathbf{x}, \mathbf{h}^{upper}(\mathbf{S}_{\alpha}^{p*}, \hat{\mathbf{h}}_{\alpha}))u_{\alpha j}^{p*+} \tag{76}$$

$$\hat{u}_{\alpha i}(\mathbf{x}) = [\hat{u}_{\alpha i}^{-}(\mathbf{x}), \hat{u}_{\alpha i}^{+}(\mathbf{x})] \tag{77}$$

If the sign vector $\mathbf{S}_{\alpha}^{\mathbf{x}}$ is unique, then we have to calculate a new interval solution

$$\hat{\mathbf{u}}_{\alpha}^{k+1*} = [\mathbf{u}(\mathbf{h}^{lower}(\mathbf{S}_{\alpha}^{\mathbf{x}}, \hat{\mathbf{h}})), \mathbf{u}(\mathbf{h}^{upper}(\mathbf{S}_{\alpha}^{\mathbf{x}}, \hat{\mathbf{h}}))] \tag{78}$$

next

$$\mathbf{S}_{\alpha}^{k+1*} = \mathbf{S}_{\alpha}^{\mathbf{x}} \tag{79}$$

$$IndSign_{\alpha} := IndSign_{\alpha} \cup \{\mathbf{S}_{\alpha}^{k+1*}\} \tag{80}$$

$$IndSoluton_{\alpha} := IndSoution_{\alpha} \cup \{\hat{\mathbf{u}}_{\alpha}^{k+1*}\} \tag{81}$$

Extreme solution can be calculated using the following formulas

$$u_{\alpha i}^{-}(\mathbf{x}) = N_{ij}(\mathbf{x}, \mathbf{h}^{lower}(\mathbf{S}_{\alpha}^{k+1*}, \hat{\mathbf{h}}_{\alpha}))u_{\alpha j}^{k+1*-} \tag{82}$$

$$u_{\alpha i}^{+}(\mathbf{x}) = N_{ij}(\mathbf{x}, \mathbf{h}^{upper}(\mathbf{S}_{\alpha}^{k+1*}, \hat{\mathbf{h}}_{\alpha}))u_{\alpha j}^{k+1*+} \tag{83}$$

$$\hat{u}_{\alpha i}(\mathbf{x}) = [\hat{u}_{\alpha i}^{-}(\mathbf{x}), \hat{u}_{\alpha i}^{+}(\mathbf{x})] \tag{84}$$

7 Calculation of the value of fuzzy function

In technical applications very often we have to calculate the value of function, which depends on the solution of fuzzy partial differential equations e.g. $y = f(\mathbf{x}, \mathbf{u}, \mathbf{h})$.
Extreme values of function f can be calculated using sensitivity analysis:

$$\mathbf{S}_{\alpha}^{\mathbf{x}} = \left[sign\left(\frac{\partial f(\mathbf{x}, \mathbf{u}(\mathbf{h}_0), \mathbf{h}_0)}{\partial h_1} \right), \ldots, sign\left(\frac{\partial f(\mathbf{x}, \mathbf{u}(\mathbf{h}_0), \mathbf{h}_0)}{\partial h_m} \right) \right] \tag{85}$$

where

$$\frac{\partial f(\mathbf{x}, \mathbf{u}(\mathbf{h}_0), \mathbf{h}_0)}{\partial h_i} = \frac{\partial f(\mathbf{x}, \mathbf{u}(\mathbf{h}_0), \mathbf{h}_0)}{\partial \mathbf{u}} \frac{\partial \mathbf{u}(\mathbf{h}_0)}{\partial h_i} + \frac{\partial f(\mathbf{x}, \mathbf{u}(\mathbf{h}_0), \mathbf{h}_0)}{\partial h_i} \tag{86}$$

Now we can apply similar procedure like in previous paragraph.
If the sign vector $\mathbf{S}_{\alpha}^{\mathbf{x}}$ is not unique, then extreme solution can be calculated using the following formulas

$$f_{\alpha}^{-}(\mathbf{x}) = f(\mathbf{x}, \mathbf{u}_{\alpha}^{p*-}, \mathbf{h}^{lower}(\mathbf{S}_{\alpha}^{p*}, \hat{\mathbf{h}}_{\alpha})) \tag{87}$$

$$f_{\alpha}^{+}(\mathbf{x}) = f(\mathbf{x}, \mathbf{u}_{\alpha}^{p*+}, \mathbf{h}^{upper}(\mathbf{S}_{\alpha}^{p*}, \hat{\mathbf{h}}_{\alpha})) \tag{88}$$

If the sign vector $\mathbf{S}_{\alpha}^{\mathbf{x}}$ is unique, then calculate a new interval solution

$$\hat{\mathbf{u}}_{\alpha}^{k+1*} = [\mathbf{u}(\mathbf{h}^{lower}(\mathbf{S}_{\alpha}^{\mathbf{x}}, \hat{\mathbf{h}})), \mathbf{u}(\mathbf{h}^{upper}(\mathbf{S}_{\alpha}^{\mathbf{x}}, \hat{\mathbf{h}}))] \tag{89}$$

next

$$\mathbf{S}_{\alpha}^{k+1*} = \mathbf{S}_{\alpha}^{\mathbf{x}} \tag{90}$$

$$IndSign_\alpha := IndSign_\alpha \cup \{\mathbf{S}_\alpha^{k+1*}\} \tag{91}$$

$$IndSoluton_\alpha := IndSoution_\alpha \cup \{\hat{\mathbf{u}}_\alpha^{k+1*}\} \tag{92}$$

Extreme values of the function *f* can be calculated using the following formulas

$$f_\alpha^-(\mathbf{x}) = f(\mathbf{x}, \mathbf{u}_\alpha^{k+1*-}, \mathbf{h}^{lower}(\mathbf{S}_\alpha^{k+1*}, \hat{\mathbf{h}}_\alpha)) \tag{93}$$

$$f_\alpha^+(\mathbf{x}) = f(\mathbf{x}, \mathbf{u}_\alpha^{k+1*+}, \mathbf{h}^{upper}(\mathbf{S}_\alpha^{k+1*}, \hat{\mathbf{h}}_\alpha)) \tag{94}$$

8 Numerical example – plane stress problem in theory of elasticity

Let us consider the following partial differential equations

$$\begin{gathered} \frac{E}{2(1+\nu)} u_{\alpha,\beta\beta} + \frac{E}{2(1-\nu)} u_{\beta,\beta\alpha} + \rho f_\alpha = 0, \quad \alpha,\beta = 1,2 \\ u_\alpha = \overset{*}{u}_\alpha, \quad x \in \partial\Omega_u \\ \sigma_{\alpha\beta} n_\beta = \overset{*}{t}_\alpha, \quad x \in \partial\Omega_\sigma \end{gathered} \tag{95}$$

where E is a module of elasticity, ν is a Poisson's ratio, u_α are displacements, ρ is a mass density, f_α are mass forces, $\sigma_{\alpha\beta}$ are stress, n_α coordinate of the unit vector which is normal to the boundary $\partial\Omega$, $\overset{*}{t}_\alpha$ are boundary traction. These are equilibrium equations of the plane stress elasticity problem. We can write these equations in the variational form

$$\int_\Omega \sigma_{ij} \delta\varepsilon_{ij} dV = \int_\Omega \rho f_i \delta u_i d\Omega + \int_{\partial\Omega} t_i \delta u_i dS \tag{96}$$

where ε_{ij} is a strain tensor. If we take into account the constitutive equations

$$\sigma_{ij} = C_{ijkl}\varepsilon_{kl} \tag{97}$$

and geometric equations

$$\varepsilon_{ij} = \frac{1}{2}(u_{i,j} + u_{j,i}) \tag{98}$$

we can define the bilinear form

$$a(\mathbf{u}, \mathbf{v}) = \int_{\Omega} \sigma_{ij}\delta\varepsilon_{ij}d\Omega = \int_{\Omega} C_{ijkl}u_{i,j}v_{k,l}d\Omega \tag{99}$$

and the linear form

$$l(\mathbf{v}) = \int_{\Omega} \rho f_i v_i d\Omega + \int_{\partial\Omega} t_i v_i d\Omega \tag{100}$$

The variational equations of the theory of elasticity can be written in the following form

$$\forall \mathbf{v} \in V, \quad a(\mathbf{u}, \mathbf{v}) = l(\mathbf{v}) \tag{101}$$

Now we can solve this equations using FEM method. The local stiffness matrix can be written in the following form

$$\mathbf{K}^e = \int_{\Omega^e} \mathbf{B}^{e^T}\mathbf{D}^e\mathbf{B}^e d\Omega \tag{102}$$

where

$$\mathbf{B}^e = \begin{bmatrix} \frac{\partial N_1^e}{\partial x_1} & 0 & \frac{\partial N_2^e}{\partial x_1} & 0 & \frac{\partial N_3^e}{\partial x_1} & 0 \\ 0 & \frac{\partial N_1^e}{\partial x_2} & 0 & \frac{\partial N_2^e}{\partial x_2} & 0 & \frac{\partial N_3^e}{\partial x_2} \\ \frac{\partial N_1^e}{\partial x_1} & \frac{\partial N_1^e}{\partial x_2} & \frac{\partial N_2^e}{\partial x_2} & \frac{\partial N_2^e}{\partial x_1} & \frac{\partial N_3^e}{\partial x_2} & \frac{\partial N_3^e}{\partial x_1} \end{bmatrix} \tag{103}$$

$$\mathbf{D}^e = \frac{E^e}{1-\nu^{e^2}}\begin{bmatrix} 1 & \nu^e & 0 \\ \nu^e & 1 & 0 \\ 0 & 0 & \frac{1-\nu^e}{2} \end{bmatrix} \tag{104}$$

and N_i are shape functions, which will be described later.
The load vector can be calculated from the following equations:

$$\mathbf{Q} = \int_{\Omega} \mathbf{N}^T \rho \mathbf{f} d\Omega + \int_{\partial\Omega} \mathbf{N}^T \mathbf{t} dS \tag{105}$$

In calculation, we will be use triangular element, which is shown in Fig. 1.

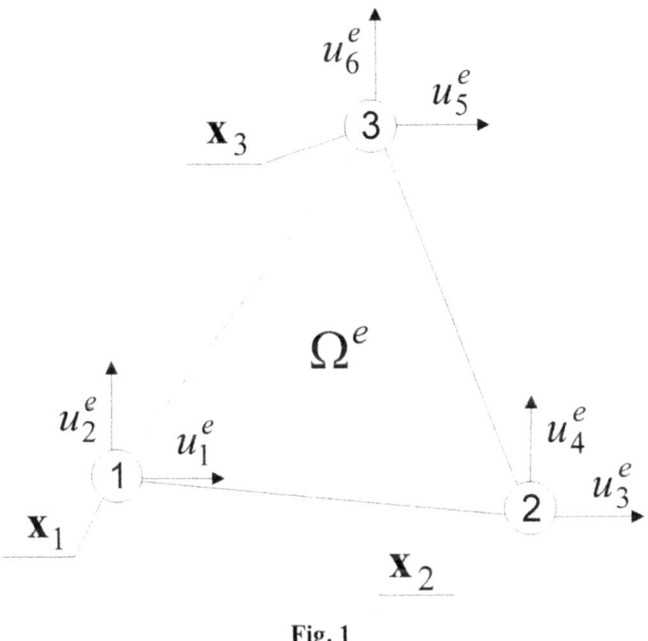

Fig. 1

Displacement in the element can be described in the following way:

$$\mathbf{u}^e(x) = \begin{bmatrix} u_1^e(x) \\ u_2^e(x) \end{bmatrix} = \begin{bmatrix} N_1^e(\mathbf{x}) & 0 \\ 0 & N_1^e(\mathbf{x}) \\ N_2^e(\mathbf{x}) & 0 \\ 0 & N_2^e(\mathbf{x}) \\ N_3^e(\mathbf{x}) & 0 \\ 0 & N_3^e(\mathbf{x}) \end{bmatrix}^T \cdot \begin{bmatrix} u_1^e \\ u_2^e \\ u_3^e \\ u_4^e \\ u_5^e \\ u_6^e \end{bmatrix} = \mathbf{N}^e(\mathbf{x})\mathbf{u}^e \quad (106)$$

Shape functions of the element "e" satisfy the following conditions:

$$N_i^e(\mathbf{x}_j) = \delta_{ij} \quad (107)$$

If we assume, that the shape functions are linear i.e.

$$N_i^e(\mathbf{x}) = a_i^e + b_i^e x_1 + c_i^e x_2 \quad (108)$$

then the function $N_1^e(\mathbf{x})$ has the following form:

$$N_1^e(\mathbf{x}) = \frac{x_1^{2e} x_2^{3e} - x_1^{3e} x_2^{2e} + (x_2^{2e} - x_2^{3e})x_1 + (x_1^{3e} - x_1^{2e})x_2}{\Delta^e} \quad (109)$$

where

$$\Delta^e = \begin{vmatrix} 1 & x_1^{1e} & x_2^{1e} \\ 1 & x_1^{2e} & x_2^{2e} \\ 1 & x_1^{3e} & x_2^{3e} \end{vmatrix} \quad (110)$$

etc.

Let us consider structure, which is shown in Fig. 2.

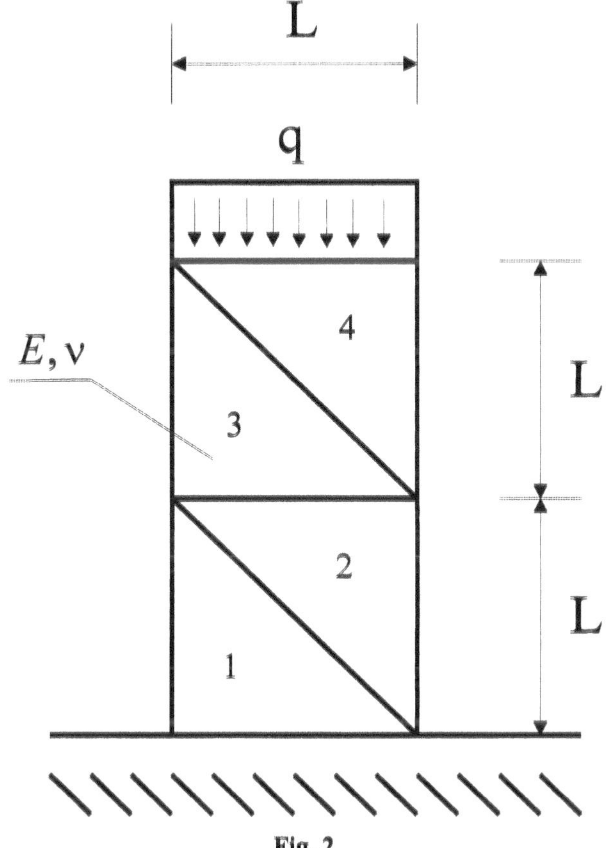

Fig. 2

In calculation we assume that L=1 $[m]$, $q = 1\left[\dfrac{kN}{m}\right]$, $\nu = 0.3$.

Table 1. Fuzzy Young's modulus

α	α=0	α=1
$\hat{E}_{\alpha}^{1}$	[189, 231][GPa]	210 [GPa]
$\hat{E}_{\alpha}^{2}$	[189, 231][GPa]	210 [GPa]
$\hat{E}_{\alpha}^{3}$	[189, 231] [GPa]	210 [GPa]
$\hat{E}_{\alpha}^{4}$	[189, 231] [GPa]	210 [GPa]

After applying algorithm 3 we get the following numerical results:

Table 2. Fuzzy stress

α	α=0	α=1
$\hat{\sigma}^1_{y\alpha}$	[0.96749, 0.974493] [kPa]	0.971063 [kPa]
$\hat{\sigma}^2_{y\alpha}$	[1.02833, 1.02955] [kPa]	1.02894 [kPa]
$\hat{\sigma}^3_{y\alpha}$	[0.98086, 1.01719] [kPa]	0.999086 [kPa]
$\hat{\sigma}^4_{y\alpha}$	[0.982807, 1.01914] [kPa]	1.00091 [kPa]

Table 3. Fuzzy displacements

No.	$\hat{u}_{i\alpha}, \alpha = 0\ [m]$
1	[0, 0]
2	[0, 0]
3	[0, 0]
4	[0, 0]
5	[3.2517e-14,7.49058e-13]
6	[3.81132e-12, 4.692e-12]
7	[-1.5243e-12,-4.9879e-13]
8	[4.4199e-12, 5.4275e-12]
9	[-1.5134e-12,1.0498e-12]
10	[8.1381e-12,9.9465e-12]
11	[-3.1758e-12,-1.7949e-13]
12	[8.7620e-12,1.0709e-11]

This problem has 8 degree of freedom.

9 Numerical example – plane stress problem in theory of elasticity

The equilibrium equations of a rod has the following form

$$\begin{cases} \dfrac{d}{dx}\left(EA\dfrac{du}{dx}\right)+n=0 \\ u\in V \end{cases}. \tag{111}$$

where E is an Young's module, A is an area of cross section, n is a load and V is some functional space. We can formulate the problem (111) in the variational form:

$$\forall v\in V,\quad a(u,v)=l(v) \tag{112}$$

where

$$a(u,v)=\int_0^L EA\frac{du}{dx}\frac{dv}{dx}dx \tag{113}$$

$$l(v)=\int_0^L nvdx+\ldots \tag{114}$$

To solution of the problem (112) we can apply finite element method. Local stiffness matrix (in local coordinate system) has the following form:

$$\overline{\mathbf{K}}^e=\int_{V^e}(\mathbf{B}^e)^T\mathbf{D}^e\mathbf{B}^e dx=\frac{E^eA^e}{L^e}\begin{bmatrix}1 & -1\\ -1 & 1\end{bmatrix} \tag{115}$$

where

$$\mathbf{B}^e=\left[-\frac{1}{L^e},\frac{1}{L^e}\right] \tag{116}$$

$$\mathbf{D}^e=[E^e] \tag{117}$$

Local stiffness matrix in global coordinate system has the following form:

$$\mathbf{K}^e = \int_{V^e} (\mathbf{C}^e)^T (\mathbf{B}^e)^T \mathbf{D}^e \mathbf{B}^e \mathbf{C}^e dx \tag{118}$$

where rotation matrix has the following form

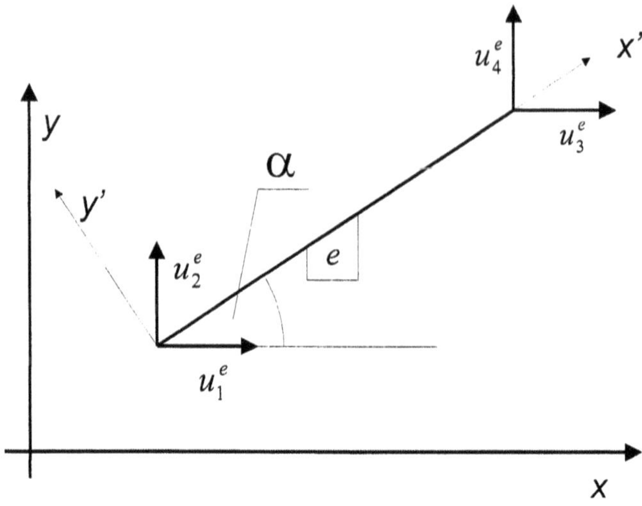

Fig. 3

$$\mathbf{C}^e = \begin{bmatrix} cos(\alpha^e) & sin(\alpha^e) & 0 & 0 \\ -sin(\alpha^e) & cos(\alpha^e) & 0 & 0 \\ 0 & 0 & cos(\alpha^e) & sin(\alpha^e) \\ 0 & 0 & -sin(\alpha^e) & cos(\alpha^e) \end{bmatrix} \tag{119}$$

Local load vector in global coordinate system:

$$\mathbf{Q}^e = \int_{V^e} (\mathbf{C}^e)^T (\mathbf{N}^e)^T \mathbf{n} dx \tag{120}$$

The structure is shown in the Fig. 4.

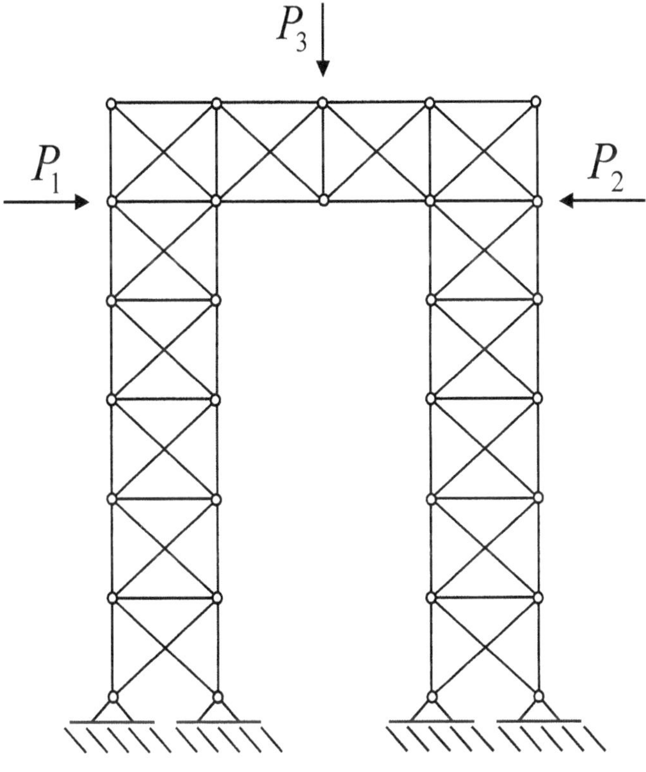

Fig. 4

Numerical data are as follows P=10 [kN], L=1 [m], $\nu = 0.3$, the Young's modulus is the same like in the previous example. Interval axial forces are shown in table 4.

Table 4. Interval axial force

No.	Axial force [N]
1	[3145.34, 4393.45]
2	[1482.48, 1914.16]
3	[-172.138, -221.845]
4	[164.454, 279.737]
5	[-958.619, -936.417]
6	[2459.35, 2536.53]
7	[1527.83, 1546.14]
8	[-343.544, -357.966]
9	[1708.72, 1617.27]

10	[-840.883, -841.035]
11	[1132.62, 1189.25]
12	[1532.73, 1547.37]
13	[-338.641, -356.736]
14	[3028.51, 2962.81]
15	[-932.071, -929.76]
16	[-278.358, -245.009]
17	[1656.79, 1671.62]
18	[-214.586, -232.489]
19	[4264.06, 4221.36]
20	[-169.222, -168.335]
21	[-751.05, -742.133]
22	[453.902, 470.55]
23	[-1417.47, -1433.55]
24	[6437.89, 6417.04]
25	[-7444.75, -7432.58]
26	[-200.408, -202.065]
27	[-2196.2, -2197.33]
28	[283.42, 285.763]
29	[4020.01, 4013.59]
30	[-200.408, -202.065]
31	[-9461.8, -9431.91]
32	[3589.87, 3583.79]
33	[-3488.96, -3478.74]
34	[713.715, 704.035]
35	[4929.89, 4924.37]
36	[720.439, 696.638]
37	[3580.36, 3594.25]
38	[-3482.95, -3485.36]
39	[-9466.06, -9427.23]
40	[4010.55, 4024]
41	[-194.644, -208.406]
42	[-2188.83, -2205.43]
43	[275.268, 294.73]
44	[-7448.38, -7428.59]
45	[-194.644, -208.406]
46	[6417.52, 6439.45]
47	[451.658, 473.02]
48	[-1419.72, -1431.08]
49	[-738.486, -755.954]
50	[-166.773, -171.028]
51	[4242.96, 4244.56]
52	[1655.57, 1672.95]

53	[-215.805, -231.149]
54	[-266.518, -258.031]
55	[-930.146, -931.887]
56	[3007.62, 2985.78]
57	[1531.23, 1549.04]
58	[-340.144, -355.068]
59	[1144.66, 1176]
60	[-839.969, -841.95]
61	[1686.62, 1641.68]
62	[1528.04, 1545.77]
63	[-343.334, -358.339]
64	[2470.18, 2524.72]
65	[-947.416, -949.597]
66	[253.654, 185.319]
67	[1683.18, 1701.27]
68	[-188.192, -202.832]
69	[3683.74, 3761.16]

10 Quasi-analytical method

Let us consider boundary value problem with fuzzy parameters

$$\mathbf{L}(\mathbf{x},\mathbf{u},\mathbf{h}) = \mathbf{f}(\mathbf{x},\mathbf{h}), \quad \mathbf{u} \in V, \quad \mathbf{h} \in F \tag{121}$$

If we differentiate the boundary value problem with respect to h_i we get a new boundary value problem

$$\mathbf{H}\left(\mathbf{x},\mathbf{u},\mathbf{h},\frac{\partial \mathbf{u}}{\partial h_i}\right) = \frac{\partial \mathbf{f}(\mathbf{x},\mathbf{h})}{\partial h_i}, \quad \mathbf{u} \in V, \quad \mathbf{h} \in F \tag{122}$$

Let us assume that we know the solution of the boundary value problem (121) for $\mathbf{h}_0^{\alpha} = mid(\hat{\mathbf{h}}_{\alpha})$ i.e. $\mathbf{u}_0^{\alpha}(\mathbf{x}) = \mathbf{u}(\mathbf{x},\mathbf{h}_0^{\alpha})$.

Now we can substitute $\mathbf{v}_i = \dfrac{\partial \mathbf{u}}{\partial h_i}$ and we get a new boundary value problem

$$\mathbf{H}\left(\mathbf{x},\mathbf{u}_{\alpha}^{0}(\mathbf{x}),\mathbf{h}_{\alpha}^{0},\mathbf{v}_{i}\right)=\frac{\partial\mathbf{f}(\mathbf{x},\mathbf{h}_{\alpha}^{0})}{\partial h_{i}},\quad \mathbf{v}_{i}\in V_{i} \tag{123}$$

Problem (123) can be solved using any method. If we know the solution of the equations (123) we can calculate extreme solution by using the algorithm 3.

Let us consider the following example

$$\frac{du}{dx}=hx,\ u(0)=u_{0}\cdot h^{2} \tag{124}$$

The exact solution is the following

$$u(x,h)=\frac{hx^{2}}{2}+u_{0}\cdot h^{2} \tag{125}$$

If we calculate derivative of the boundary value problem (124) with respect to h we get

$$\frac{d}{dx}\left(\frac{\partial u}{\partial h}\right)=x,\ \frac{\partial u(0)}{\partial h}=2u_{0}h \tag{126}$$

or

$$\frac{dv}{dx}=x\cdot h,\ v(0,h)=2u_{0}h \tag{127}$$

where

$$v(x,h)=\frac{\partial u(x,h)}{\partial h} \tag{128}$$

the solution of the equation (127) is the following

$$v(x,h)=\frac{x^{2}}{2}+2\cdot u_{0}\cdot h \tag{129}$$

If $h>0$, then $v(x,h)>0$ and the function $u=u(x,h)$ is monotone (for fixed x) and extreme solution can be calculated in the following way:

$$\hat{u}_\alpha(x) = [u_\alpha^-(x), u_\alpha^+(x)] \tag{130}$$

where

$$u_\alpha^-(x) = u(x, h_\alpha^-) = \frac{h_\alpha^- x^2}{2} + u_0 \cdot (h_\alpha^-)^2 \tag{131}$$

$$u_\alpha^+(x) = u(x, h_\alpha^+) = \frac{h_\alpha^+ x^2}{2} + u_0 \cdot (h_\alpha^+)^2 \tag{132}$$

It should be noted that if we know the numerical solution $\mathbf{u}_\alpha^0$ of the problem (121) we could not calculate $\frac{\partial \mathbf{u}}{\partial h_i}$ directly.

11 Finite difference method

The derivative $\frac{\partial u_i}{\partial h_j}$ can be calculated directly using the finite difference

$$\frac{\partial u_i(\mathbf{x}, \mathbf{h}_\alpha^0)}{\partial h_j} \approx \approx \frac{u_i(\mathbf{x}, h_{1\alpha}^0, \ldots, h_{j\alpha}^0 + \Delta h_j, \ldots, h_{1\alpha}^0) - u_i(\mathbf{x}, h_{1\alpha}^0, \ldots, h_{j\alpha}^0, \ldots, h_{1\alpha}^0)}{\Delta h_j} \tag{133}$$

The value $u_i(\mathbf{x}, \mathbf{h}_\alpha^0)$ is a solution of the boundary value problem

$$\mathbf{L}(\mathbf{x}, \mathbf{u}, \mathbf{h}_\alpha^0) = \mathbf{f}(\mathbf{x}, \mathbf{h}_\alpha^0), \quad \mathbf{u} \in V. \tag{134}$$

Other methods of calculation of the sensitivity can be find in the book (Kleiber 1997).

12 Point monotonicity tests

12.1 First order monotonicity tests

If derivative of the function $u = u(\mathbf{h})$ has constant sign, then we can assume that the function $u = u(\mathbf{h})$ is monotone. The value of the function $\frac{\partial u}{\partial h_i}$ can be approximated by using the linear function:

$$\frac{\partial u_{(1)}(\mathbf{h})}{\partial h_i} = \frac{\partial u(\mathbf{h}_0)}{\partial h_i} + \sum_{j=1}^{m} \frac{\partial^2 u(\mathbf{h}_0)}{\partial h_i \partial h_j}(h_j - h_j^0) \tag{135}$$

An interval function is an interval-valued function of one or more interval arguments. Consider a real-valued function f of real variables $x_1,...,x_n$ and an interval function $\hat{f}$ of interval variables $\hat{x}_1,...,\hat{x}_n$. The interval function $\hat{f}$ is said to be an interval extension of f if

$$\forall (x_1,...,x_n) \in D_f, \quad \hat{f}(x_1,...,x_n) = f(x_1,...,x_n) \tag{136}$$

where D_f is a domain of the function f. That is, if the arguments of $\hat{f}$ are degenerate intervals (i.e. $\hat{x}_i = x_i$), then $\hat{f}(\hat{x}_1,...,\hat{x}_n)$ is a degenerate interval equal to $f(x_1,...,x_n)$.
From properties of interval extensions (Neumaier 1990) arise that

$$\text{if } 0 \notin \frac{\partial \hat{u}_{(1)}(\hat{\mathbf{h}}_\alpha)}{\partial h_i}, \text{ then } \forall \mathbf{h} \in \hat{\mathbf{h}}_\alpha, \quad \frac{\partial u_{(1)}(\mathbf{h})}{\partial h_i} \neq 0. \tag{137}$$

and we can assume that the function $u = u(\mathbf{h})$ is monotone.

12.2 High order monotonicity tests

We can also approximate derivative of the function $u = u(\mathbf{h})$ using high order polynomials

$$\frac{\partial u_{(2)}(\mathbf{h})}{\partial h_i} = \frac{\partial u(\mathbf{h}_0)}{\partial h_i} + \sum_{j=1}^{m} \frac{\partial^2 u(\mathbf{h}_0)}{\partial h_i \partial h_j}(h_j - h_j^0) + \\ + \frac{1}{2}\sum_{j=1}^{m}\sum_{k=1}^{m} \frac{\partial^3 u(\mathbf{h}_0)}{\partial h_i \partial h_j \partial h_k}(h_j - h_j^0)(h_k - h_k^0) \tag{138}$$

$$\text{If } 0 \notin \frac{\partial \hat{u}_{(p)}(\hat{\mathbf{h}}_\alpha)}{\partial h_i}, \text{ then } \forall \mathbf{h} \in \hat{\mathbf{h}}_\alpha, \ \frac{\partial u_{(p)}(\mathbf{h})}{\partial h_i} \neq 0. \tag{139}$$

and we can assume that the function $u = u(\mathbf{h})$ is monotone.

13 Numerical example - displacement of the shell structure

The equilibrium equations of shell structures can be written in the following form:

$$\begin{aligned} & T^{\beta\alpha}|_\beta - b_\gamma^\alpha M^{\beta\gamma}|_\beta + b^\alpha = 0 \\ & T^{\beta\alpha} b_{\alpha\beta} + M^{\alpha\beta}|_{\beta\beta} + b^3 = 0 \\ & T^{\beta\alpha} n_\beta + b_\gamma^\alpha M^{\beta\gamma} n_\beta = p^\alpha, \quad x \in \partial\Omega \\ & M^{\beta\alpha}|_\beta n_\alpha + \frac{d}{ds}(M^{\alpha\beta}\tau_\alpha n_\beta) = p^3, \quad x \in \partial\Omega \end{aligned} \tag{140}$$

where

$$u^\alpha|_\beta = u^\alpha{}_{,\beta} + \left\{ \begin{matrix} \alpha \\ \beta\gamma \end{matrix} \right\} u^\gamma, \quad u^\alpha{}_{,\beta} = \frac{\partial u^\alpha}{\partial x^\beta}, \quad \alpha, \beta = 1,2 \tag{141}$$

$$\left\{ \begin{matrix} i \\ jk \end{matrix} \right\} = \frac{1}{2} g^{il} (g_{jl,k} + g_{kl,j} - g_{jk,l}) \tag{142}$$

g_{ij} is a metric tensor.

Let us consider shell structure, which is shown in Fig. 5.

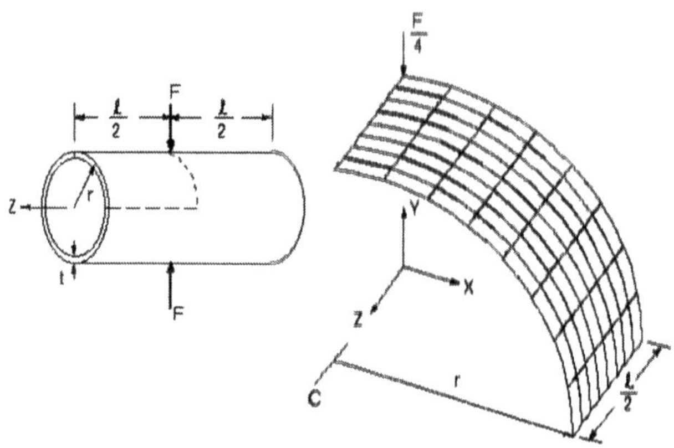

Fig. 5

In calculation we assume the following numerical data $E \in [2.0 \cdot 10^5, 2.2 \cdot 10^5][MPa]$, $\nu \in [0.2, 0.3]$, L=0.263 [m], r=0.126 [m], F=444.8 [N], $t = 2.38 \cdot 10^{-3} [m]$. We will be looking for an interval displacement in direction of the force F. Usign first order monotonicity test we can check monotonicity of the solution. Because the function $u = u(E, \nu)$ is monotone, then extreme values of the solution can be found using only the endpoints of given intervals. The interval solution is as follows:

$$\alpha = 0: \; u \in [-0.043514, -0.03748] \; [m], \tag{143}$$

$$\alpha = 1: \; u = -0.04102 \; [m]. \tag{144}$$

Using point monotonicity test we can calculate the interval solution only in some selected points. In this example professional FEM program Ansys was applied.

14 Taylor model of the solution

If the solution is sufficiently smooth, then we can approximate them by using Taylor series

$$u_{i\alpha}(\mathbf{h}) = u_i(\mathbf{h}_\alpha^0) + \sum_{j=1}^{m} \frac{\partial u_i(\mathbf{h}_\alpha^0)}{\partial h_j}(h_i - h_\alpha^0). \tag{145}$$

Extreme values of the solution can be approximated directly by using equation (145) and interval arithmetic

$$\hat{u}_{i\alpha} = \hat{u}_{i\alpha}(\hat{\mathbf{h}}_\alpha) = u_i(\mathbf{h}_\alpha^0) + \sum_{j=1}^{m} \frac{\partial u_i(\mathbf{h}_\alpha^0)}{\partial h_j}(\hat{h}_{i\alpha} - h_\alpha^0). \tag{146}$$

15 This method has very low computational complexity (*m*+1 system of equations) (Akapan et all 2001). Unfortunately, the equation (146) gives only approximate solution. Interval monotonicity tests

15.1 Linear equations

Let us consider the problem (36) and assume that we know the solution of the following systems of linear interval equation

$$\hat{\mathbf{K}}(\hat{\mathbf{h}}_\alpha)\mathbf{u} = \hat{\mathbf{Q}}(\hat{\mathbf{h}}_\alpha). \tag{147}$$

$$\hat{\mathbf{K}}(\hat{\mathbf{h}}_\alpha)\frac{\partial \mathbf{u}}{\partial h_i} = \frac{\partial \hat{\mathbf{Q}}(\hat{\mathbf{h}}_\alpha)}{\partial h_i} - \frac{\partial \hat{\mathbf{K}}(\hat{\mathbf{h}}_\alpha)}{\partial h_i}\hat{\mathbf{u}}(\hat{\mathbf{h}}_\alpha). \tag{148}$$

where

$$\hat{\mathbf{u}}(\hat{\mathbf{h}}_\alpha) = hull\sum(\hat{\mathbf{K}}(\hat{\mathbf{h}}_\alpha), \hat{\mathbf{Q}}(\hat{\mathbf{h}}_\alpha)). \tag{149}$$

If

$$0 \notin \frac{\partial \hat{u}_i(\hat{\mathbf{h}}_\alpha)}{\partial h_j} = hull \sum \left(\hat{\mathbf{K}}(\hat{\mathbf{h}}_\alpha), \frac{\partial \hat{\mathbf{Q}}(\hat{\mathbf{h}}_\alpha)}{\partial h_j} - \frac{\partial \hat{\mathbf{K}}(\hat{\mathbf{h}}_\alpha)}{\partial h_j} \hat{\mathbf{u}}_\alpha(\hat{\mathbf{h}}_\alpha) \right), \quad (150)$$

15.2 then the solution of the problem (36) is monotone (with guaranteed accuracy). Numerical example – heat transfer

Let's us consider heat transfer problem

$$\begin{cases} R_1 < r < R_2 : & \frac{1}{r}\frac{d}{dr}\left(r\lambda \frac{dT(r)}{dr} \right) + Q = 0 \\ r = R_1 & : -\lambda \frac{dT(r)}{dr} = \alpha\left(T(r) - T_b\right) \\ r = R_2 & : T(r) = T_t \end{cases} . \quad (151)$$

In calculation we assume the following numerical data $R_1 = 0.0005\ [m]$, $R_2 = 10 \cdot R_1$, $\alpha = 2000\left[\frac{W}{m^2 \cdot K}\right]$, $T_b = 32^o C$, $T_t = 37^o C$, $Q = 10245\left[\frac{W}{m^3}\right]$, $\lambda \in [0.21, 0.23]\left[\frac{W}{m \cdot K}\right]$.

Numerical solutions are shown in the table 5.

Table 5. Interval temperature

No	$T_i^-[^oC]$	$T_i^+[^oC]$
1	36.586	36.619
2	35.470	35.494
3	34.782	34.800
4	34.282	34.298
5	33.894	33.582
6	33.302	33.308

7	33.065	33.070
8	32.857	32.859
9	32.669	32.671
10	32.500	32.500

15.3 Nonlinear equations

Sometimes system of algebraic equations is nonlinear

$$\mathbf{F}(\mathbf{x},\mathbf{u},\mathbf{h}) = \mathbf{0}, \quad \mathbf{h} \in \hat{\mathbf{h}}_{\alpha}. \tag{152}$$

From implicit function theorem arise that

$$\frac{\partial \mathbf{F}}{\partial \mathbf{u}} \frac{\partial \mathbf{u}}{\partial h_i} = -\frac{\partial \mathbf{F}}{\partial h_i}, \quad i=1,...,m. \tag{153}$$

Equation (153) is a system of linear equation with unknown $\frac{\partial \mathbf{u}}{\partial h_i}$, because of that

$$\frac{\partial u_i}{\partial h_j} = -\frac{\begin{vmatrix} \frac{\partial F_1}{\partial u_1} & \cdots & \frac{\partial F_1}{\partial u_{i-1}} & \frac{\partial F_1}{\partial h_j} & \frac{\partial F_1}{\partial u_{i-1}} & \cdots & \frac{\partial F_1}{\partial u_n} \\ \cdots & \cdots & \cdots & \cdots & \cdots & \cdots & \cdots \\ \frac{\partial F_n}{\partial u_{11}} & \cdots & \frac{\partial F_n}{\partial u_{i-1}} & \frac{\partial F_n}{\partial h_j} & \frac{\partial F_n}{\partial u_{i+1}} & \cdots & \frac{\partial F_n}{\partial u_n} \end{vmatrix}}{\left| \frac{\partial \mathbf{F}}{\partial \mathbf{u}} \right|}, \tag{154}$$

$$\frac{\partial u_i}{\partial h_j} = -\frac{\left| \frac{\partial \mathbf{F}}{\partial (u_1,...,u_{i-1},h_j,u_{i+1},...,u_n)} \right|}{\left| \frac{\partial \mathbf{F}}{\partial \mathbf{u}} \right|}. \tag{155}$$

From equation (155) arrays, that if the following determinates

$$\left|\frac{\partial \mathbf{F}}{\partial \mathbf{u}}\right|, \quad \left|\frac{\partial \mathbf{F}}{\partial(u_1,\ldots,u_{i-1},h_j,u_{i+1},\ldots,u_n)}\right|. \tag{156}$$

have constant sign, then the derivative $\frac{\partial u_i}{\partial h_j}$ has also constant sign and the functions $u_i = u_i(\ldots,h_j,\ldots)$ are monotone.

From properties of the determinates and Darbox theorem arise, that if

$$\forall \mathbf{h} \in \hat{\mathbf{h}}_\alpha, \left|\frac{\partial \mathbf{F}(\mathbf{x},\mathbf{u}(\mathbf{h}),\mathbf{h})}{\partial(u_1,\ldots,u_{i-1},h_j,u_{i+1},\ldots,u_n)}\right| \neq 0, \, , \tag{157}$$

$$\forall \mathbf{h} \in \hat{\mathbf{h}}_\alpha, \left|\frac{\partial \mathbf{F}(\mathbf{x},\mathbf{u}(\mathbf{h}),\mathbf{h})}{\partial \mathbf{u}}\right| \neq 0. \, . \tag{158}$$

i.e. the Jacobean matrix are regular, then the functions $u_i = u_i(\ldots,h_j,\ldots)$ are monotone.

From properties of interval arithmetic, arise that

$$\forall \mathbf{h} \in \hat{\mathbf{h}}_\alpha, \quad \left|\frac{\partial \mathbf{F}(\mathbf{x},\mathbf{u}(\mathbf{h}),\mathbf{h})}{\partial \mathbf{u}}\right| \in \left|\frac{\partial \hat{\mathbf{F}}(\mathbf{x},\mathbf{u}(\hat{\mathbf{h}}_\alpha),\hat{\mathbf{h}}_\alpha)}{\partial \mathbf{u}}\right| \tag{159}$$

and

$$\begin{aligned} &\forall \mathbf{h} \in \hat{\mathbf{h}}_\alpha, \quad \left|\frac{\partial \mathbf{F}(\mathbf{x},\mathbf{u}(\mathbf{h}),\mathbf{h})}{\partial(u_1,\ldots,u_{i-1},h_j,u_{i+1},\ldots,u_n)}\right| \in \\ &\in \left|\frac{\partial \hat{\mathbf{F}}(\mathbf{x},\mathbf{u}(\hat{\mathbf{h}}_\alpha),\hat{\mathbf{h}}_\alpha)}{\partial(u_1,\ldots,u_{i-1},h_j,u_{i+1},\ldots,u_n)}\right|. \end{aligned} \tag{160}$$

We can see that, if the interval Jacobean matrices (159, 160) are regular, then the functions $u_i = u_i(\mathbf{h})$ are monotone.

15.4 Numerical example – frame structure

The equilibrium equations of beam is as follows:

$$\frac{d^2}{dx^2}\left(EJ\frac{d^2u}{dx^2}\right)=q, \quad u\in V. \tag{161}$$

If we apply the finite element, we get equilibrium equations in the following form:

$$\mathbf{K}(\mathbf{h})\mathbf{u}=\mathbf{Q}. \tag{162}$$

Let us consider a structure, which is shown in the Fig. 6.

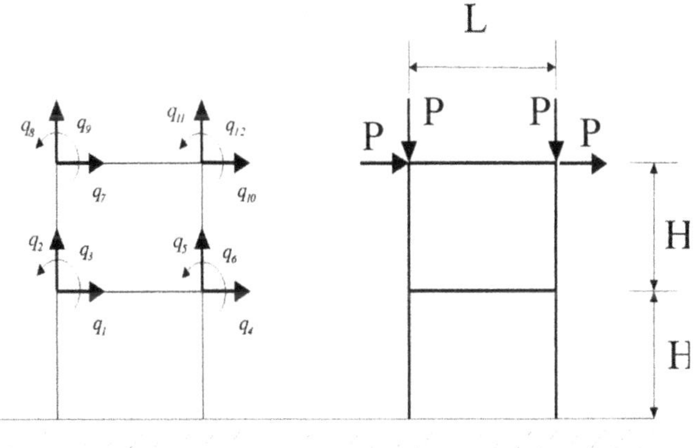

Fig. 6

In calculation we assume the following data $E\in[210,220][GPa]$, $J\in\left[\frac{0.05^4}{12},\frac{0.055^4}{12}\right][m^4]$, $A\in[0.05^2,0.055^2][m^2]$, L=H=1 $[m]$, P=1 $[kN]$.

Table 6. Interval displacements

No.	q_i^- $[m]$	q_i^+ $[m]$
1	0.035716	0.037414
2	0.000008	0.000009
3	-0.011230	-0.010718
4	0.035716	0.037414
5	-0.000021	-0.000017
6	-0.011230	-0.010718
7	0.082163	0.086067
8	0.00009	0.000010
9	-0.007494	-0.007151
10	0.082163	0.086067
11	-0.000033	-0.000026
12	-0.007494	-0.007151

15.5 Subdivision

16 The interval extension of the Jacobean matrix may become singular even for very narrow intervals $\hat{\mathbf{h}}_\alpha$. In this case, we can divide these intervals and repeat procedure again. Optimization methods

16.1 Description of the algorithm

If the intervals $\hat{\mathbf{h}}_\alpha$ are very wide, then we cannot apply methods, which were described below. In such situation, optimization methods can be applied.

$$u_{i\alpha}^- \Leftarrow \begin{cases} \min u_i \\ \mathbf{L}(\mathbf{x},\mathbf{u},\mathbf{h}) = \mathbf{f}(\mathbf{h}) \\ \mathbf{h} \in \hat{\mathbf{h}}_\alpha, \mathbf{u} \in V \end{cases} \quad u_{i\alpha}^+ \Leftarrow \begin{cases} \max u_i \\ \mathbf{L}(\mathbf{x},\mathbf{u},\mathbf{h}) = \mathbf{f}(\mathbf{h}) \\ \mathbf{h} \in \hat{\mathbf{h}}_\alpha, \mathbf{u} \in V \end{cases}. \quad (163)$$

Approximate solution can be defined as follows:

$$u_{i\alpha}^{-} \Leftarrow \begin{cases} \min u_i \\ \mathbf{K}(\mathbf{h})\mathbf{u} = \mathbf{Q}(\mathbf{h}) \\ \mathbf{h} \in \hat{\mathbf{h}}_\alpha \end{cases} \quad u_{i\alpha}^{+} \Leftarrow \begin{cases} \max u_i \\ \mathbf{K}(\mathbf{h})\mathbf{u} = \mathbf{Q}(\mathbf{h}) \\ \mathbf{h} \in \hat{\mathbf{h}}_\alpha \end{cases}. \tag{164}$$

16.2 Numerical example – displacements of beam

Let us consider beam structure, which is shown in Fig. 7.

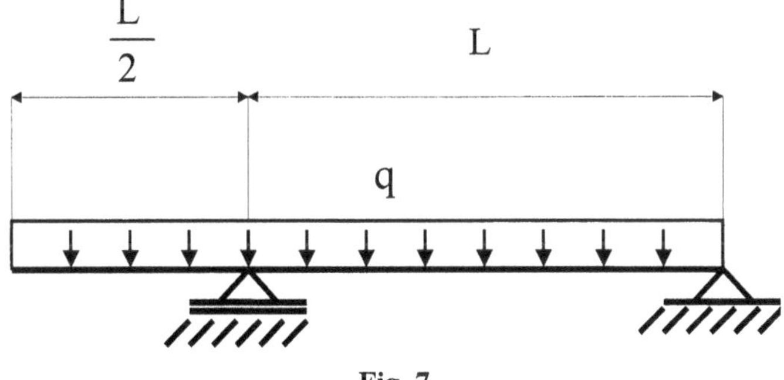

Fig. 7

The equilibrium equation has the following form

$$\begin{cases} \dfrac{d^2}{dx^2}\left(EJ\dfrac{d^2u}{dx^2}\right) = q(x), \\ u\left(\dfrac{L}{2}\right) = 0, \ u\left(\dfrac{3L}{2}\right) = 0, \ \dfrac{d^2u(0)}{dx^2} = 0, \ \dfrac{d^2}{dx^2}u\left(\dfrac{3L}{2}\right) = 0 \end{cases}. \tag{165}$$

In calculations we assume that $E \in [2.0 \cdot 10^5, 2.2 \cdot 10^5][MPa]$, $J \in \left[\dfrac{0.049^4}{12}, \dfrac{0.051^4}{12}\right][m^4]$, $L = [0.999, 1.001][m]$, $q \in [9.9, 10.1][kN]$. Numerical results are shown in Fig. 8.

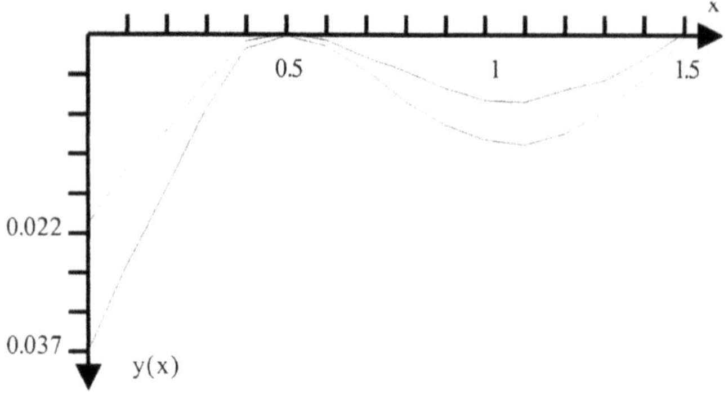

Fig. 8

17 Conclusions

1) Calculation of the solutions of the fuzzy partial differential equations is in general very difficult (NP-hard).
2) In engineering applications the relation between the solution and the uncertain parameters is usually monotone.
3) Using methods which are based on sensitivity analysis we can solve very complicated problems of computational mechanics (even with thousands degree of freedom).
4) If we apply the point monotonicity tests we can use results, which were generated by the existing engineering software.
5) Reliable methods of solution of the fuzzy partial differential equations are based on the interval arithmetic. These methods have high computational complexity.
6) In some cases (e.g. if we know analytical solution) the optimization method can be applied.
7) In some special cases we can predict the solution of the fuzzy partial differential equations.
8) The fuzzy partial differential equation can be applied to modeling of mechanical systems (structures) with uncertain parameters.

References

[1] Akapan U.O., Koko T.S., Orisamolu I.R., Gallant B.K. (2001) Practical fuzzy finite element analysis of structures, Finite Element in Analysis and Design, vol. 38, pp. 93-111
[2] Buckley J.J., Qy Y. (1990) On using α-cuts to evaluate fuzzy equations, Fuzzy Sets and Systems, vol. 38, pp.309-312

[3] Buckley J.J., Feuring T. (2000) Fuzzy differential equations, Fuzzy Sets and Systems, vol. 110, pp. 43-54
[4] Ciarlet P.G. (1978), The Finite Element Method for Elliptic Problems, North-Holland, New York
[5] Kleiber M. (1997) Parameter Sensitivity in Nonlinear Mechanics, Theory and Finite Element Computations, John Willey and Sons, New York
[6] Kreinovich V., Lakeyev A., Rohn J., Kahl P. (1998) Computational Complexity Feasibility of Data Processing and Interval Computations, Kluwer Academic Publishers, Dordrecht
[7] Kulpa Z., Pownuk A., Skalna I. (1998) Analysis of linear mechanical structures with uncertainties by means of interval methods, Computer Assisted Mechanics and Engineering Sciences, vol. 5, pp.443-477
[8] McWilliam S. (2000) Anti-optimization of uncertain structures using interval analysis. Computers and Structures, vol. 79, pp.421-430
[9] Neumaier A. (1990) Interval methods for systems of equations, Cambridge University Press, New York, 1990
[10] Noor A.K., Starnes J.H., Peters J.M. (2000) Uncertainty analysis of composite structures, Computer methods in applied mechanics and engineering, vol. 185, pp.413-432
[11] Pownuk A. (2001) Application of fuzzy sets theory to assessment of reliability of civil engineering structures (in Polish), Ph.D. Dissertation, Silesian University of Technology